Fundamentals of

SOLAR ASTRONOMY

WORLD SCIENTIFIC SERIES IN ASTRONOMY AND ASTROPHYSICS

Editor: Jayant V. Narlikar
Inter-University Centre for Astronomy and Astrophysics, Pune, India

---

Published:

Volume 1: Lectures on Cosmology and Action at a Distance Electrodynamics
F. Hoyle and J. V. Narlikar

Volume 2: Physics of Comets (2nd Ed.)
K. S. Krishna Swamy

Volume 3: Catastrophes and Comets*
V. Clube and B. Napier

Volume 4: From Black Clouds to Black Holes (2nd Ed.)
J. V. Narlikar

Volume 5: Solar and Interplanetary Disturbances
S. K. Alurkar

Volume 6: Fundamentals of Solar Astronomy
A. Bhatnagar and W. Livingston

Volume 7: Dust in the Universe: Similarities and Differences
K. S. Krishna Swamy

Volume 8: An Invitation to Astrophysics
T. Padmanabhan

Volume 9: Stardust from Meteorites: An Introduction to Presolar Grains
M. Lugaro

Volume 10: Rotation and Accretion Powered Pulsars
P. Ghosh

Volume 11: Find a Hotter Place!: A History of Nuclear Astrophysics
L. M. Celnikier

Volume 12: Physics of Comets (3rd Edition)
K. S. Krishna Swamy

Volume 13: From Black Clouds to Black Holes (3rd Edition)
J. V. Narlikar

*Publication cancelled.

World Scientific Series in Astronomy and Astrophysics – Vol. 6

Fundamentals of SOLAR ASTRONOMY

"The point of living is to study the Sun"

– Anaxagoras (499–428 BC)

Arvind Bhatnagar
Udaipur Solar Observatory, India

William Livingston
National Solar Observatory, Arizona, USA

NEW JERSEY • LONDON • SINGAPORE • BEIJING • SHANGHAI • HONG KONG • TAIPEI • CHENNAI

Published by

World Scientific Publishing Co. Pte. Ltd.
5 Toh Tuck Link, Singapore 596224
USA office: 27 Warren Street, Suite 401-402, Hackensack, NJ 07601
UK office: 57 Shelton Street, Covent Garden, London WC2H 9HE

British Library Cataloguing-in-Publication Data
A catalogue record for this book is available from the British Library.

Cover page: Displaying the icon of Surya — the Sun god; ancient emblem of the dynasty of Mewar (Udaipur, Rajasthan, India) kings, along with solar Coronal Mass Ejections and prominences.

World Scientific Series in Astronomy and Astrophysics — Vol. 6
FUNDAMENTALS OF SOLAR ASTRONOMY

ISBN 978-981-238-244-3
ISBN 978-981-256-357-6 (pbk)

Printed in Singapore

Professor M. K. Vainu Bappu
(1927 – 1982)

This book is dedicated to our friend and mentor
Professor M. K. Vainu Bappu.

Arvind Bhatnagar
William Livingston

Preface

The aim of this book is to inculcate, motivate and inspire readers to take up the study and observations of our nearest star – the Sun, and enjoy its beauty and glory. Our Sun is the only star in the Universe which presents its surface details, as there is no other star near enough to show features of the order of a few hundred kilometers. Extending from deep inside the Sun to the solar surface and beyond, the Sun manifests a variety of phenomena, ranging from a few hundred kilometers to thousands of kilometers in size, temperature ranging from a few thousands to several million degrees and in temporal domain from a few seconds to several decades and dynamical events with speeds from a few tenths of km/sec to thousands of km/sec. Thus why not make use of this unique celestial laboratory to study the physical characteristics of matter and to understand other celestial bodies in the Universe.

The remarkable dynamical phenomena occurring on the Sun, such as mass ejections in the form of eruptive prominences, filaments, surges, sprays, Coronal Mass Ejections, transient events; like solar flares, ephemeral regions, sunspots, granulations etc., make the study of the Sun extremely fascinating and interesting. One can see various solar phenomena occurring *right in front of one's eyes* and follow them for hours and days. The purpose of this book is to present some of these fascinating phenomena, in their full glory to the readers through ample number of illustrations, sketches and photographs.

This book is mainly addressed to those who are starting to study the Sun and want to pursue an advance course in solar physics, but lack the basic knowledge of solar astronomy. To encourage young people, especially the budding amateur solar astronomers, we have pointed out to the high quality early **visual** solar observations made in the seventeenth and eighteenth centuries, through small telescopes by Father Secchi, Langley, Captain Tupman, Professor Fernley and many others of solar granulations, sunspots, prominences, spicules, solar corona etc. We have emphasized that keen, persistent and careful visual observation through small telescopes can, not only provide extremely useful scientific data, but also gives great joy and fun, and one can always think of serendipity discovery that is just awaiting to be made about the Sun. Thus people with limited means in terms of equipment need not be discouraged, but follow the example of early observers and take up observing the Sun and contribute to its global watch.

From time immemorial Sun has occupied a central stage in all ancient cultures. It had been and still worshiped in many cultures, countries and civilizations. Our ancestors had considered the Sun as god and goddess, because it gave them light, warmth, seasons and the very existence of life on this planet. To perpetuate its glory and might, all the ancient cultures created mythological stories about the Sun, and interwoven them in their daily cultural life and rituals. To give an idea about these ancient myths, in Chapter 1 we have briefly described them, as these are not readily available in the standard texts on solar astronomy. The readers will note that actual solar observations during the year were very important activity in ancient times, to mark the solstices and equinoxes, which were part of the cultural, religious and agricultural life. For this purpose huge structures like the ***Stonehenge*** were built more than 5000 years ago.

We have tried to keep to bare minimum the use and derivation of mathematical equations, only some basic knowledge of physics and mathematics is required to understand the text. There is some amount of repetition also, which to some extent it is intentional so that various Chapters could be read independently too.

In Chapter 2, we have given a brief description of some of the operating solar optical and radio observatories, ranging from very small observatories with 10-15 cm aperture telescopes to the state-of-the–art observatories, such as the New Swedish Solar Observatory, the Dutch Open telescope, the German Vacuum Solar Observatory, the THEMIS and the National Solar Observatory at Kitt Peak, USA. It has been pointed out that even small observatories have and are significantly contributing to synoptic solar observations, in spite of available highly sophisticated solar telescopes and space missions. A brief description of some of the operating and planned space solar missions has also been given. We feel guilty of the fact that in this list of solar observatories we have not been able to include the enormous contribution being made by amateur solar astronomers.

In Chapter 3, some of the basics of solar structure, energy generation, transport, irradiance, solar rotation and the neutrino puzzle are discussed. Chapter 4 gives a description of the *Quiet Sun*, although the Sun is never quiet, it is in *action* all the time. In Chapter 5, we present the *Active* aspect of our Sun, covering activities in the photosphere, chromosphere and the corona. To appreciate and enjoy the *Sun in Action,* the readers are advised to see time lapse movies now available from Websites of several solar observatories.

In Chapter 6, we have given methods and techniques to determine basic solar parameters, such as the Solar Parallax, mass, distance, temperature, heliographic coordinates of solar features. These may be found useful for those initiated in solar astronomy from other disciplines. Chapter 7 covers in some detail, description of solar optical instruments, especially the various types of light feeds (solar telescope), spectrographs, imaging equipment, like narrow band filters and spectroheliographs. In this Chapter we have gone to great length in discussing the principle and working of birefringent filters, this is essentially because narrow band filters are the heart of any solar observational investigation, and description of such filters is not readily available in standard textbooks. We hope that the discussion given on the birefringent filter will familiarize the readers enough, not to consider it as a 'black box'.

The fascinating phenomenon of the total solar eclipse is discussed in Chapter 8, emphasizing the importance of eclipse observations, and what we have learnt and what more can be learnt. Since the early days in nineteenth century, enormous scientific data and results are now available, but still some unsolved problems persist. However, we urge the newcomers to this field, that there is nothing like watching the whole event of the total solar eclipse with naked eyes (of course after taking due care), and suggest the readers of this book to witness at least once, one of the nature's most beautiful and fantastic phenomenon.

In Chapter 9, we take the readers to the solar interior and introduce the new subject of *helioseismology*. We have not dealt the topic in great detail, but have simply discussed the basic principles of helioseismology and given the latest results obtained through this technique.

In Chapter 10, is a description of personal experience of a solar observer to share his joy of observing the Sun.

The authors feel apologetic that it was impossible to mention all the references to the enormous wonderful work that is being carried out in solar astronomy and mentioned in this book. Actually, the scientific literature in all sciences, especially in Astronomy is inflating at an exponential rate and it has become almost impossible to keep track of all the research papers in spite of the 'Information highway' and Internet access etc. In this book we have tried to mention at number of places older references on the subject, which had been often forgotten or left out or people are simply not aware of them. Lately, it has been noticed that younger people hardly refer to literature earlier than 10 years, and in this process either miss the earlier findings or '*re-discover*' the same phenomenon. We found it interesting to look in the past literature and were amused to note that how the concept about our Sun has changed. In the eighteenth century, even the great astronomer Sir William Herschel thought that the Sun could be inhabited and that sunspots are windows to the interior! Now after 300-350 years, we talk of resolving features on the order of 70-100 kilometers, know precisely the physical condition of even the solar interior and are preparing to see the Sun in 3-dimension (STEREO mission).

We strongly feel that to bring out the real beauty of our Sun and its activities, it is most essential to display it through high quality illustration and pictures, which are now available from modern ground-based solar telescopes and solar space missions such as YOHKOH, TRACE, EIT and SXI. Therefore, in this book we have tried to present as many good pictures and illustrations as possible, those would inspire and motivate the beginners to take up Solar Astronomy, as a subject for study, enjoyment and fun. We believe that one good illustration is equivalent to thousand words; hence this book contains a fairly large number of illustrations, which is the crux to manifest the beauty of our Sun. The

Publisher, World Scientific Publishing Company, Singapore (WSPC) had been generous to allow large number of high quality illustrations appearing in this book.

We wish to acknowledge several authors, publishers and individuals who have provided and permitted us to make use of photographs, illustrations for this educational book. We would like to record our gratitude to Ashok Ambastha, Nandita Srivastava, Sushant Tripathi, Kiran Jain of the Udaipur Solar Observatory for their help in preparing this manuscript. Jingxiu Wang and Li Ting sent us photographs of Chinese observatories and Sun's myth in China. Takeo Kosugi and his student K. Yaji helped to process YOHKOH images of January 14, 1993 event. Pam Gilman and Steve Padilla sent us a latest picture of the 150-foot Solar tower telescope at Mount Wilson Observatory and K. Sundararaman of Kodaikanal Observatory sent us pictures of the Observatory and spectroheliograms. Major typing and computer setting of diagrams, text etc., was done very devotedly by Ms. Anita Jain and helped by her husband, Naresh Jain, the authors feel indebted for their help. One of the authors (AB) acknowledges receipt of partial financial grant from the Indian Space Research Organization (ISRO) for this project, and wishes to thank Professor U. R. Rao, Chairman, Physical Research Laboratory's Governing Council and Dr. K. Kasturirangan, former Chairman, ISRO for their interest in this project.

We have taken great care that no mistake has crept into the text, but if any, we shall be responsible. The Publisher, WSPC has taken great care to perfectly reproduce the large number of color and black and white illustrations presented in this book.

Arvind Bhatnagar,
Udaipur, Rajasthan, India.
William Livingston,
Tucson, Arizona, USA.

Contents

Chapter 1

Ancient Solar Astronomy

1.1 Mythologies about the Sun

Among cultures of antiquity, the Sun has always occupied a central position. It caught the imagination of early man because the Sun gave him warmth, light, life, and acted as his clock. Because of this, he made the Sun his god and goddess, and worshipped it. Even today, in modern times, the Sun is worshipped in many countries and religions. Number of temples dedicated to the Sun god had been built. Many of the great cities of the ancient world were known as "The City of the Sun", such as Baalbec, Rhodes, and Heliopolis. More then just cult centers, scientists and astronomers of the day who lived in these cities studied the Sun, Moon and planets, in an effort to devise accurate calendar systems. What are the folklore and mythological stories about our Sun from these civilizations? It is of interest to note that many of these stories originated at different times in history, and in far off places, yet they still possess meaning to us.

1.1.1 *In Early Europe*

In early Europe generally the Sun was considered as a male god, but among the Indo-Europeans it was a female goddess, and the Moon was a male god. In German, and Gaelic languages the word for Sun is still female. In many other languages a common solar association is still reflected, for example: in Sanskrit, the Sun is called 'Surya' and Savitra or Savita, in Gaul 'Sulis', in Lithuanian 'Saule', and in Latin and

German ‘Sol’. In addition, in Sanskrit the solar year is called ‘Sama’, which is similar in modern English to the word ‘summer’, and Celtic words such as ‘Samhain’ mean summer's end. Commonality is found in the names of the Sun among various cultures.

1.1.1.1 *Norse*

Europe has a long history with celestial deities. It was, in fact, named after the goddess Europa. Long ago a tribe known as Tautens colonized Europe or what is now called the European countries. Tauten people stemmed from an even older people known to us as Indo-Europeans. Early Tautens believed in a Sun goddess, Sol, and a Moon god Mani. Today in the German language, Sun is addressed as Die Sonne, a female noun, and the Moon as Der Mond, male. Like the dawn goddesses of the Greeks, Hindus and Egyptians, the early Germans propitiated a dawn goddess known as Ostara, or Eoster. It is this goddess from which the Christians incorporated a ceremony known as Easter, and her season, lencten in Anglo-Saxon, or literally "spring", became the Christian "Lent", leading to the Easter holiday. This reasoning leads to the medieval belief that the Sun "danced" on Easter day. Yet Eoster's most dominant symbol remains the ‘egg’, which symbolizes birth and renewal.

Celestial knowledge of the Norse is seeped in symbols and myth. For thousands of years, the most sacred and important symbol was the ‘Wheel of the Year’, represented by a 6 or 8 spoke wheel, or by a solar cross within a wheel. Such wheels are depicted on the famed silver cauldron of Gundestrup, which shows a horned deity touching a wheel. The Norse people, who lived in what is now known as Yorkshire, often cut out a solar wheel and placed it on the tops of mounds, inserting a pole or pillar to make a solar compass or a sundial. As in many other ancient cultures, the solstices played a key role in their lives, customs, and religious traditions. Solstices refer to the most northern and southern positions of the Sun in the sky. The modern word "solstice" stems from the Latin "sol stetit", or literally meaning that the "Sun stands still", and the official modern name of the Sun. Sol also finds it's origins in Latin, where sol is a feminine noun meaning ‘Sun’.

Norse people devised their calendar taking into consideration the midsummer solstice. Among the Norse, the god Balder is the most closely associated with the solstices. In a myth that explains the actions of the midsummer and midwinter Sun, Balder, the son of the god Odin, was said to die at the hands of his evil brother who, wielded a mistletoe stake each summer solstice. He was reborn at the winter solstice, or what is still known in Germany as Mother Night (the 'mother' in question being the goddess who brings the new born Sun back into existence).

There are a large number of Norse myths about the Sun. In the epic of Sigmund, also known as Sigurd or Siefried, the Sun's magic sword is named Balmung, which means 'Sun beam'. In this tale, the hero comes across a valkrie surrounded in a ring of fire. It is a lovely Brunhild, who symbolizes a dawn maiden. The Saxon god, Saxnot (sax-sword) also had a magic sword, and one was said to have hung in his temple in such a way as to reflect the dawn's first light. Even Odin was associated with the Sun. The tale explains that Odin, in search of wisdom, once went to the well of Mimir (memory) to drink deeply and gain knowledge in the process. The guardian of the well asked one eye as a price for the act. Odin plucked the eye and threw it into the well where it became the Sun.

Presently in Scandinavia, on the eve of the summer solstice, thousands of people flock to the hillsides to light bonfires and to watch the Sun set, following a tradition started in the dawn of time. Though originally a tribute to the Sun, the event has since been assimilated by the Christians and transferred to honor St. John. Another notable, and still living midsummer tradition is the construction of large wheels made of wood or straw which are set on fire and rolled down hills to represent the Sun's journey toward the winter.

1.1.1.2 *England/Ireland/Scotland*

Norse tribes such as the Angles, Saxons, and native people of areas such as the Celts and Picts, invaded and influenced the English-Scottish people. This explains the Irish name for the Sun goddess Grian, a female noun. It indicates a close relationship with the Celts culture and their Indo-European descendents. The Irish concept of the 'solar cross' was prevalent and the 'central mound cosmology' was considered sacred centers known as 'Tara'. They were constructed in such a way that from a central station extended four divisions or provinces. On holidays such as 'Samhain' (meaning Sun's end), to mark the end of summer, large bonfires were lit in these sacred centers, Tara, on the tops of mounds across the countryside. Another Irish deity is the spring goddess, 'Bride' (bright), who has much in common with the Norse's Ostara. A special temple complex in Kildare, originally known as Cill Dara, was dedicated in her honor. In this temple there was a circular building with an eternal flame burning in it, stoked with sacred oak wood. A holiday in her honor on February 2, known as Imbolc is often associated with the fertility of sheep. However the most important aspect of Bride's reign is the New Year's returning Sun. To mark this event, the modern day Catholic nuns admirably absorbed not only the goddess and her shrine, but follow also the old customs. Once a year, followers of St. Bride still go to the spiritual center where they circle a central pillar with a candle, visibly re-enacting the yearly journey of the Sun. There is another Irish Sun goddess, Aine. In her honor there was an annual festival on each summer solstice day. The legend says that Aine had the ability to transform into a horse, perhaps referring to an ancient memory of the 'horse fetter', the *Analemma* of the Sun. Lugh, a Celtic Sun god, was said to be honored each year at the harvest festival of Lughnasad. His temple site gave a name to what is today called London.

In many Irish passage graves, carvings of the Sun's symbol are seen which support the idea that the ancient Irish associated the dead with the Sun. A multitude of other structures, such as megaliths, stone circles, graves and religious sites, seem to be aligned with solar events, for example with solstices and equinoxes. The famous passage grave is at Newgrange. Liamh Greine, or 'The Cave of the Sun', is aligned such that on the winter solstice day a beam of sunlight at dawn illuminates the inside of the structure for approximately 17 minutes. Such associations have given rise to modern day superstitions in Ireland that those carrying the deceased past a graveyard, or sometimes a standing stone, had to circle it '*sun-wise*' (clockwise) two or there times to avert ill. Otherwise a sunbeam falling on someone at a funeral would foretell of his or her death!

In 17th century Scotland there was a very similar concept of tying life with the Sun. When a child was born, a ceremony called 'saining' was done. An attendant would carry a candle *sun-wise* around both the mother and baby. Like most other pagan customs, Christianity later absorbed this and the meaning converted from receiving a blessing by the Sun to warding off the devil. In Gaelic we also find the source of the modern day word used by Wiccans when casting a circle. 'Deosil', which means *'sun-wise'*, meant to walk in the clockwise direction of the Sun.

To this day pagans are still tracing the Sun's path. Owing to its diverse history, not much is found in England of ancient pagan sites, culture, and traditions. The Romans left behind some sites, however, such as the Chanctonbury Ring, called 'Mother Goring' by the locals. Archaeologists believe that this site is actually the remains of a Romano-British temple, and the rituals are re-enactment of the hero's quest around the celestial circle and the final victory of reaching the Sun. In Dorset, Cerne Abbas, a giant-Sun deity is carved on a hillside. Some say it is the Saxon god - Heil, Hayle, or Helis, equivalent to the Greek Helios and the Norse Hel. St. Michael's Mont in Cornwall was originally called Dinsul, meaning, 'mount of the Sun'. The mountain is an island, and legend has it that it is the sole remnant of a lost culture called Lyonesse, which many associate with the Celtic Isle of Avalon, or the Norse Summerland. Cornwall has a large number of such historic sites, including the standing stone called the Men-an-Tol, a large circular stone with a central hole. It retains the tradition that to gain health, one must crawl through the hole towards the Sun! At the stone circle Long Meg and Her Daughters, there is an alignment to the winter solstice sunset, and the site of Castlerigg aligns to both the midsummer sunset and February 1, the ancient Imbolc.

1.1.2 *North America*

1.1.2.1 *Among the Navajo Indians*

Tsohanoai is considered the Sun god for the Navajo Indians of North America. As the story goes he is supposed to have a human form and carries the Sun on his back everyday across the sky. At night, the Sun god rests, hanging on a peg on the west wall of Tsohanoai's house.

Tsohanoai had two children, Nayenezgani (killer of enemies) and Tobadzistsini (child of water). They lived separated from their father in their mother's house in the far west. Once adults, they decided to find their father and seek his help in fighting the evil spirits that were tormenting mankind. After many adventures they met Spider Woman, who told them where they could find the Sun god and provided them two magic feathers to keep them safe. Finally, they arrived at Tsohanoai's house where Tsohanoai gave them magic arrows to overcome Anaye, the evil monsters that devoured men.

1.1.2.2 *Among the Pueblo American Indians*

Among the Pueblo tribe, which is the descendents of Chacoan people, the Sun is depicted as carrying a bow and arrow. The bow-and-arrow and arrows are associated with the Sun in the cosmology of the historic Pueblo peoples. In certain Pueblo traditions the arrow is seen as a vertical axis and may refer to nadir and the zenith, or the world below and above. In a version of a Zuni story, the father, the Sun gives his sons bows and arrows and directs them to lift with an arrow the Sky-father to the zenith. In another story, the Sun directs his sons to use their bows and arrows to open the way to the world below so that the Pueblo people can emerge to the Earth's surface and receive the Sun's light. At the solstices the Pueblos give offerings of miniature bows and arrows to the Sun. The Pueblo people had developed an accurate calendar. It has been described as a synchronization of the monthly lunar cycles with the annual solar cycle.

1.1.2.3 *Among the Anasazi Indians*

Anasazi Indian tribes occupied Chaco Canyon from about 400 to 1300 AD. In this arid region these early inhabitants left evidence of a skilled and highly organized society that displayed interest in astronomy. The famous rock painting of the supernova of 1054 AD, several other petroglyphs, and many solar alignment sites found in Native American Indian regions testify to their astronomical awareness. Details of some of these are given later in this section. Through precise observation of the Sun and the recurrence of solstices and equinoxes, the Anasazi developed an accurate calendar for agricultural and ceremonial purposes. This astronomical knowledge was also commemorated in design and alignment of major buildings of their time.

1.1.3 *South America*

1.1.3.1 *In Aztec Culture*

The Aztec people belong to the most evolved civilized culture of their time in Latin America. They developed astronomy, mathematics, along with a solar calendar, and also they were Sun worshippers. The Aztecs considered Huitzilopochtli as their god of the Sun and of war. He was pictured as a blue man, fully armed with humming bird feathers on his head. His mother was called Coatlicue. It is believed that Aztecs used to offer human sacrifices to propitiate the Sun god, Huitzilopochtli. The victims were usually prisoners captured in the frequent wars that Aztecs fought against their neighbors. The sacrifices were intended to secure rain, harvests and further success in war.

1.1.3.2 *In Mayan Civilization*

More than 2000 years ago the Mayan people lived in the present day Yucatan peninsula of Mexico and Guatemala. They had a very rich scientific and cultural life. Mayan life literally centered on astronomy, mathematics, and the calculation of time and calendars. Mayans based their calendar not only on the Sun and the Moon, but also on rising and setting of the planet Venus. This is mentioned in one of the surviving Mayan books, the Dresden Codex, written more than 1000 years ago. The worship of the Sun figured in their rituals, too. It is believed that many Mayan kings ascended to the throne on May 1, the date when Pleiades and the Sun are in conjunction. The kings were often depicted holding an upright staff, perhaps a sacred gnomon or sundial in their hands, to demonstrate their connection with and understanding of the Sun. On the solstices days in Mayan regions (latitude $\pm$ 23.5^{o}), at noon the Sun does not cast shadow from upright sticks. This was considered to be one of the most sacred events of the year. Many of their ancient Sun rituals were Christianized by Spanish missionaries and are still followed under that guise today. A good example is in the modern religious holiday called the 'Passions'. This festival is celebrated in honor of the Sun's influence on growing corn (Sun beams from gods). Modern Mayans refer to Jesus as the 'Lord Sun', and demonstrate the annual battle between the summer Sun and the darkness of winter. Several Mayan buildings and cities are aligned keeping astronomical phenomena in mind. At the time of winter solstice, a beam of sunlight falls directly into the famed pyramid of Lord Pascal to light up a carved sarcophagus cover. The most famous of all Mayan cities is Chichen Itzá in Yucatán. This was founded over 1000 years ago and the famous pyramid called 'Pyramid of Kukulcan' was built here. Each year at the spring and autumnal equinox, around 4 p.m., the Sun casts a shadow on the stepped structure. As if by magic, this shadow appears to form a slithering serpent which slides down the face of the pyramid to the Earth below. It is believed that this serpent represents the deity, Quezalcoatl, the "feathered serpent", also known as Kukulcan. Sometimes it is assumed to be a male god. Natives of the area have a long held belief in this divine serpent goddess, who has fostered life on Earth and delivered mankind from evils.

1.1.3.3 *Among the Inca in Peru*

Ancestors of the Inca people lived in Peru, South America. Incas were highly sophisticated in mathematics, astronomy and agriculture. The ancient Incas were known to have had a string "computer", called a *quipu*, and had mastered the science of hydrodynamics through their extensive canal building and experience in irrigation. Incas had abundant gold but made no ornamental use of it. They considered gold as 'tears wept by the Sun'. Incas utilized the metal for its sacred connection to the Sun rather than for its monetary value. In this way gold was prized. The sacred Sun temple in Cuzco was literally covered with gold; it is referred to as Coricancha, the 'Place of Gold'. The western wall of the temple contained an idol made of gold and is positioned so as to catch the western sunlight. Nearby this temple, the Inca raised special pillars to serve as sundials. At the time of solstices, when no shadow was cast, they declared that 'god sat with all his light upon the column'.

Incas considered Inti as their Sun god. It was believed that Inti and his wife called Pachamama, the Earth goddess, were benevolent deities. According to an ancient Inca myth, Inti taught his son Manco Capac, and his daughter Mama Ocollo, the arts of civilization, and sent them to the Earth to instruct mankind about what they had learned. Even today Inti is worshiped in Peru during the Festival of Inti Raimi in Cuzco.

1.1.3.4 *Among the Mamaiuran Amazon Indian Tribe*

Mamaiuran, an Amazon Indian tribe that lives along the banks of the Xingu River in Brazil, named their Sun god as *Kuat*. According to a Mamaiuran legend, at the beginning of time it was continuously night and the Indian tribes were forced to live in perpetual fear of attack from wild animals. Light could not reach the Mamaiurans because the wings of birds blocked the sky. Kuat and his brother Iae decided to steal some light from the vulture god, Urubutsin, king of the birds. The two brothers hid themselves in a corpse, and waited until the birds approached. As soon as Urubutsin landed on the corpse to eat the maggots, Kuat grasped the vulture god's legs. Unable to get away and deserted by his followers, Urubutsin was obliged to agree that he would share daylight with the two brothers. To make the light last for long time, it was established that day should alternate with night. As a result, Kuat became associated with the Sun and Iae with the Moon.

1.1.4 ***Egypt and the Middle East***

1.1.4.1 *In Egypt*

In ancient Egypt, *Re or Ra* was known as the Sun god and the creator of our world. He took many forms, each depending on where he was. Usually Re was portrayed with a hawk's head, wearing a fiery disk like the Sun on this head. Surrounding the disk was a cobra-goddess, representing his power to bring death. In the Underworld, the Sun god took the form of a ram-head. In this guise, Re even had power over Osiris, the ruler of the Underworld. It is said that in the beginning of time, an egg rose from the primeval waters and from it emerged the Sun god, Re. Once out of its shell, Re had two children, Shu and Tefnut, who became the air and clouds. They had two more children, Geb and the goddess, Nut, who became the Earth and the stars. They in turn had two sons, Seth and Osiris, the father of Horus. It is said that Re wept one day, which lead to the creation of humans from his tears. He also created the four seasons for the Nile, the heart of Egypt. Re combines with Horus to form Re-Harakhte, god of the Sun and the heavens.

According to an Egyptian legend, Het-Heru, or Hathor, the mother of the Sun god Re, was considered as solar dawn. Each morning she gave birth to the Sun and carried it from the east to west, wearing the Sun disk between her horns. She was depicted in prehistoric Egypt by a cow head on a pole. In the 'Hall of the Cycle of the Gods', a temple dedicated to the precession of the equinoxes, serious rituals took place dedicated to Hathor. Here she reigned as a Sun goddess, while mirrors on each side represented the solstices. Another tale involving Hathor says that, as Re grew old and paranoid, Hathor was sent on a mission to destroy her human enemies. Her rage was so great, that she was only stopped by the reflection at dawn of her own face in a makeshift mirror consisting of a pool of beer.

From a detailed study of ancient Egyptian symbols related to solar eclipses, Robin Edgar supports E. Walter Maunder's theory that the ancient Egyptian "winged solar disk" symbol was in all probability inspired by ancient observations of total solar eclipses. A photograph of the winged Sun is shown in Figure 1.1. It is seen in many ancient Egyptian hieroglyphic inscriptions and on the royal seals and cartouches of Egyptian pharaohs.

Egyptian religious symbols are also found carved on obelisks, capstones of pyramids, alabaster bas-reliefs, and painted onto now fading ancient wall murals. A winged Sun disk symbol graces the stone lintels of the entrances to many temples and palaces in Egypt. Numerous Egyptian pharaohs employed this ubiquitous religious symbol of their Sun gods as a royal sign of their divine status. It seems that the early

Figure 1.1 Egyptian symbol of a 'Winged Sun'.

Egyptians were inspired from observations to have adapted as a symbol the depiction of a solar eclipse, with equatorial streamers of the solar corona stretching out on either side of a 'black Sun'. This image bears a striking resemblance to the outspread wings of a glorious celestial bird. At the same time the plume-like polar rays distinctly resemble the fanned-out tail-feathers of a gigantic cosmic bird.

1.1.4.2 *Middle East*

One of the greatest and most advanced cultures of all time was that of ancient Sumer. This early civilization had scientists, school teachers, universities, pharmacies, and lawyers who were very advanced in natural sciences. Sumer formed the foundation for the Babylonians and Hebrews. In Sumer it was thought that immortals lived on the primordial mountain of heaven, and the Earth was 'the place where the Sun rose' at the dawn of creation. This later turns into the Garden of Eden concept of the Hebrews, with the Adam and Eve story. The Sun and Moon; Utu and Nanna, were considered to be two special deities that saw everything humans did. This was from their lofty positions in the sky and attended their judgment to give witness to each person's deeds. Babylon took this idea farther, saying that the Sun and Moon, Shamash and Sin, are themselves born from the great mother serpent in the sky, called Tiamat, which is now identified as the constellation Draco. Many thousands of years ago this constellation once contained the pole star. The Babylonian king Hammurabi credited Shamash for creating the famed Hammurabi Code of laws. In Phoenicia, the Sun god was also known as *Bel* or *Baal,* and had the combination name of '*Baal-Samin*', the great god, the god of light and the heavens, the creator and the rejuvenator. The root word Samin, for Sun later occurs in the Hebrew tale of Samson, the blind and super strong hero, who pulls down the beams of a mill-house. It is known that this tale is a metaphor for the Sun god, Samin, and the precessional cycle of the Earth's axis by which the 'old' Sun cycle comes to an end, with the movement of the celestial pole. The 12 signs of the zodiac are also represented in Hebrew culture by the 12 tribes of Israel, and by the 12 stones that are worn on priest's chest plates. The stones represent the order of the Sun's progress through the zodiac signs, starting with jasper and ending with amethyst. The two sardonyxes on the shoulders represented the two "eyes" of the gods, the Sun and Moon.

1.1.5 *Greek and Roman Mythology*

It has often been said that Greece was the seat of western civilization. The Greeks may be considered as cousins to both the Indian Hindu and the Norse cultures. All three seem to have been much alike in myth, science and practice, perhaps due to their common Indo-European heritage. In Greek mythology, *Apollo* was considered as the god of the Sun, logic, and reason. He was also a fine musician and healer, a son of Jupiter (in Greek, Zeus) and Leto (Letona). The story goes that Apollo's mother, Leto, traveled all over Greece to find a place to give birth to Apollo. She finally came upon an island named Delos. The island agreed to allow the birth of Apollo if she in turn founded a temple on the island. Leto agreed and when Apollo grew up he changed Delos into a beautiful island.

Very early texts of the Greeks mention a goddess associated with the Sun or dawn. It was said that Circe, was the daughter of the Sun, who lived on the sacred primordial island, guarding a magic cup of the gods. Whoever tasted from this cup lost his upright shape and fell downward into a groveling swine. One Deomocritus regarded this ambrosia as the vapour by which the Sun is nourished. Aurora, the dawn goddess, was said to arise each morning and open the gates of the sky for the Sun god, Apollo. Another Sun deity in Greek mythology is mentioned as a male god, *Helios*, the Sun. Evidence also points, however, to an earlier female Sun goddess under the name of *Helice*, which seems to closely link with the Norse goddess Hel. Two modern words remind us of Helios, whether male or female. The element Helium was named after Helios when it was first discovered on the Sun. And the Christian "halo" comes from the Greek word "halos", or "helos", meaning the circular disk of the Sun or Moon, often depicted over the heads of celestial deities.

1.1.6 *In Asia*

1.1.6.1 *In India*

Sun worshipping existed in India from the most ancient times down to the present day. In the Vedic period (about 3000 BC) the Sun, known in Sanskrit as *Surya*, was worshipped under various names, of which the chief were *Surya* and *Savitra* or *Savita*. The other Sun deities are *Mitra, Pushan* and *Vishnu*. The Sun god, Surya, is described as far seeing, all seeing, a spy of the whole world, he who beholds all beings, and the good and bad deeds of mortals. He is considered as the preserver and the soul of all things, both stationary and moving. Enlivened by him, men pursue and perform their work. The god Surya shines for the entire world, for all men and all gods. He dispels the darkness with his light. He rolls up the darkness as a skin. His beams throw off the darkness as a skin onto water. Even today a religious Hindu is suppose to recite the *"Gayatri mantra"* (prayer), every morning in praise of the Sun god and offer water to it. The Gayatri mantra literally means that, "May we receive the glorious brightness of this (Surya), the creator, the God, who shall prosper our works". Several Indian communities believe that they are the descendents of the Sun god - Surya.

In a popular version, Surya is considered as an anthropomorphic figure, a the son of *Dyaus*, the wide spreading sky, and is described as 'all creating' and 'all seeing'. In this aspect his most ancient and significant name is *Prajapati*, 'the lord of Creation'. He traverses the heavens in his golden chariot drawn by seven horses, and *Usha*, or the dawn, is the charioteer, with *Asvin*, the twin gods of the morning, his children. By his power, he drives away the demons of sickness and expels diseases and all the subtle and dreaded influence of darkness.

There are several ancient temples in India dedicated to the Sun god, Surya. The best known are at Konark (Orrisa), Gaya (Bihar), Varansi (Uttar Pradesh), Modhera (Gujarat), and Srinagar (Kashmir). Some of these temples are aligned to solar phenomena, such as solstices and equinoxes, and also to Sunrise or at mid day. On certain astronomically important days, the sanctum sanctorum of the temple is illuminated by a sunbeam. In India today, the winter solstice called the *'Uttrayan'*, when the Sun starts its northward journey in the sky, is considered a very auspicious celestial event and is celebrated throughout the country.

1.1.6.2 *In China, Japan and Korea*

Although no records could be found in olden Chinese and Korean literature, wherein the early people of these countries considered the Sun as a god, as in the case of other ancient civilizations of the world, Japan is an exception. According to an ancient belief, the Japanese royal family is descended from the Sun goddess, *Amaterasu*. Even today the Japanese maintain a most sacred Shinto shrine to the Sun goddess at Ise, a city with a 2000 year history. Even the Japanese national flag represents the disk of the Sun. Ancient Chinese were known to have well developed astronomical (and other) sciences. They knew how to calculate the circumstances of solar eclipses, and could predict their occurrence, as indicated by the famous apocryphal story of the two Chinese court astronomers, Hsi and Ho. It is said that, being too drunk, they did not predict the occurrence of the solar eclipse of October 22, 2134 BC, and hence the Emperor beheaded them. To what extent this story is correct is not known, but the idea it brings to mind is that the early Chinese astronomers were well verse in observations and the calculation of eclipses. There is another side of this story that Hsi-Ho was perhaps the name of the Sun deity, who had the responsibility of preventing eclipses. Thus the ancient Chinese also had a Sun deity. There is an interesting story in Chinese folklore about the Sun. That in the very early days, there were nine suns in the sky, due to which the Earth became terribly hot, and the farmers were extremely miserable as their plants in the fields started dying. The hero Hou-Yi, a very brave and strong man, shot off eight suns by his arrows, leaving only one Sun in the sky. Thus the Earth became safe for life and the farmers got rich harvests.

1.2 Major Ancient Solar Observing Sites

As man evolved through time, he started looking at the Sun and its movement across the sky during the day, weeks, months and year, and discovered that the movement and position of the Sun is related to many phenomena around him. He watched the Sun rise daily in the east and set in the west, he noticed that the Sun does not rise or set at the same place in the horizon, but seems to shift its position from day to day, during the year. Perhaps he also discovered that its position in the sky repeats after about 365 days. Sometimes the Sun appeared quite high towards the north side and sometimes in the south. These positions of the Sun coincided with the seasons. When the Sun was towards the north, it was summer (in the northern hemisphere), quite warm even hot, but when it was towards the south side it was winter, quite cold. As man developed agriculture, it required tiling of fields, sowing and harvesting of crops etc., and he soon realized that the Sun has a profound influence on agriculture and on his daily needs. Thus the early man put the Sun at a pedestal and considered it as his god or goddess and coined folklore and stories to explain many unexplained phenomena. As the Sun was so important for him, he started making observations of the Sun so that he could keep a watch on its movement during the day, during the year, and also help him to make predictions on its position and solar events that may occur in time. For this purpose, he either constructed equipment suitable at that time, or used the natural configurations of rocks, buildings etc., to keep track of the Sun. Let us now take a look at ancient observatories or observing sites built and used by various civilizations around the world, beginning from the early Neolithic period to almost the eighteenth century AD.

1.2.1 *In Europe*

1.2.1.1 *The Stonehenge*

Perhaps the earliest observatory or site for observing the Sun was built around 2950 – 2900 BC, in the Middle Neolithic period on the Salisbury Plain in southern England. It is now known as Stonehenge. In the 1940s and 1950s, Richard Atkinson indicated that the Stonehenge was built over period of many centuries and had three distinct phases of development. The first and the oldest phase was Stonehenge I, dating back to 2950 – 2900 BC, then followed the Stonehenge II period dating from 2900 to 2400 BC, and the third was the Stonehenge III phase from 2550 to 1600 BC.

The earliest portion of the complex was built during Phase I. It consists of a circular bank, or ditch, and a counter-scarp bank of about 100 meters in diameter. Just inside the earthen bank is a circle of 56 'Aubrey' holes.

After 2900 BC and for the next 500 years and until 2400 BC, during the Phase II, the Aubrey post-holes were perhaps used as indicated by the timber settings in the centre of the monument and at the north-eastern entrance. However, the Aubrey holes no longer hold posts and are partially filled. The numerous post-holes around the monument indicate that timber was used for the structures, but no clear patterns or configurations are discernible that would suggest their shape or form. Perhaps these were used for sighting celestial objects.

During the Phase III period, from 2550 to 1600 BC, the monument underwent a complicated sequence of settings with large stones. The first stone setting was comprised of a series of Bluestones placed in what are known as the Q and R holes. This originally had set of 30 stones but now has only 17. These are neatly trimmed upright sets of massive sand-stones blocks, each weighing more than 25 tons. These stones form a circle of more than 33 m in diameter and 4 m in height. They form two horseshoe shaped patterns. Some of the pairs of stones have massive stone lintels, raised four meters above the ground, as shown in Figure 1.2.

Considering that many stones were brought from Marlborough Downs, some 32 km from the present site near the city of Salisbury, enormous work must have gone in building such huge structures. *The question is for what purpose?*

Figure 1.2 (Left) - Stonehenge, showing massive stones arranged in a semi- circular horseshoe shape with cross member lintels. (Right) - Sunrise seen over Heelstone from center of the circle.

Around 1771 AD, it was realized that Stonehenge, in the Neolithic (New Stone Age) period, was used to observe the Sun and to mark the time and day of the summer solstice (at present on June 21/22, when the Sun is in its northern-most position in the sky). When seen from the center of the Stonehenge circle, the Sun rose directly at a particular stone called the 'Heelstone', which is approximately 16 feet high (4.88 m) with another 4 feet (1.22 m) buried below the ground. One of its most misunderstood aspects, however, concerns this Heelstone. For decades it has been debated why it was so named. It is believed that the 'Heel' stone is a corruption of Welsh 'hayil', or Norse Hel, both of which mean Sun. Likewise in the English town of Helston, a stone once stood called the Hel Stone, though it has long since been removed. Alignment of the Sun with less prominent pairs of stones perhaps referred to the sunrise at other significant times of the year, such as the equinoxes, which fall 6 months before and after solstices.

Sir Norman Lockyer got interested in Stonehenge in the 1890s. He worked on the presumption that the midsummer Sun rose originally over the Heelstone at the time of its construction. He calculated back from the point where the Sun now rose at midsummer dawn (in 1901) precisely over the Heelstone, and thereby established the date Stonehenge might have been built. This turned out to around 1680 BC. However, Lockyer's calculations were flawed, because there were considerable error in his sightings and he used the wrong tables. His results are now usually dismissed. Later in 1950, Gerald Hawkins and Sir Fred Hoyle studied Stonehenge in great detail and proposed that besides being used as an observing site for astronomical sighting for solstices, it was also used to predict the solar and lunar eclipses. From these inferences it seems that the Neolithic people had a good knowledge of astronomy, including the movement of the Sun during the year and also of a calendar, which was required for the timing of agricultural, social and religious activities.

1.2.1.2 *In Ireland at Newgrange*

In Europe, indications of early solar observations came from the burial tombs and similar structures. In Ireland, there is a Megalithic passage tomb at Newgrange. This dates back to about 3200 BC. A 19-m long inner passage leads to a cruciform chamber with a corbelled roof that is surrounded by 97 kerb stones arranged in a circle. The direction of the entrance to the tomb is such that the passage and the chamber at Newgrange are illuminated by the winter solstice sunrise. At dawn, on solstice just after 9 am the Sun begins to rise across the Boyne Valley from Newgrange over a hill known as Red Mountain. For the following seventeen minutes, between 19 and 23 December, the sunbeam stretches into the narrow passage of Newgrange tomb and on into the central chamber. In Neolithic times it illuminated the rear stone of the central recess of the chamber.

With simple stone technology, the Neolithic people captured a very significant astronomical and calendric moment. This tomb at Newgrange was precisely built so that at the time of winter solstice at sunrise, the first Sun's rays would strike the burial chamber at the end of the tomb (Figure 1.3). The timing and location of solstices and other astronomical events were important for the early Irish people for their day-to-day needs in agriculture, calendar, social and religious activities. Such burial chambers have been also found elsewhere in Ireland.

Figure 1.3 (Left) - Light beam shining through the tunnel. (Right) - Entrance to the Newgrange passage tomb.

1.2.1.3 *In Ancient Germany*

In the December 2003 issue of the Scientific American magazine, Madhusree Mukerjee has shown that a vast shadowy circle of 75- meter wide has been also found in a flat field near Goseck, Germany. He suggests that this circle represents the remains of perhaps the world's oldest observatory, dating back 7,000 years. From an etched disk recovered at the site, archaeologists reason that the observatory was used by Neolithic and Bronze age people to measure the heavens. Originally it consisted of four concentric circles, a mound, a ditch and two wooden palisades. In the middle stood three sets of gates facing Southeast, Southwest and North. On the winter solstice day, someone at the center of the circles would see the Sun rise and set through the southern gates. Aerial surveys have identified 200-odd such circles scattered across Europe, but the Goseck structure is the oldest and best preserved of the 20 excavated thus far. This is now called the German Stonehenge; it precedes Stonehenge by at least two millennia.

1.2.2 *In Ancient Egypt*

According to Egyptian mythology, the sky goddess Nut supported (Figure 1.4) the vault of heavens, and the Sun god performed its daily journey across the sky in a chariot. The Sun spent the night hours going through the underworld from west to east.

As we have said, the Egyptians considered the Sun as one of their gods. Several temples dedicated to the Sun god Re and god Thebes had been built within the modern city of Luxor at Karnak. The main axis of the temple at Karnak of Ammon, Re, is aligned within 3 arc minutes of east-west direction. Even the sides of the some of the pyramids of Gizeb, dating back to 2000 BC, run east-west. Facing east the guarding statue of Sphinx receives the first of the Sun's rays on the vernal equinox.

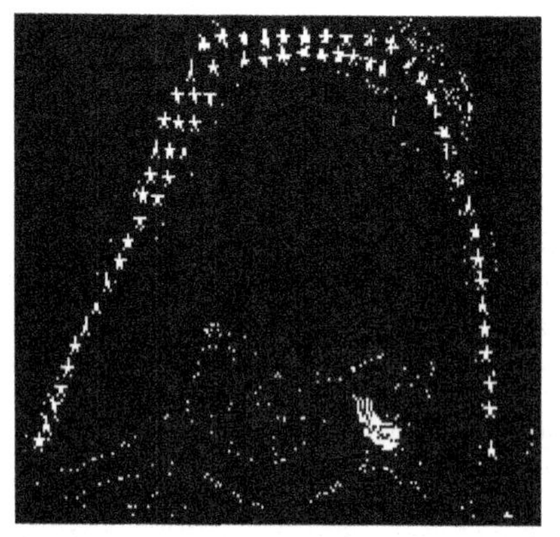

Figure 1.4 Goddess-Nut supporting the vault of heavens and the Sun performing its daily journey across the sky.

Figure 1.5 Great Sun temple of Abu Simbal.

Similarly, there are several other Sun temples in Egypt, like the great temple of Abu Simbal (Figure1.5). It stands as the world's greatest Sun sculpture. Twice a year in this temple, on the equinox days in March and September, the first rays of the Sun illuminates the inner sanctuary, where it lights up a statue of Pharaoh Ramses II, flanked by two Sun-gods. In the Edfu temple, sunlight comes through a carefully executed opening in the ceiling to illuminate pictures of each of the 12 hours on the wall.

Recently, McKim Malville et al., (1998) have shown that during the Megalithic and Neolithic periods (about 3000 BC) astronomy had flourished to great heights in the Nabta Playa region of the southern

Egypt. This is evidenced by the megalithic stone alignments to cardinal and solstice directions. From these findings, it appears that the Stonehenge and many similar Neolithic solar observing sites in Europe were not the only astronomical sites, but that early man had also built such structures in other far off places. Another early site in Egypt has alignments of stone in a circle dating back to perhaps 7000 years ago. Thus Egypt is perhaps the oldest site in the world to have made astronomical observations of the Sun and other celestial events.

1.2.3 *In Ancient Babylon*

The Babylonians surpassed Egyptians in their astronomical knowledge. They confined their observations to the Moon, instead of the Sun. Mesopotamia, now Iraq, in 3000 BC was a great civilization for astronomical studies. They had built observatories, or watch-towers, called *Ziggurats*. The tower of Babel is the best known example. Babylonian astronomer-cum-priests made observations of the planets, Moon and Sun. They kept astronomical diaries by noting down the positions of heavenly bodies using cuneiform writing on soft tablets of clay that were later baked. From these tablets they were able to predict the future positions of these bodies. They knew that solar eclipses occur in cycles, one of these lasts for 135 months, during which there were 23 'dangerous' periods when eclipses were likely to occur. Babylonian astronomer-priests were also aware of the *Metonic* cycle of 19 years, named after Greek Meton of Athens.

1.2.4 *In the Early Americas*

1.2.4.1 *Solar Astronomy among Native American Indians*

As in any other ancient civilization, the native American Indians were not behind in their pursuit of astronomical observations. The well known petroglyph (rock engraving/paintings) by the Native American Indian tribes shows the crescent Moon and a star shape to its left. This petroglyph has been interpreted as a depiction of the Crab nebula explosion in late June 1054, which reached its maximum brightness on 4-5 July 1054 AD. This petroglyph is located on an east facing cliff and about five hundred metres northeast of the ruins of Peñnasco Blanco in Chaco Canyon. Calculations of the Moon's orbit back to 5 July 1054 have shown that the Moon was waning, just entering the fourth quarter. These calculations also indicate that at dawn on 5 July 1054 in the American southwest, the Moon was within 3 degrees of the supernova, and its crescent was oriented as seen on the pictograph.

Another petroglyph, on the South side of a large boulder near the Una Vida ruins in Chaco Canyon, shows a solid round disk surrounded by elongated features of about the length of the circular disk's diameter and distributed all around it (Figure 1.6). This has been interpreted as a schematic depiction of the solar corona seen at the time of a total solar eclipse. Actually, there were four total solar eclipses visible in the San Juan basin between 700 and 1300 A.D; one on 13 April 804, 2nd on 11 July 1097, 3rd on 13 June 1257, and 4th on 17 October 1259. Scientists believe that this petroglyph may refer to observations of the total solar eclipse of perhaps 11 July 1097. This petroglyph is a reasonably good depiction of the solar corona as seen close to the solar maximum period when the 'helmet' streamers are found at all heliocentric latitudes.

A boulder seen at this site appears to have been used as an ancient solar observing station for anticipating the coming of the summer solstice. This and many other alignments of rocks, windows and buildings for sighting and predicting solstices and equinoxes were used and built by the early American Indian tribes. Among the famous ones is the 'Sun Dagger' (Figure 1.7) on the top of Fajada Butte also in Chaco Canyon. It was discovered by Anna Sofaer *et al.*, (1979) in 1977, through an exercise called 'Solstice project'. It is called 'Sun Dagger' because the sunbeam entering the cave through a set of three rock slabs placed accidentally or intentionally positioned in just the right direction, appears like a dagger. These mark the summer and winter solstices, the Vernal (spring) and Autumnal (fall) equinoxes, and helped to make a calendar.

The 'Sun Dagger' was probably created or conceived by the Anasazi Indian community in the12th century. There were a pair of patterns, one with 9 and half spirals and another nearby smaller one with 2 and half spirals. These are carved on a flat vertical rock wall, oriented north-south and facing east on top of a bluff near Pueblo Bonito. Leaning against the rock wall, are 3 large stone slabs, which may have been moved there or fortuitously available at the site.

Figure 1.6 Petroglyph showing round circle depicting the Sun and outer lines are coronal streamers, perhaps this refers to the total solar eclipse of 11 July 1097 as seen by Native American Indians.

The dagger-shaped bright light pattern is formed by the Sun's rays, passing through the openings between the stone slabs, and descends vertically through the center of the large engraved spiral at the time close to the summer solstice midday, when the Sun is at its highest point in the sky. On the two equinoxes days, when the sun's altitude is lower, the vertical path of the light dagger shifts well right to the center of the large spiral. A second, smaller light pattern passes through the center of the smaller spiral. On the winter solstice, when the Sun is at the lowest possible midday altitude, the two daggers shift to the right to "frame" the large spiral, as shown in Figure 1.8 and in Figure 1.9.

A number of major buildings in Native American settlements are oriented to the cardinal directions, and also some unusual doorways

Figure 1.7 Showing the 'Sun dagger' as a bright beam of sunlight shining vertically down on a rock which has carvings of two spirals.

for observations to anticipate important celestial events likes solstices and equinoxes. Similar spiral clock-calendars are found as petroglyphs elsewhere in New Mexico and Arizona.

Since discovering the 'Sun dagger', Sofaer, Sinclair and the Solstice Project research team have documented numerous other solar markings on Fajada Butte. In addition, the Solstice Project's survey of the large Chaco buildings revealed that they are all oriented to the Sun and Moon. A possible Sun-watching station has been identified at the South end of Cliff Palace. Looking from this location the south-western horizon is featureless, except for the *Sun Temple* standing some 300 meters away on the mesa top across Cliff Canyon. At winter solstice, seen from the observing station, the setting Sun touches the horizon between the Sun Temple's two main towers. It has also been suggested that the smallest tower in the Temple's West end might have served as the horizon marker to *anticipate* the winter solstice by some 20 days.

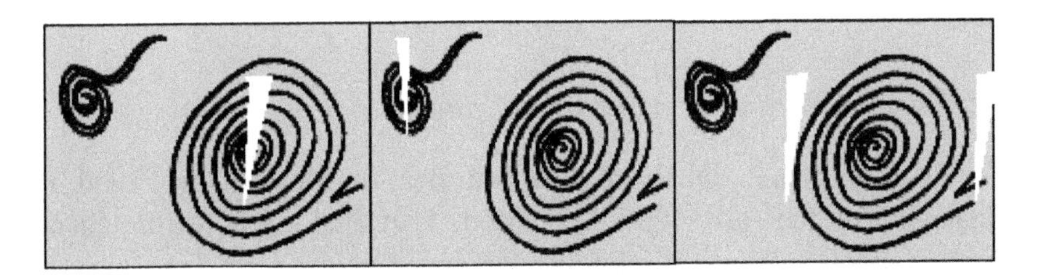

Figure 1.8 Showing drawings of the two spirals and position of 'Sun dragger' on summer solstice, equinox and winter solstice.

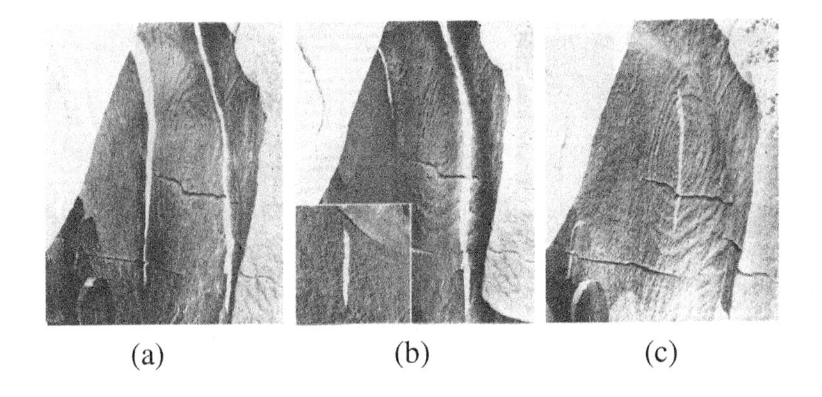

(a) (b) (c)

Figure 1.9 Actual photographs of the petroglyph, displaying position of the Sun dagger taken on (a) at near midday summer solstice, on 26 June 1978, at 11:13:15 a.m., (b) at equinoxes midday on 21 September 1978, 10:50:50 a.m., the inset shows the bisection of the smaller spiral by the left light formation and (c) near winter solstice midday on 22 December 1978, 10:19 a. m.

1.2.4.2 *Solar Astronomy among Aztecs*

Most of the astronomical beliefs of the Aztec community of central Mexico are known from the existing literature, in particular from a work written at the time of the Spanish conquest called *Codex Mendoza*. There are many Aztec monuments that prove there Aztec people made observations of the movement of the Sun in the sky, especially in coordination with Venus. Pillar doors and windows in their stone monuments are clearly seen aligned at sunrise and sunset times on solstices.

1.2.4.3 *Solar Astronomy in Maya Civilization*

The Mayan people inhabited Chichen Itzá in Mexico and Tikal in Guatemala in the early 1000 AD period. They had a rich astronomical, mathematical and scientific knowledge and developed a calendar based on the Sun and Moon. They also used the helical rising and setting of the planet Venus. Many features of their cities and buildings seem to be aligned to astronomical directions and events.

1.2.5 *In Far East – Asia*

1.2.5.1 *In Ancient China*

Astronomy was especially important in the spiritual and academic life of ancient China. There it was truly a science. In fact, mankind's first record of an eclipse of the Sun was made in China in 2136 BC. Early Chinese astronomers made very systematic naked eye observations of the Sun. Sunspot records date back to 28 BC and a few from even earlier times. Rulers in China encouraged court astronomers to make such observations of heavenly objects and to keep precise records. The Chinese observation of the Crab supernova in 1054 AD is a brilliant example. Emperor Zhengtong, a Ming dynasty ruler from 1436 to 1449 AD, had built the ancient Beijing observatory at the south corner of the old city wall. A 46-foot high platform holds 8 Ming dynasty bronze astronomical instruments, two were built in 1439 and another six in 1673. Astronomical alignments figured even in the dwellings of the Chinese emperors. They built a nine square plan of the 'Hall of Light', with four square walls around it to mark the four seasons of the year. The alignment of the walls of the hall was along the cardinal east-west and north-south directions. The observation and location of the Sun played an important role in early Chinese culture. Chinese court astronomers knew about the precession of the earth's axis and also the position and dates of solstices and equinoxes.

1.2.5.2 *In Early Japan*

In Japan, the Asuka region south of Nara is one of the most historically and culturally rich regions. Reliable historical evidence indicates that as early as the mid-6th century exchanges between the Chinese scholars and the more aristocratic members of the Japanese court occurred in this region. These exchanges not only gave rise to the infusion of technology, religion, and other aspects of Chinese culture, but also helped astronomical calendar reckoning and astrology. From the archeo-astronomy perspective, one of the most interesting of these is the tomb at Takamatsu Zuka Kofun. While the exact date of construction is unknown, this tomb discovered in 1972, provides one of the earliest and most definitive examples of Chinese and Korean astronomical influence on Japan in the 7th century. The two particular stones found in this region and worth mentioning are the *Sakafune Ishi* (literally meaning 'liquor ship rock') and the *Masuda Iwafune* (Masuda was the name of a lake, thought to have been near this stone, hence the translation would literally mean "Masuda ship rock"). These might have been used for alignments with solstices or the cardinal directions. In the 1980 April/June issue of ***Archeoastronomy***, Kunitomo Sakurai claims that the central 'trough' of *Sakafune Ishi* is "well aligned along the true east-west direction, which coincides with the Sun's path at the vernal and autumnal equinoxes." Through triangulation and sightings along the 'troughs' toward distant mountain passes to the west, Sakurai concluded that the primary function of *Sakafune Ishi* was that of a sunset observing station for determining both winter and summer solstices. Sakurai also mentions that Emperor Temmu in 675 AD built an astronomical observing platform there.

1.2.5.3 *In Early Korea*

The golden age of science and astronomy in Korea was during the reign of King Sejong (1412–1450 AD), the fourth monarch of the Choson dynasty. Perhaps the most noteworthy achievements of this period were the invention of many ingenious instruments for astronomy and horology, as described in *The Hall of Heavenly Records*, compiled by Joseph Needham and other scholars (Cambridge University Press, 1986). During his reign King Sejong also built a Royal observatory in the main palace of Seoul. He arranged a series of astronomical and horological devices around the Kyonghoeru Pond in Kyongbok Palace. These included a simplified armillary sphere, a self-striking clock, a "jade clock", and a 40-foot high bronze gnomon to measure the exact altitude of the Sun. At least four kinds of sundials were invented under King Sejong's reign. The most distinguished is a sundial, shaped like a bowl. None of the original sundials have survived. The peak of astronomical and calendarial advances made during this period was the compilation in 1442 AD of a Korean version of the traditional calendar, called Ch'ilchongsan (on the calculations of the Luminaries). This work made it possible for scientists to calculate and accurately predict major heavenly phenomena, such as solar eclipses and other stellar movements. No mention has been found in the Korean literature of sighting 'stone' structures, like the one at Stonehenge, or the temples in Egypt and or in India, or the *Sakafune Ishi* in Japan.

1.2.6 *In Ancient India*

The Indian contributions to astronomy and mathematics date back to the *Vedic period* that is before 1500 BC. Descriptions of planetary motion, Sun, Moon, equinoxes, solstices and calendar are found in many old Indian treatises available from the *Vedanga Jyotish* period 1500 BC to 500 BC, from the *Jain Puranic period* 500 BC to 400 AD, and from the *Siddhantic period* of 400 AD to 1900 AD. Although the knowledge and importance of equinoxes and solstices can be traced back to the earliest periods in Indian astronomy, no records of actual observations of the Sun, or sunspots, or solar eclipses is found. In India several temples were built to propitiate *Surya*, the Sun god.

For example, the Modhera Sun temple (Figure 1.10) in Gujarat was built around 1026 AD, at the latitude of 23°.6. Another one; the famous Konark Sun temple is on the eastern coast near Bhubneshwar in Orrisa.

Figure 1.10 Temple dedicated to the Sun god Surya was built in 1086 AD, at Modhera, Gujarat, India to propitiate the Sun god and monitor the movement of the Sun.

The orientation of the Modhera Sun temple is such that the first rays of the rising Sun illuminates directly an idol on equinox days; on the summer solstice day, the Sun shines directly overhead at noon, casting no shadow. These temples were constructed keeping in mind the movement of the Sun during the year and for timing of the passage of solstices and equinoxes for religious, agriculture and calendar purposes.

1.2.7 *Solar Astronomy in the Medieval Period*

1.2.7.1 *Solar Observatories at Maraga, Iran and Samarkand*

In the medieval period in Europe, Middle Eastern and Asian countries, solar astronomy was a part of the larger discipline to study the Sun, stars, Moon and the planets. During this time several astronomers had built observatories for making precise astronomical observations to record the position of celestial bodies. Nasir-ul-din al tusi, a Persian astronomer, built an observatory at Maraga in 1259 AD and published his laborious work in the form of tables, cataloguing the position of stars and the

Moon. This is known as '*Ilkhamic tables*'. Ulugbek, the grandson of *Timur the lame*, king of Persia, devoted at Samarkand in about 1425 AD. From his numerous observations of the Moon, planets and 1018 stars made with amazing precision, he compiled a set of useful tables which

Figure 1.11 Ulugbek's Observatory in Samarkand, a sextant built in 1425 AD.

superseded those of Ptolemy's tables known as 'Syntaxes'. Its Arabic translation is known, as '*Al Mayista*'. These tables were the principal source for Arabic astronomy for centuries. It reappeared later under the name of '*The Almagest*', which means in Arabic – *greatest*. At present the only structure left at Ulugbek's observatory in Samarkand, is the underground giant marble sextant, as shown in Figure 1.11.

1.2.7.2 *Solar Observatories in India*

Around the beginning of the 18th century, Sawai Raja Jai Singh II,

Maharaja of Jaipur built several observatories. He was well versed in the Indian astronomy and the astronomical treatise known as the '*Surya Sidhanta*' (Solar principle), composed by the famous Indian astronomer Aryabhatta in the 4th century AD. Jai Singh II was also aware of the Arabic astronomy of that period and was much impressed by Nasir-ul-din's observatory built in the thirteen century, and by Ulugbek's observatory built in the fifteenth century at Samarkand. Based on Ulugbek's astronomical instruments, Jai Singh II built much larger and massive masonry instruments at five places in India. Their purpose was to measure precisely the position of stars, planets, Sun, Moon and the zodiacal signs. Although Jai Singh II also extensively used a small metallic Astro-lab, but he was convinced that the brass instruments being small in size and the divisions marked on them being very small they could never give high accuracy. So he built huge instruments of stable masonry structures. The first Jai Singh's major observatory was built in Delhi in 1724 AD, and another one in Jaipur in 1734 AD. Three more observatories were built at Ujjain, Benaras (now known as Varansi), and Mathura. The last one does not exist now. For stellar, planetary and solar observations, he constructed 12 or 13 astronomical instruments at each of these five observatories. For solar observations there are four main instruments; *Samrat yantra* (instruments), *Jaiprakash yantra*, *Ram yantra* and *Shasthanisa yantra*. *Samrat yantra* is shown in Figure 1.12. It is essentially a huge equinoctial sundial with a 90-foot high gnomon to determine the local solar time, to an accuracy of about 1 second. The *Jaiprakash yantra* is the most versatile instrument. It is a hemispherical bowl of 27 feet in diameter at the Delhi observatory, and 24 feet at Jaipur. This instrument was used to locate the particular zodiac sign in which the Sun appears at the moment of observation. *Jaiprakash yantra* can be said to be an elaborate version of the "bowl of Berossus," the Babylonian who flourished in about the 3rd century BC. The *Ram yantra* at Delhi has a diameter of 55 feet and is 11 feet high. It was used to determine the azimuth of the Sun and stars. *Shasthanisa Yantra* or the sextant instrument is a huge concave arc of 60 degrees and 28 feet 4 inches of radius, lying in the meridian. There are 2 pairs of graduated arches built into the masonry that supports the east and west end of *Samrat* quadrant. Small holes in roof of each structure allow the sunlight to fall on the graduated arcs at local noon, giving an image of the Sun of about 75 mm in diameter, acting like a pin hole camera. Large sunspots can be seen on the solar image, but there are no past records of sunspot sightings, if made from this instrument. The records of the altitude and declination of the Sun on each day were maintained. This instrument gives the altitude of the Sun at local noon, or the declination of the Sun at that moment. The instrument is capable of giving accurate results of the altitude, but the readings for transit time are said to be an error to about 4 minutes.

Figure 1.12 Samrat yantra, the largest sundial in the world, at Jaipur Observatory.

These massive 18th century masonry astronomical instruments were regularly used until the early twentieth century, but are not used now. For more details readers are referred to the monograph by M F. Soonawala on *Maharaja Sawai Raja Jai Singh II of Jaipur and his observations,* Published by the Jaipur Astronomical Society, Jaipur, 1952 and to *'The Astronomical Observatories of Jai Singh'* by G. R. Kaye's classic treatise published by the Government Press, Calcutta in 1918.

Chapter 2

Modern Solar Observatories

2.0 Introduction

During the last four decades, solar astronomy has seen enormous progress on both the experimental and theoretical side. All over the world new and the state-of-the-art solar observing facilities have been created and every day new concepts and ideas are coming up to improve the spatial resolution and spectral coverage. In this Chapter we give brief description of instruments and scientific programs being carried out at some of the major and typical solar observatories around the globe. The interest in solar ultraviolet, X-rays and gamma rays is steadily growing, since the early days of U-2 rocket exploration of the solar ultraviolet spectrum in late forties. Exploration of the Sun from space-based platforms has enormously increased since the days of Skylab mission in early seventies. There are several new space initiatives in the pipeline. We shall briefly give details of the some of the space based solar experiments also.

2.1 Ground-based Solar Optical Facilities

The earth's atmosphere allows us to see the electromagnetic radiation in visible and infrared spectrum from about 3000Å to about 22 microns and in radio wavelengths. Ground-based telescopes are confined to these spectral bands. Of course the neutrinos generated in the Sun's center penetrate our atmosphere, but to detect them special techniques are used. The main objectives of all the ground-based optical solar facilities had been to achieve the highest feasible angular resolution of solar features and cover wide spectral bandwidths. To achieve this, astronomers take great trouble in selecting proper sites where the 'solar seeing' is best over a large fraction of the time. Numerous site surveys have been and are being conducted to locate the very best site for solar optical and infrared observations. New sophisticated techniques to minimize or completely remove the effect of 'seeing' have been recently developed and remarkable results are being obtained to improve the image quality, through the use of image restoration techniques.

2.1.1 *Optical Solar Observatories in North and South America*

1. Solar Tower telescopes at Mount Wilson, California

At present two telescopes are operative; the 60-foot and the 150-foot tower telescopes. Both of these telescopes were designed and built by George Hale in the early 1900s. These telescope systems have a long standing history and are known for many pioneering discoveries in solar physics. Here we shall not go into these details. The 60-foot solar tower is now operated by the University of Southern California (USC). It is a part of a world-wide network for helioseismology studies using velocity field measurements made in the sodium D line through a Magneto-Optical filter. Daily full disk white light photographs are also regularly taken with this telescope to maintain the long series of images started here almost 100 years ago. The 150-foot tower telescope is now operated by the University of California, Los Angles (UCLA), see Figure 6.2a. Continuing a long tradition, daily sunspot drawings and visual measurements of sunspot magnetic fields are made. Consistent data from such sunspot drawings and magnetic field measures are available for nearly 100 years, and these are of enormous importance for long term synoptic studies. Daily full disk longitudinal magnetic and velocity field observations are also taken using the magnetograph in magnetic and Doppler modes. The velocity data have been central to solar rotation studies.

2. The Big Bear Solar Observatory (BBSO)

This observatory is located on the North shore of the Big Bear

Lake, California at an altitude of 2000 metres, originally built by the California Institute of Technology in 1969, under the guidance of Professor Harold Zirin, and it is perhaps the first serious effort to harness the good solar 'seeing' known to be available from a lake site. Since 1997, it is being operated by the New Jersey Institute of Technology. Several telescopes ranging from a 65-cm aperture vacuum telescope to 25-cm telescopes are mounted on a single equatorial mount. Being located in a superb site in the middle of a lake, sub-arc second solar observations are frequently made. The observatory's main scientific objectives are to obtain high resolution chromospheric, photospheric and vector magnetic field observations in several wavelengths. There are also several other instruments for general solar astronomy research. In Figure 2.1 is shown a general view of BBSO.

Figure 2.1 Big Bear Solar Observatory in the lake.

3. The San Fernando Solar Observatory (SFO)

This observatory was established by the Aerospace Corporation in 1969-70 near an artificial water reservoir. The main objective is to obtain high spatial resolution spectroheliograms and longitudinal field magnetograms with a 65-cm vacuum telescope. At present this observatory is operated by the Department of Physics and Astronomy of

Figure 2.2 General view of the San Fernando Solar Observatory.

the California State University at Northridge. In Figure 2.2 we see the unique retractable 4-petal dome high up on a tower, housing the vacuum telescope and the spectroheliograph.

4. John Wilcox Solar Observatory

This observatory was established in 1975 near the Stanford University campus. The solar telescopes and associated equipment were designed for taking low-angular resolution synoptic magnetic and velocity field observations and to measure the magnetic field of the Sun seen as a star. In Figure 2.3 is shown the exterior view of this observatory.

5. Helio Research Solar Observatory, California

This is a unique private 'one person' effort by Sara Martin to establish a simple observing facility for taking chromospheric observations of solar mass ejections, filament eruptions, and other solar activities. A 25-cm refracting objective is mounted on a peculiar equatorial mount, followed by a narrow band Lithium Niobate ($LiNibO_3$) etalon filter and a CCD camera. Valuable scientific research data on solar mass ejections are being collected to understand the mechanism of filament eruption, CMEs (Coronal Mass Ejections) and solar geomagnetic effects. In Figure 2.4 is shown the equatorial solar spar telescope of Helio Research observatory, designed and built by Dong Martin.

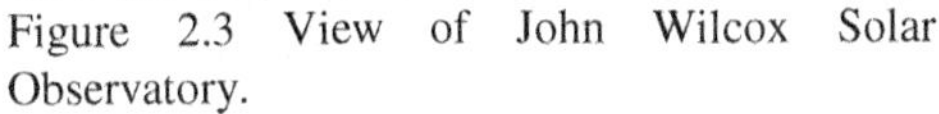

Figure 2.3 View of John Wilcox Solar Observatory.

Figure 2.4 Solar spar of the Helio Research Observatory.

6. National Solar Observatory at Kitt Peak, Arizona.

This observatory was established in 1962-63 by Robert R. McMath for the purpose of high quality solar observations. Providing a huge photon flux from its 1.5-m aperture objective, this all reflecting telescope is the largest solar telescope in the world. A schematic of the optical layout of this telescope is shown in Figure 6.4. Due care was taken to optimize seeing by cooling the exterior of the telescope enclosure through circulating liquids. Together with the main 1.5-m telescope there are two additional east and west auxiliary telescopes of 60-cm aperture each, and a 13.5-m general purpose spectrograph. This facility is now known as McMath-Pierce Solar Telescope in recognition of Keith Pierce's role in the design. Its light gathering power and extended wavelength coverage from near UV to far IR, together with the high resolution spectrograph and a 1-m Fourier Transform Spectrometer, make this McMath-Pierce Solar Telescope a unique facility. Recently, a low order solar adaptive optics system has been incorporated which has greatly improved solar observations and is yielding almost diffraction limited images.

For nearly 30 years a 70-cm aperture vacuum solar telescope, fed by a

Figure 2.5 Picture of SOLIS telescope on Kitt Peak, Arizona

2-mirror coelostat mounted on a separate tower, was operational at Kitt Peak. Its purpose was to make full disk longitudinal magnetic field maps together with spectroheliograms in the HeI line at 10830 Å, and in other wavelengths. In September 2003 this telescope was de-commissioned and a new SOLIS (Synoptic Optical Long term Investigations of the Sun) instrument, designed by Jack Harvey, has been placed on the tower. It will be largely a remote-controlled system. There are three instrument components on a single equatorial mount. One instrument is a Vector-Spectro-Magnetograph (VSM) fed by a 50-cm helium filled telescope with an active mirror image tracking system. The VSM is capable of recording full disk vector magnetograms with 1 arc sec pixels in the FeI 6301.5 and 6302.5Å lines, deep longitudinal magnetograms in the same lines for very weak magnitude fields, longitudinal magnetograms in CaII 8542Å, and also intensity images in HeI 10830Å. The second instrument is a Full-Disk-Patrol (FDP) telescope with a 14-cm aperture refractor for taking filtergrams through a narrowband (0.25Å) tunable birefringent filter from CaII K line to H-α, and in 10830 Å through a separate filter. The third instrument is an Integrated Sunlight Spectrometer (ISS) which will very precisely measure selected spectrum lines in integrated sunlight. SOLIS began operations in 2004 is shown in Figure 2.5.

7. National Solar Observatory at Sacramento Peak, New Mexico.

The Sac Peak Observatory, as it is familiarly known, was established in 1948 by Donald Menzel and Walter Orr Roberts as a high altitude coronagraphic site (2800 meters above the sea level). It is known for long uninterrupted intervals of clear sky. Several telescopes are now employed here for synoptic and research work. For example, there is the 40-cm coronagraph in the Evans Solar Facility, which takes daily scans around the solar disk at various limb distances in the green, yellow and red coronal lines. The Hilltop dome also provides real-time H-α images along with coronagraphic and spectrographic observations.

In 1968 Dick Dunn (Dunn 1969) designed and built a unique domeless vacuum telescope. Now this telescope is named as Dunn Solar Telescope (DST). The design of the DST is quite simple, consisting basically of three mirrors, two windows and an evacuated optical path. Sunlight enters the tower through a 76-cm fused silica window located 41 meters above the ground. By placing the window so high up, image distortion by ground heating is minimized. A pair of movable 1.1-m mirrors directs the sunlight down to the 1.2-m diameter vacuum tube

Figure 2.6 Exterior view of Dunn Solar Tower telescope at Sac Peak, USA.

that runs vertically to the center of the observing room. The sunlight is reflected from the concave 1.6-m diameter main mirror of the telescope and then back up to the observing room, producing a 51-cm diameter image of the Sun. Three vertical 1.5-m (5 foot) diameter tubes, clustered around the central tube, extend upward through the ceiling and contain spectrographs. By slightly tilting the main mirror at the bottom of the central tube by a computer control, the Sun's image can be focused on any of the spectrographs or at three additional viewing ports. In Figure 2.6 is shown an exterior view of the DST. It has played a key role in high-resolution solar physics. Now coupled with adaptive optics, the DST is currently providing some of the best high resolution solar images.

8. Solar Observing Facility at Marshall Space Flight Center, Huntsville.

This facility was built in 1973 to measure vector magnetic fields in support the Skylab mission. This was the first effort to obtain on routine basis vector magnetic field maps of active regions. The telescope has a 20-cm aperture telescope with a 6x6 arc minutes field of view.

9. Mees Solar Observatory on Haleakala, Maui, Hawaii.

This observatory is operated by the University of Hawaii. There are several telescopes to observe the corona, chromosphere and magnetic fields. Two coronagraphs take observations in H-α and Fe XIV lines. The Haleakala Stokes Polarimeter is available which produces vector magnetograms through the Imaging Vector Magnetograph (IVM). This uses a Fabry-Perot filter to scan the Fe I 6302 Å line in all four Stokes parameters (polarization states). Along with these instruments, there is the Mees CCD Imaging spectrograph (MCCD) which repeatedly records the spectra of an active region. The Mees white light telescope provides full disk solar images and the Mees video telescope takes H-α pictures of active regions.

10. Mauna Loa Solar Observatory.

Located on the big island of Hawaii, it is operated by the High Altitude Observatory, Boulder, Colorado. Here the main instruments include: 1) A 23-cm refracting Mark IV K coronameter that produces images in white light for polarization measurements, 2) A polarimeter for inner coronal studies with a removable occulting disk that is often used for disk observations in H-α, 3) A chromospheric Helium I Imaging photometer that records solar images in the 10830Å line, as well as at a number of other nearby wavelengths, using a liquid crystal variable retardation Lyot filter.

11. Stull Observatory at Alfred, N.Y., U.S.A., has a 20-cm Schmidt-Cassegrain telescope to observe in H-α active region over a field of view of 10x10 arc min. This telescope is used mainly for educational purposes.

12. Prairie View Solar Observatory, Houston, has a 35-cm Gregorian

Vacuum telescope for high resolution and full disk H-α images to study pre-flare and energy build-up processes. The observatory is involved in the Max Millennium Project for flare research. A Magneto-Optical Filter is under construction to observe Doppler and magnetic fields in the potassium line of 7699 Å.

13. Space Environment Laboratory at Boulder, Co, USA. A 20-cm refractor is used for taking full disk H-α, white light, and CaK line observations.

14. Solar observations at Universidad de Sonora, Hermosillo, Sonora, Mexico.

The University of Sonora, Mexico operates two solar observatories; one the Estacion de Obseracon Solar and other Observatorio Carl Sagan. At both these observatories, observations of active regions in the continuum, H-α, and the CaII K line are obtained using two heliostats and a 15-cm refractor. Real time H-α pictures are distributed through the World Wide Web.

15. Solar observations in Argentina.

At El Leoncito in the Argentina Cordillera de Los Andes, at an altitude of 2552 m, an H-α solar telescope for Argentina (HASTA) is operational to provide daily full disk H-α pictures. Equipment is a 10-cm aperture telescope, a 0.3A Lyot passband filter, and 1x1K square pixel CCD camera. At the same location a Mirror Coronagraph for Argentina (MICA), having a 6-cm aperture and pixel resolution of 8 arc sec, takes observations of the solar corona from 1.05 to 2.0 solar radii in the green and red coronal lines, and in H-α. This instrument provides high temporal resolution observations of transient phenomena such as the evolution of solar prominences and coronal streamers. MICA is almost identical to the LASCO-C1 coronagraph on SOHO and is meant as a ground support for the space-based LASCO instrument.

2.1.2 *Solar Observatories on the European Continent*

1. The French-Italian THEMIS (Telescope Heliographi Que por

Figure 2.7 General view of the THEMIS French-Italian Telescope building.

Du Magnetisme et des Instalilities Solaries) Observatory. At the Observatories del Teide, Tenerife, Spain, is a compact and axially symmetrical 90-cm aperture telescope known as THEMIS, designed to have zero or minimum instrumental polarization. It is a new generation solar telescope designed and built by INSU/CNRS (France) and CNR (Italy) agencies. The telescope is installed at the International Observatory of the Canary Islands (Tenerife, Spain), which is operated by the Instituto de Astrofisica de Canaries. The main scientific goals of THEMIS are to measure very accurately the polarization and vector magnetic fields. In Figure 2.7 is shown a general view of THEMIS.

2. Kiepenheuer Solar Observatory

This observatory on Tenerife Island, Spain, is operated by the Kiepenheuer Institute of Solar Physics. Here the following main instruments are operational:

1) A 70-cm aperture Vacuum Tower Telescope (VTT), which is to a

Figure 2.8 View of the especially designed THEMIS dome which minimizes internal dome seeing for high resolution observations.

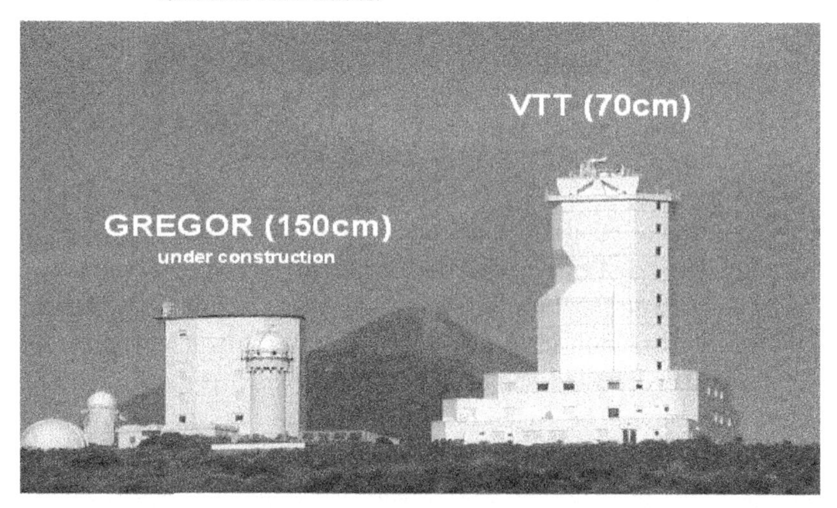

Figure 2.9 The German solar observing facilities at Tenerife.

large extent is a copy of the Kitt Peak Vacuum Telescope. There is a large 2-mirror coelostat located atop a high tower. The solar beam is sent vertically down through a vacuum chamber. Post focus instruments include a vertical Echelle spectrograph, a filter device for simultaneous observations in several wavelengths, and Fabry-Perot interferometer,

2) A 45-cm vacuum Gregory-coudè telescope (see Kneer & Wiehr 1989) which is an equatorial mounted telescope feeding a horizontal Echelle spectrograph, and
3) A 40-cm aperture equatorial Vacuum Newton Telescope is also available (see Schröter *et al.* 1985).

A new major project of building a 1.5m aperture GREGOR telescope is underway. A general view of the German solar observing facilities is shown in Figure 2.9.

3. Dutch Open Telescope (DOT)

Figure 2.10 Dutch Open Telescope at La Palma, Canaries.

At the Roque de los Muchachos Observatory, La Palma, Spain (see Rutten *et al.* 1999), a novel telescope based on an open telescope concept by Zwaan and Hammersschlg is operational. The idea of DOT is that the optics are completely open to the air and high winds flush away the heat generated at the primary mirror surface. This keeps the telescope's structure cool (isothermal) so that any air turbulence in the telescope is minimized. To flush out the heat from the telescope mirrors, it is necessary to have in fact fairly high winds. As reported by Rutten, a minimum wind speed of 5m/s is required, but optimum would be about 10m/s. Under such a wind pressure, the mechanical stability of an 'open' telescope has to be very strong. An advantage of the 'open' telescope concept is that it can in principle be scaled to much larger apertures. Such scaling is not feasible for a traditional evacuated or helium filled telescope since it would require impossibly large entrance windows. Windows also make it impossible to reach far infrared and near ultra violet wavelengths. Except for the one DOT of 45-cm aperture, at present there is no other similar solar telescope in the world.

Extremely high resolution solar granulation and sunspot observations have been obtained by this telescope, showing diffraction limited details of 0.2-0.3 arc sec. A general view of this telescope is shown in Figure 2.10.

4. Meudon Observatory, Paris

A 2-mirror coelostat feeds a 60-cm aperture tower telescope. A large spectrograph with a double pass system is available to cover wide spectral regions, simultaneously over a field of view of 1x8 arc min. Full disk spectroheliograms in H-α and the CaII K lines are also regularly obtained.

5. New Swedish Solar Telescope (NSST) at La Palma.

The NSST consists of a 97-cm aperture fused silica singlet objective lens located on a high tower. This lens serves as an entrance window to a vacuum tube through which the sunlight travels vertically down to the observing room. Various secondary optical systems are available for particular requirements. In Figure 2.11 is shown a general view of the NSST. The telescope is coupled to an adaptive optics system which gives diffraction limited solar images. As an example, a high spatial resolution picture of a sunspot and granulation displaying details of better than 0.1 arc sec. This telescope along with image restoring techniques has produced remarkable high resolution solar images.

Figure 2.11. Exterior view of New Swedish Solar Telescope.

6. The Astronomical Institute of Wroclau University, Poland, operates a 53-cm coronagraph that provides filtergrams and feeds a multi-channel subtractive double-pass spectrograph. There are two smaller instruments for studying H-α on the disk and limb activity.

7. Ulugh Beg Astronomical Institute, Uzbekistan. The Solar division of this institute has a 9 cm aperture reflector for solar observations and hosts IRIS and TON helioseismological instruments at Parkent & Tashkent respectively.

8. Einsteinturm Solar Observatory, Potsdam, Germany.

This original observatory has been operational since mid-1920. It contains a 2-mirror coelostat that feeds a 60-cm objective lens of 14-m focal length, followed by a Littrow spectrograph. Solar research topics involve vector and spectro-polarimetry studies of sunspots. In Figure 2.12 we see this architecturally attractive building.

Figure 2.12 View of the Einsteinturm observatory in Potsdam, Germany

9. Kharkov Astronomical Observatory, Ukraine, provides digital full disk images in HeI 10830 Å, H-α and CaII K, using a spectroheliograph with a 2 arc sec pixel detector.

10. Ondrejov Observatory, Czech Republic provides spectrograms using a 23-cm aperture horizontal telescope and a large multichannel flare spectrograph. Slit jaw images in H-α and simultaneous spectra of active regions are obtained in the CaII 8542 A and CaII K lines. White light images of a limited field of view are also available from an old 20-cm Clark refractor dating from 1858 and H-α images from a 20.5-cm refractor. It also has 2 patrol refracting telescopes of 7.5 and 11-cm

apertures for full disk white light and H-α images. There is a 13-cm aperture coronagraph for H-α prominence observation, 2 horizontal telescopes of 23- and 50-cm aperture for high resolution white light and H-α images for magnetic field and Dopplergrams.

11. At the Baikal Solar Observatory, Russia, the main solar telescope has a 76-cm aperture lens serving as an entrance window. This Large Solar Vacuum Telescope (LSVT) was constructed in 1980 and is based on the McMath-Pierce telescope design. It is coupled to a high dispersion spectrograph. A 25.5-cm refractor is also available for high resolution H-α images, along with two 18-cm refractors for H-α and white light observations.

Figure 2.13 Heliostat of Baikal Solar Observatory.

12. The Kiepenheuer Institute, Tenerife, also provides full disk 1024x1024 pixel digital H-α images from a 15-cm aperture telescope located on the building of VTT.

13. The Debrecen Observatory, Hungary, is famous for its long records of daily sunspot positions from white light images. It has 13-cm and 15-cm aperture telescopes for full disk white-light observations. It also has

a 53-aperture coronagraph, which is used to take pictures in H-α with a Lyot filter. However the location of the observatory does not permit it to actually observe the corona.

14. Kanzelhöhe Solar Observatory, Austria, has two refractors of 11- and 10-cm apertures for full disk and H-α synoptic images.

15. The Instituto Ricerche Solari, Locarno, Switzerland was founded by the Universities–Sternerarte, Gottingen, and was operational until 1984. Thereafter most of the instruments were moved to Tenerife. In 1993 the instrumentation was rebuilt. Now the observatory contains an evacuated Gregory-coudè telescope with a 45-cm primary mirror and a Czerny-Turner type spectrograph.

16. Abastumani Astrophysical Observatory, Tbilisi, Georgia, operates a 53-cm multi-purpose coronagraph built in 1973. It cannot be used to observe the solar corona because of scattered light. This instrument is used for chromospheric and photospheric observations. There is also a 44-cm general purpose horizontal solar telescope.

17. Astronomical institute of Wroclaw University, Poland has a 53-cm coronagraph which provides filtergrams and feeds a multi-channel double-pass spectrograph. Two smaller telescopes are also available for H-α disk and limb observations.

18. Sayan Solar Observatory, Izmiran, Russia. Full disk longitudinal magnetograms of 10 arc sec resolution and vector magnetograms of 4 arc sec resolutions are obtained. Full disk H-α alpha filtergrams are taken with an 18-cm aperture refractor.

19. Alma-Ata Observatory, Kazakhstan at an altitude of 1450 meters, has a coronagraph for synoptic observations of limb activity and H-α and CaII K-line filtergrams. Originally this observatory was associated with the Sternberg Institute in Moscow.

20. Pulkovo Observatory station, Kislovodsk, Russia, has 3 coronagraphs

of 53-, 20- & 10-cm aperture for taking synoptic observations in 5303Å, 5694Å, 6374Å, 10742Å, 10798Å and H-α lines, along with a 30-cm aperture tower telescope and a chromospheric telescope.

21. Kharkov Astronomical Observatory, Ukraine, has a spectroheliograph for taking full disk CaII K, H-α and HeI observations.

22. Kandilli Observatory, Turkey has two refractors of 15- and 10-cm aperture for full disk H-α and CaII K-line observations.

23. Bucuresti, Romania, has two refractors of 8- and 13-cm aperture for full disk white light and H-α images.

24. National Astronomical Observatory, Rozhen, Bulgaria, has one 20-cm aperture coronagraph and a chromospheric telescope for H-α images.

25. Lvov, Ukraine, has a photoheliograph for both full disk white light and H-α observations.

26. Stara Lesna Observtory, Tatranska Lomnica, Slovak Republic. This observatory has a horizontal solar telescope of 50-cm aperture coupled with a spectrograph, and a double solar telescope of 20- and 15-cm aperture for sunspots observations.

27. Lomnicky Stit Observatory, Slovak Republic, located at an altitude of 2632 m, has a 20-cm coronagraph for FeX, FeXV, and CaXV coronal line photometry, and H-α limb prominence observation.

28. Hvar Observatory, Croatia has a refractor for visual observations, regular observations are available.

29. Catania Astronomical Observatory, Italy has a 15-cm aperture telescope for full disk white light and H-α images.

30. Haute Provence observatory, France has a heliograph and takes full disk H-α images.

31. Royal Observatory, Belgium, operates 12-cm and 6-cm refractors for taking full disk H-α and white light synoptic observations.

32. Pic du Midi Observatory, France, operates a coronagraph for 5303Å, H-α, and HeI 10830Å lines synoptic observations. High resolution white light and H-α images are taken with a 50-cm refractor housed in a dome especially designed by Rösch, shown in Figure 2.14. Altitude is 2861 m.

Figure 2.14 View of the 'turret' dome at Pic du Midi. The telescope tube with its 50-cm objective lens extends out of the dome, the telescope tube is isolated from the dome structure and sealed to prevent exchange of air through the dome opening.

33. Bordeaux Observatory, France, has an equator refractor for velocity field observations using a Magneto-Optical filter.

34. Tashkent Observatory, Uzbekistan, has a 9-cm refractor to take white light images of the Sun. There is also is a TON (Taiwan Oscillation Network) site for helioseismology.

35. Crimean Astrophysical Observatory, Nauchny, Ukraine, has a solar tower telescope (TST-2) of 45-cm aperture for making HeI 10830Å maps. A larger system (TST-1) has a 1.2-m coelostat feeding a 90-cm aperture cervit mirror for vector magnetograms and Dopplergrams using a NaD Magneto-Optic filter.

36. Capodimonte Astronomical Observatory, Naples, Italy, has a telescope for full disk intensity and longitudinal magnetic and velocity field observations in the Potassium 7699 Å line using a Magneto Optic filter.

37. Roma Astronomical Observatory, Monte Porzio, Italy, has a telescope for full disk Ca II K and continuum observations.

38. Ebre Observatory, Roquetas, Spain, has a small refractor for full disk white light solar images.

39. Observatório Astronómico da Universida-de-Coimbra, Portugal, started in 1926. It has a spectroheliograph for full disk H-α, CaII K and continuum observations.

2.1.3 *Solar Observatories in Asia-Australia*

1. Culgoora Solar Observatory, Australia, runs a 12-cm refractor for full-disk H-α patrol and has a radio spectroheliograph to observe between the 18 MHz and 1.86 MHz frequency range.

2. Hiraiso Solar-Terrestrial Research Centre, Japan, operates a 15-cm refractor for full disk white light and high resolution H-α observations, along with radio spectrograph in the frequency range of 25-2500 MHz.

3. Norikura Solar Observatory, Japan, has three coronagraphs, one of 25-cm aperture and two 10-cm aperture telescopes to take limb observations in 5303Å, H-α, He D3, He 10830Å lines, and in the continuum. The altitude is 2876 m.

4. National Astronomical Observatory, Mitaka, Japan, (NAOJ). This observatory has two 15-cm aperture flare telescopes and two 20-cm refractors for high resolution vector magnetic fields, velocity fields, white light and H-α observations. Full disk magnetic and velocity field observations are also made using a STEP 6.5-cm heliostat refractor. 10- and 4-cm refractor telescopes take full disk, white light and H-α flare

patrol observations. In Figure 2.15 is shown NAOJ's flare telescopes.

Figure 2.15 Solar flare telescope consisting of 4 telescopes at the NAOJ.

Figure 2.16 The Domeless Solar telescope at Hida Observatory, Japan.

5. Okayama Astrophysical observatory, Japan, has a 65-cm aperture coudè refractor for vector magnetic field observations.

6. Hida Observatory, Japan is operated by the University of Kyoto. It has a 60-cm aperture domeless vacuum solar telescope, located on a high tower, as shown in Figure 2.16. This telescope is used to make high resolution white light, H-α, and vector magnetic field observations. For flare monitoring a full disk telescope is also available.

7. Yunnan Astronomical Observatory, Kumning, operates a 26-cm vacuum telescope for high resolution H-α observations and a 50-cm solar spectral telescope for spectroscopic and 2-D observations of solar active regions using a 2048 x 2048 pixel CCD camera.

8 Huairou Solar Observing station of the Beijing Astronomical Observatory, China.

Figure 2.17 Huairou Solar Observing station on the north shore in Lake Huairou, China.

To obtain the best seeing, this observatory is located on the north shore of Lake Huairou, north of Beijing. At this observatory there are several medium and small size telescopes available to make a large variety of solar observations. The 60-cm aperture 9-channel Gregorian reflector takes observations in CaII, HeII, MgI, four lines of FeI, HeI and H-α over a field of view for 5x4 arc min. A 35-cm vacuum refractor telescope is also available for taking longitudinal magnetic field observations in Fe I and H-β lines over a field of view of 3.5x5.4 arc min. 10-cm, 14-cm and 8-cm aperture telescopes are available for taking full disk vector magnetograms, full disk and high resolution H-α and full disk Ca II filtergrams respectively. In Figure 2.17 is shown the high tower with its retractable dome. This is also one of the TON sites for helioseismology.

9. Bohyunsan Optical Astronomical Observatory, South Korea, operates two 20-cm and two 15-cm aperture telescopes for white light, H-α, and vector magnetic field observations as shown in Figure 2.18.

Figure 2.18 View of Korean Solar Telescope with roll - off shed.

10. Udaipur Solar Observatory, Udaipur, India.

This observatory was started in 1975 on an island in a large lake for high resolution solar observations. It has 25-cm and 15-cm refractors for small field high resolution observations in H-α, mounted on the old CSIRO, Australia's 10-foot spar. A digital video magnetograph, using a Lithium Niobate etalon filter in the 6122 Å CaI line was operational until 1999 for longitudinal magnetic field measurements. A full disk 12.5-cm aperture Razdow telescope is available for full disk H-α synoptic observations and a 15-cm Zeiss coudè telescope is also available for taking filtergrams and spectrographic observations. A multi-slit H-α spectrograph was operational during 1988 -1994 for simultaneous spectroscopic observations over an active region. This is also a GONG (Global Oscillation Network Group) site for helioseismological studies. See Figure 2.19.

Figure 2.19 The lake bound Udaipur Solar Observatory with 3 domes housing the 10-foot spar, the Razdow and 15-cm Zeiss coudè telescopes.

11. Kodaikanal observatory, Kodaikanal, India.

This observatory was started in 1898 at the present site, earlier it was known as the Madras observatory, located at Madras, which was established by the East India Company in 1782. Dating back from 1900, the observatory has a 30-cm siderostat coupled to a spectroheliograph and a 20-cm photoheliograph. In 1958-59, a 60-cm aperture two mirror coelostat along with a 35-cm aperture horizontal telescope and a 60-foot Littrow spectrograph were added to solar instruments for high spatial and spectral resolution observations. The observatory is engaged in

Figure 2.20. Solar Tower Telescope of the Kodaikanal Observatory.

taking daily spectroheliograms in CaII and H-α, and white light photoheliograms. The observatory maintains a continuous record of spectroheliograms and photoheliograms since 1901. In Figure 2.20 is shown the solar tower telescope of the Kodaikanal Observatory.

12. State Observatory, Nainital, India.

A 15-cm Zeiss Coudè telescope along with an H-α filter is used to take observations of active regions for flare and other transitory events.

2.2 Solar Radio Observatories

With the invent of solar radio astronomy soon after the World War II, enormous progress was made in the field of solar radio physics, in Australia, England and America. Several new radio astronomy observatories were started in Europe, Soviet Union, America, Asia and

Australia.

1. Nobeyama Solar Radio Observatory, Japan

This observatory was started in 1992, with an interferrometric radio-heliograph dedicated to solar observations. The Radio heliograph has 84 antennas with 80–cm diameter dishes placed in a T-shaped base lines, 490 m in the E-W and 220 m long in the NS directions. The antennas are densely populated near the intersection of EW and NS arms and are sparsely placed away from the intersection. This design aims to image

Figure 2.21 Showing Nobeyama Radio Observatory with 84-80 cm aperture dishes arranged in EW and NS directions.

the extended solar disk and compact active regions and bursts. The field of view (FOV) of the interferometer is 35 minutes of arc, providing a full view of the Sun. The observing frequency until November 1995 was only 17 GHz but since then observations are being made in 34 GHz also. The spatial resolution 17 GHz is 10 arc sec and in 34 GHz is 5 arc sec, with a temporal resolution of 1 second and in active mode it could be as high as 50 ms. Measurements are made in both Right handed and Left handed circular polarizations, daily from 23 UT to 07 UT. This is a very powerful instrument to monitor in 2-dimensions the radio emissions in the low chromosphere and the transition region. The data are available in the form of contour and gray level displays, as shown in Figure 4.24. Total radio flux and circular polarization are also measured in the frequency band of 1-80 GHz range, with a separate telescope, polarimeter and spectrograph.

2. Cracow Solar Radio Observatory, Poland.

This radio solar observatory of the University of Jagellonian, Cracow, Poland was started in 1957, with a 7-meter parabolic radio telescope, which was replaced in 1995 with a new 8-meter polar mounted parabolic antenna, for systematic daily radio monitoring of the Sun in the 810 MHz band covering a band width from 275 MHz to 1775 MHz, at an approximate interval of 135MHz. The solar flux is recorded every 11 seconds. Observations are taken every day from 6.30 to 14.00 UT in the winter and from 5.00 to 15.00 UT during the summer.

3. Izmiran Solar Radio Laboratory, Russia.

Systematic daily solar radiometric observations are made at 169 MHz, 204 MHz, 3000 MHz fixed frequencies in patrol mode, in the 25 - 270 MHz range digital radio spectrograms are made in patrol mode and during active periods, spectrograms are taken at high time resolution between the frequency range of 220-260 MHz.

4. Culgoora Solar Observatory, Australia.

This observatory is part of the IPS Radio and Space Services of Australia, and is mainly used for daily monitoring of solar activity in optical and radio wavelengths. Optical full disk H-alpha solar observations are made using a 12-cm aperture refractor. For monitoring of the radio emission in the frequency band of 18 MHz - 1.98 GHz wave band a radio spectrograph is used.

5. Hiraiso Solar Terrestrial Research Center, Japan.

This center is engaged in taking synoptic solar observations in optical and radio wavelengths. A 15-cm aperture telescope takes full disk H-alpha and white light observations and a radio spectrograph takes daily radio flux measurements in waveband of 25 MHz -2500 MHz.

6. Learmonth Solar Observatory, Australia.

This observatory is also run by the IPS Radio and Space Services of Australia. At this observatory monitoring of both optical and radio emissions is made on regular synoptic basis. This is also a world wide SOON (Solar Observing Optical Network) site for providing full disk and high-resolution H-alpha, white-light and magnetic field observations, for Space-Weather forecast purposes. Solar Radio flux measurements are also made in the frequency range of 0.2, 0.4, 1.4, 2.7, 5.0, 8.8, 15 GHz and a sweep frequency interferometer is also operational in the frequency range of 25-75 MHz.

7. Radio-astrophysical Observatory, Siberia, Russia.

This Radio observatory is run by the Institute of Solar-Terrestrial Physics, Siberian Division, Russian Academy of Science. Daily images

Figure 2.22(a) General view of the Sebrian Solar Radio Interferrometric Telescope, with 128, 2.5-meter diameter antennas in EW and NS directions to almost make full disk images in 5.7 GHz.

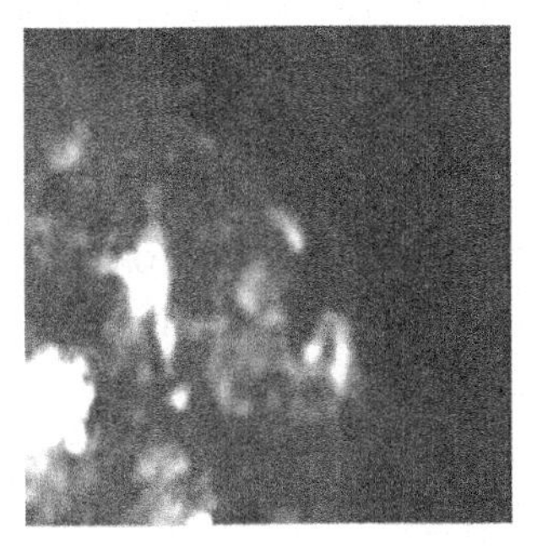

Figure 2.22(b) Solar image in 5.7 GHz taken from SSRT showing eruptive prominence on September 4, 2000 at 04:52UT and bright radio emissions on the disk.

of the Sun are taken at wavelength of 5.2 GHz, through the Siberian Solar Radio Telescope (SSRT), which is a crossed interferometer, consisting of two arrays of 128 x 128 parabolic antennas 2.5 meters in diameter each, spaced equidistantly at 4.9 meters and oriented in the E-W and N-S directions. The telescope field of view is slightly more than the diameter of the radio Sun at this wavelength. The length of each linear baselines of the interferometer is 622.3 meters. In Figure 2.22(a) is shown a general view of the Siberian Solar Radio Telescope, and in Figure 2.22(b) is shown a solar image taken in 5.7 GHz displaying prominence and bright emission on the disk.

2.3. Current Solar Space Missions

The need to take solar observations from above the terrestrial atmosphere was felt long back, because the short wavelengths of the electromagnetic spectrum are not accessible from the ground, due to the atmospheric absorption. In this section we shall describe only those solar space missions which are currently in operation (except for YOHKOH) or being planned for the near future.

1. Ulysses

It was a keen desire of many scientists to study the unexplored heliosphere around the Sun in 3 dimensions, which could be achieved only if a spacecraft flies over the poles of the Sun. This was made possible by the launch of Ulysses spacecraft on 6 October 1990. The trajectory of the spacecraft was so designed that it swings over the poles of the Sun, taking the advantage of Jupiter's gravity. The Ulysses passed over the southern solar pole in September 2000- January 2001 and from September 2001–December 2001, it passed over the northern pole. The main objective of Ulysses mission is to study as a function of latitude; the properties of solar wind, heliospheric magnetic field, solar radio bursts, plasma waves, solar X-rays, solar and galactic cosmic rays and interplanetary/interstellar neutral gas and dust. There are nine instruments on board to measure these parameters. The total weight of the space craft is 367 kg while the scientific payload weighs only 55 kg.

2. YOHKOH

One of the most productive solar space mission was YOHKOH meaning 'Sun Beam' in Japanese, was a project of the Institute for Space and Astronautical Sciences (ISAS) Japan with collaboration with American and British scientists. From October 1991until it got lost in December 2001 produced enormous high quality valuable data for more than 10 years. The main objectives of this mission was to obtain soft and hard X-ray images of the full disk and high resolution images of active regions, wide band spectral studies, through soft and hard X-ray spectrometers, Gamma ray spectral studies. And through a Bragg Crystal spectrometer (BCS) to study the resonance lines complexes of H-like Fe XXVI, Ca XIX and S XV lines.

YOHKOH was launched on 31 August 1991from Kagoshima Space Center, into a slightly elliptical orbit of about 600 km altitude at 31° inclination with a period of 97 minutes. The space craft weighed about 400 kg and had four instruments; a soft X-ray telescope (SXT) which was a grazing incident telescope with diameter of 23 cm and focal length of 1.53 m, this formed X-ray images in the range of 0.25 to 4.0 keV (3-60Å). Both full disk and small field observations were possible with SXT. A set of metallic filters located near the focal plane was used to separate different X-ray energies for plasma temperature diagnostics.

For hard X-ray observations a Fourier synthesis telescope (HXT) in four energy bands (15, 24, 35, 57, and 100 keV) with a temporal resolution of 0.5 sec was available, with full disk field-of-view (FOV), thus HXT could detect flares any where on the Sun.

A Wide Band Spectrometer (WBS) was also available to take soft and hard X-ray and Gamma ray spectrograms.

A Bragg Crystal Spectrometer (BCS) was also provided to study the H-like Fe XXIV, Ca XIX and SXV lines.

3. SOHO

Solar and Heliospheric Observatory (SOHO) is perhaps the most comprehensive and sophisticated space mission launched since Skylab in 1971, for the study of the Sun from its deep core to the outer corona and solar wind. The three main scientific objectives of the SOHO mission are:-

1. To study the solar interior, using the technique of helioseismology,
2. To study the heating mechanisms in the solar corona, and
3. To study the solar wind and the acceleration processes.

There are three experiments for helioseismic studies, namely GOLF, for global sun velocity oscillations (l = 0-3), VIRGO for Low degree (l = 0-7) irradiance oscillations and solar constant, MDI for velocity oscillations (l up to 4500) and full disk high resolution magnetograms.

For atmospheric remote sensing there are 6 experiments, namely SUMER for plasma flow characteristics from chromosphere through corona, CDS for temperature and density measurements in corona and transition region, UVCSfor Ion temperature, velocities and abundances in corona, LASCO for observations of solar corona extending from 2.5 to 30 solar radii. SWAN for measurement of solar wind mass flux anisotropies and its temporal variations.

Figure 2.23 An artistic view of SOHO spacecraft in fight, with its battery of instruments.

There are three experiments for 'IN SITU' measurements for solar wind; CELIAS for energy distribution and composition, COSTEP for energy distribution of ions and electrons (p, He) 0.04-53 Me V/n and electrons 0.04-5 Me V, and the third experiment is ERNE for energy distribution and isotopic composition of ions and electrons 5-60 Me V.

Observations for helioseismic, Coronal Mass Ejections (CME), EUV, solar wind studies have already provided wealth of information and new perspective of the Sun has emerged.

SOHO was launched on board Atlas II- AS on December 2 1995 and is placed in 'halo' orbit on 14 February 1996, around the L1 Lagrangian point, between the gravity of the Earth and the Sun. From this vantage location SOHO is able to observe the Sun continuously for 24 hours a day, 7 days a week, accept for a brief period of 3 months during summer of 1998, due to ground telemetry error. In Figure 2.23 is shown an artistic view of SOHO in flight. This is an International co-operative project between NASA and ESA, for which 12 sophisticated instruments were developed by 39 consortia of institutions from 15 countries.

4. WIND & Polar

Wind and Polar are 'sister spacecrafts', their primary objective is to measure the mass, momentum and energy flow and the time variability, throughout the solar wind-magnetosphere-ionosphere system near the Earth. Wind was launched on 1 November 1994 on board Delta II rocket from Cape Canaveral air force station. There are 8 instruments to measure these parameters.

5. ACE

The Advance Composition Explorer (ACE) was launched on 25 August 1997 into an orbit around the L1 Lagrangian point at 240 R_e between the Sun and the Earth. The main objective of ACE is to determine the elemental and isotropic composition of several samples of matter in the solar corona, interplanetary and local interstellar medium and galactic matter. To achieve this objective, ACE carries a set of 9 *in situ* instruments to measure the charge state composition of the nuclei from Hydrogen (H) to Nickel (Ni) ($1 \leq Z \leq 28$) from the solar wind energies (~1 keV/nuc) to galactic cosmic rays (~500 MeV/nuc), it also provides real time solar wind data to NOAA for Space Weather forecast.

6. TRACE

TRACE (**T**ransition **R**egion and **C**oronal **E**xplorer) mission is aimed to achieve the highest spatial resolution images hitherto obtained from any space mission (0.5arc sec pixel) of the solar photosphere, transition region and the corona with high temporal resolution also.

Figure 2.24 Photograph of TRACE spacecraft

The TRACE satellite was launched on 2 April 1998 from the Vandenberg Air Force Base on a Pegasus –XL launch vehicle, into a sun-synchronous polar orbit at an altitude of 625 km and 97°.8 inclination. TRACE carries a single 30-cm aperture telescope, using 4 normal incidence coating for EUV & UV on quadrants of the primary and secondary mirrors. The segmented coatings on the solid mirrors form identical sized and perfectly co-aligned images at selected temperature range from 6000 to 10 million degrees (Fe IX/X at 171 Å, Fe XII/XXIV at 195 Å, FE XV at 284 Å, Lyman α, CIV at 1550 Å, UV continuum at 1600 Å, and at 1700 Å and white light continuum at 5000 Å). The field of view is 8.5x8.5 arc minutes, obtained over 1024x1024 CCD detector. The pointing is stabilized to 0.1 arc sec against the spacecrafts jitter.

Thousands of beautiful solar images taken at these wavelengths are available daily from TRACE. In Figure 2.24 is shown a photograph of TRACE spacecraft.

7. GOES / Solar X-ray Imager (SXI)

The next generation of US weather satellites GOES have a small X-ray imager mounted on the solar array yoke, called the Solar X-ray Imager (SXI) gives continuous monitoring of the Sun in X-rays. It provides 512x512 pixel images in 10 wavebands from 6 to 60 Å, at 1 minute intervals. The first satellite equipped with SXT was launched in summer of 2001.

8. GENESIS

GENESIS mission of NASA-JPL is to capture an integrated sample of solar wind material and return it to Earth for isotropic and elemental abundance. Genesis spacecraft was launched on 8 August 2001, it spent about 2.5 years at the L1 Lagrangian point, collecting solar wind material before returning on 8 September 2004. The capsule containing the solar wind material crash landed and got destroyed, at the time of writing this it was not known whether useful data was possible to retrieve. Its key scientific objective was to precisely determine the solar isotropic abundances which would greatly improve the elemental solar abundance also. Genesis was to measure the oxygen at 0.1% level which is at present known to 10% accuracy. It was hoped that this would allow to distinguish between the two important theories for the solar system heterogeneity and thus would provide crucial information on the origin and formation of the solar system.

9. CORONAS- F

CORONAS-F (Complex **OR**bital Near- Earth **O**bservations of the **S**olar **A**ctivity) is a Russian solar space mission with the main objective to understand better the solar activity, in particular flares and the solar interior. CORONAS-F was launched on 31 July 2001 on Cyclone-3 rocket from Russia's Northern Cosmodrome in Plesetsk, in a polar orbit at an altitude of 500 km and inclination of 83 deg. The CORONAS-F is second in series of CORONAS, the first was launched on 21 March 1994 and provided data until 1995. CORONAS-F carries 15 instruments, among them are 10 X-ray spectrometers and imagers, 2 are UV instruments, a radiometer, a coronagraph and several full disk photometers.

10. RHESSI

The **H**igh **E**nergy **S**olar **S**pectroscopic **I**mager (HESSI) is NASA's mission to explore the basic physics of particle acceleration and explosive energy release in flares. This mission has been renamed as Ramaty High Energy Solar Spectroscopic Imager, in honor of the famous solar scientist Dr. Reuven Ramaty. It was launched in 2003. RHESSI has a Fourier synthesis hard X-ray and Gamma ray imaging spectrometer, and has the capability to provide high resolution images of the flares over a broad spectral range, from soft X-rays (3 keV) to gamma rays (20 MeV) and also with high spectral resolution. It takes full disk solar image with spatial resolution varying from 2 arc sec at 100 keV to 36 arc sec at 1 MeV, at a temporal resolution of 10 ms for the basic image.

2.3.1 *Planned Solar Space Missions*

1. SOLAR –B

The proposed SOLAR-B is a Japanese mission of ISAS with American and British participation, and will be launched in August 2005. This is infact a follow up of the highly successful YOHKOH mission. The main scientific objectives are to study the generation and transport of magnetic fields and their role in heating the chromosphere and the corona and in eruptive solar events, like flares, mass ejections etc. The SOLAR-B will consist of mainly the following three instruments:-

A diffraction limited 50-cm aperture Solar Optical Telescope (SOT) with focal plane package instruments to take high resolution photospheric and chromospheric images and spectro-photometry. It will obtain diffraction limited images (0.2 arc sec ~ 150 km) from 3880 to 6600Å, with the help of tunable narrow band birefringent filter and a spectro-photometer. It will also make vector magnetic field measurements from Stokes spectra of Fe I lines at 6301 and 6302 Å, with 0.16 arc sec /pixel over a field of view of 164x328 arc sec. Broad band system will take diffraction limited images with 0.05 arc sec/pixel in the H line of Ca II, CN and g-band heads and in the continuum. The narrow band system will make filtergrams, magnetograms, Dopplergrams and Stokes images in several photospheric lines, Mgb and Hα, at 0.08 arc sec per pixel resolution.

There will be a XRT (X-Ray Telescope), a more advanced version of the Yohkoh's SXT, it will provide images from 2 to 60 Å in the soft X-ray range. The image scale will be 1 arc sec/pixel, which will be 2.5 times better than the SXT. XRT is a grazing incident modified Wolter - I X-ray telescope of 35-cm inner diameter and focal length of 2.7 meters. Full disk images will be available on a 2048x2048 back illuminated CCD chip yielding a spatial resolution of 1 arc sec. There will be a small optical telescope using the same CCD for co- alignment with the SOT.

The third instrument on board Solar-B will be the Extreme Ultraviolet Imaging Spectrometer (EIS), which consists of multi-layer coated off axis telescope mirror of 15-cm aperture and a multi-layer coated toroidal grating spectrometer. The telescope and spectrograph combination will have spatial resolution of 2 arc sec. The spectral resolution will be good enough to measure velocity to 3 km/s. Half of each optics is coated to optimize reflectance at 170-210 Å and the other half to 250-290 Å range.

2. STEREO

The main scientific objective of NASA's STEREO (**S**olar **T**errestrial **Re**lations **O**bservatory) is to understand the origin and consequences of the Coronal Mss Ejections (CMEs) through 3-D pictures of CMEs. This could be achieved by using identical instruments on two spacecrafts, one flying ahead of the Earth and other behind, and then combining the observations made simultaneously from the two spacecraft, a 3-D *stereo* image of CME can be created. In Figure 2.25 is shown an artistic view of the proposed STEREO mission. It will also study the propagation of disturbances through the heliospheric and their effects at 1 AU.

The two spacecrafts will carry the following identical instruments:-

i. SECCHI (**S**un **E**arth **C**onnection **C**oronal and **H**eliospheric **I**nvestigation) which will include 4 instruments, two white light coronagraphs, covering the range from 1.15 to 4.0R_0 and 2 to 15 R_0, an EUV full disk imager with 1 arc sec /pixel and a heliospheric imager that will image the heliosphere from 12R_0 to beyond the Earth's orbit.

ii. STEREO/WAVES (SWAVES) an interplanetary radio burst tracker that will trace the generation and development of traveling radio disturbances from the Sun to the Earth's orbit.

Figure 2.25 Artistic view of Stereo mission.

iii. IMPACT (**In** *situ* **M**easurement of **P**articles and **C**ME **T**ransits) will measure the 3-D distribution and the plasma characteristics of the solar energetic particles and the local vector magnetic fields.
iv. PLASTIC (**Pl**asma and **S**uper **T**hermal **I**on **C**omposition) experiment will provide plasma characteristics of protons, alpha particles and heavy ions.

The STEREO mission spacecrafts are expected to be launched in February 2006 and hopefully will last at least for a period of 2 years and may continue for five years.

3. SDO

SDO (**S**olar **D**ynamics **O**bservatory) is the first initiative mission in NASA's "Living with a Star" program, to develop the scientific understanding necessary to effectively address those aspects of the Sun-

Earth system that directly affect life and society. The main questions to be addressed by SDO are:-

What mechanisms drive the 11- year solar cycle?

How magnetic flux is generated, concentrated and dispersed across the solar surface?

Where do the observed variation in the Sun's total and spectral irradiance arise and how they are related to the magnetic cycle?

Is it possible to make accurate and reliable forecast of Space Weather and climate?

Several groups are working towards formulating suitable instrumentation, and most probably it may be launched in 2007.

4. Solar Orbiter

The Solar Orbiter mission of the European Space Agency (ESA) envisages to observe the Sun from a distance of only 45 solar radii or 0.21 AU. The Solar Orbiter will be in a Sun synchronous orbit in order to examine Sun's surface from a co-rotating vantage point with extremely high spatial resolution and will also be able to obtain images of the sun's polar region. It is considered for 2008-2013 time frame.

Chapter 3

Structure of Solar Atmosphere

3.1 From the Solar Interior to the Photosphere

What makes the Sun shine? Energy generating nuclear reactions in the Sun's core are explained. How does this energy travel from the core to the surface? How do we know all this? How do we construct a model of the solar interior? We begin by observations of the solar atmosphere. How do we judge the model is correct? These are the subjects of this chapter.

3.1.1 *Hydrostatic Equilibrium in Solar Interior*

The center of the Sun lies some 700,000 km below the visible surface of the photosphere. No light ray can emerge directly from the core of the Sun to us. Then what chances do we have to do more than just speculate about the temperature, pressure, or its chemical composition? What are the parameters available to us on which we can base a theory of the solar interior?

There are three basic measurements available from astronomical observations that bear on the structure of the Sun as a whole:-

First, we know that the total mass of the Sun is $m_{\odot} = (1.9891 \pm 0.0012) \times 10^{30}$ kg. To construct a theoretical model of the Sun, we should know exactly how much material is available to build the Sun.

Second, we know the radius of the Sun which is $r_{\odot} = (6.9626 \pm 0.0007) \times 10^{5}$ km. This tells us how large a space our model must fill.

The third parameter we know is the luminosity of the Sun, i.e., how much the energy the Sun puts out in all directions every second. This "candle power" of the Sun is 0.9 x 10^{26} calories per second, or $L_{\odot}$= (3.845 ± 0.006) x10^{26} watts. Most of this energy is in the form of light.

These are the three corner stones on which the theory of the solar interior is based. However, still further information has been gained from other observations, namely that of the solar surface. This gives us important starting points for a solar interior model.

From the identification of lines in the solar spectrum, Russell in 1929, found the approximate relative abundance of carbon, nitrogen, oxygen, manganese, silicon and iron on its surface layers. At that time he could not estimate very well the abundance of the two lightest elements; hydrogen and helium. The observations also indicated that the outer layers of the Sun are gaseous with a temperature of 6000 Kelvin. The pressure is about $1/10^{th}$ that of the Earth's atmosphere. On the basis of these observational data set, can one build a reasonably accurate picture of the solar interior? The answer to this question is **no**, if it were not for another basic fact: geological evidence points out that the Sun has changed very little for at least a billion years. It is this amazing stability that gives us the clue as to the construction of solar interior. What conditions must be fulfilled so that our 'model' Sun comes out as stable as the real one?

There are two conditions, one dynamic and another thermodynamic. The first one states that if we want to avoid having our model Sun collapse or expand, we must arrange that all the forces acting on any small volume of gas inside the Sun exactly balance each other. The major forces acting inside the Sun are gravity and gas pressure. Of these two big forces, gravity will always pull any volume element towards the Sun's center. To balance gravity we must arrange that gas presses outwards with a force exactly equal to the gravity. How can we provide this?

Consider a particular cylindrical volume element, with its axis towards the center of the Sun, of 1 km thickness somewhere inside the Sun (Figure 3.1). If the gas pressure P was the same throughout the Sun then it would press downwards on the top surface of this element equally as it presses upwards on the bottom. As a result gas pressure would not move the element in any direction. But to balance the gravity ρg (where ρ is the density of the material and g the gravitational acceleration), we must make the gas pressure (P+dP) greater at the bottom than the top. Thus our cylindrical element will stay in equilibrium without any motion arising from unbalanced forces. Hydrostatic equilibrium is maintained between the force of gravity and the difference in the gas pressure dP.

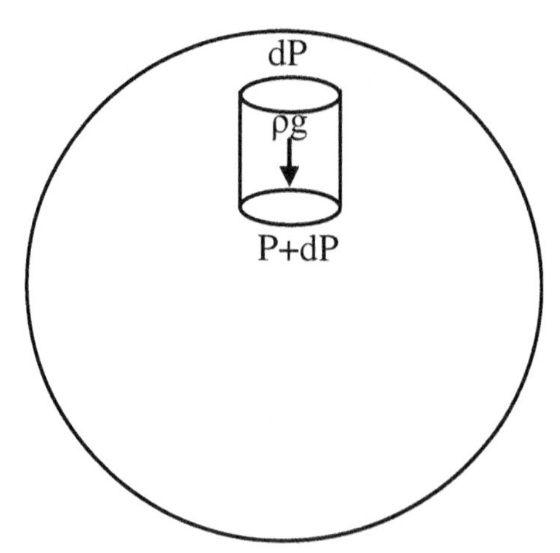

Figure 3.1 Showing a hypothetical cylindrical volume element in the Sun with its axis pointed towards the Sun's center parallel to gravity.

Since for every element the pressure must be higher at the bottom than at the top, we can conclude that the gas pressure steadily increases from the surface towards the center, and this increase of pressure per kilometer must be exactly as large as is necessary to balance gravity. The balance between the gas pressure differences dP across the cubic volume element of opposite faces and the gravitation force is then

$$dP = -\rho g dr, \tag{3.1}$$

where g is the gravitational acceleration at a distance r from the center of the Sun. Since g is positive, the inward acceleration of gravity balances an outward decrease of pressure with increasing r. A body of large mass such as the Sun is held together by its self-gravity. The acceleration g is determined by the mass M(r) lying within the sphere of radius r, so that its magnitude is:-

$$g = \frac{GM(r)}{r^2}, \tag{3.2}$$

where G is the universal gravitational constant which is G = $6.67x10^{-8}$ dyne cm^2/g^2 . Substituting G from this expression in equation (3.1) we obtain

$$\frac{dP(r)}{dr} = -\frac{GM(r)\rho}{r^2}, \tag{3.3}$$

where ρ is the density in a spherical shell of radius r. Mass M(r) can be obtained by integrating over the radial density distribution of the star, as

$$M(r) = \int_0^r 4\pi r^2 \rho(r) dr. \tag{3.4}$$

Equation 3.3 governs the variation of pressure through a star. This can be related to the density by the gas law: -

$$Pg = nKT = \frac{K\rho T}{\mu m_H} = 0.825x10^8 \frac{\rho}{\mu},$$

where K is the Boltzmann constant, m_H is the mass of a hydrogen atom and μ the mean molecular weight, or the average weight of each individual particle in atomic mass units. For example, ionized hydrogen contributes 2 particles, proton + electron, hence μ = 0.5.

To have a stable solar model, we have to simultaneously consider a second condition from thermodynamics. Consider the surface layers which emit light and heat in the form of radiation. We call this the luminosity of the Sun ($L_\odot$). But how can a body of gas, such as the solar surface layers, steadily emit heat without cooling? This can only happen if the surface layers receive, every second, just as much heat as it emits. From where can this compensation come? This can only come from below; from the solar interior. Hence, we conclude that the layer just below the surface sends a flow of heat energy upwards into the outer layer, and that this flow of heat must be exactly equal to the solar luminosity. But not only the surface layer, rather every layer within the Sun must be thermally stable. That means every layer must emit heat flux upwards just as large as it receives from below. There must be a heat flow from the very core of the Sun all the way to the surface passing through all the layers and finally appearing as the Sun's luminosity. Only then can the temperature remain constant with time at every point in the Sun.

The luminosity or the total net flux of energy L(r) outward from a sphere of radius r and mean density ρ is given as:-

$$L(r) = 4\pi \int_0^r r^2 \rho \varepsilon dr \,, \tag{3.5}$$

where ε is the rate of nuclear energy production per gram, ρ the density of the material, and r the radius of the Sun. Since most of the nuclear energy generation occurs in the central regions, at the solar surface we can take L (r) = $L_\odot$, the solar luminosity.

3.1.2 *Energy Generation*

The source and mechanism of energy generation in the Sun had been a matter of great concern for the early scientists. Even until the nineteenth century, people tried to explain the solar energy generation through chemical combustion of hydrogen and oxygen. Knowing the mass of the Sun and the rate of heat produced in this reaction, the life of the Sun turned out to be only 3000 years! Another idea, proposed by J.R. Mayer in 1842, was that in-falling meteorites on the Sun can produce sufficient heat through friction and impact to raise the temperature to 28 million degrees. The Sun's energy in this case would last for 22 million years at the present day expenditure rate. Lord Kelvin (William Thomson) and Hermann von Helmholtz objected to this hypothesis on the basis that the mass of the Sun should increase so rapidly that the consequences would have been revealed by an accelerated motion of the planets. Alternately they proposed that if the Sun contracted annually by 20 meters enough heat energy would be generated by this extra compression to account for the measured solar luminosity. This slow contraction too small to be measured, would keep the Sun shining for about 50 million years. Due to the uncertain knowledge of geological time scales in mid-nineteenth century, this seems a long enough time period for the Sun's life. But by the early twentieth century it become abundantly clear that fossil and geological records show evidence of the Earth's existence for hundreds of millions of years. It was quite embarrassing for astronomers to have a Sun 'younger' than the Earth! To solve this problem, Sir James Jeans came with an idea in the 1920s that radioactivity could provide the vital heat energy. Today we know that this hypothesis was incorrect, but it provided guide lines for scientists to think in the right direction; that the transformation of atomic nuclei must be the energy source in the Sun.

In 1929, R. d' E. Atkinson and Houtermanns (1929) suggested that solar energy might arise from the capturing of extra protons by the nuclei of atoms, forming heavier nuclei, and that this union is accompanied by a release of nuclear energy. In early 1930s, scientists believed that this reaction of capturing of a proton was not possible because the positive electrical charges on protons cause them to repel each other too strongly- the Coulomb repulsive force. At this point George Gamow came to rescue and proposed that in the strange world of atomic particles this simple argument was unsound. Using the new science of Quantum Mechanics he showed that the protons can 'tunnel' into each other and get near enough together for nuclear glue to make them stick.

By the late thirties it became apparent that the sub-atomic processes were at work in generating energy in the Sun. Suddenly, in 1939, the impasse was resolved by Hans Bethe (1939) in America and Carl von Weizsäcker (1938) in Germany. They independently extended the idea of Gamow and proposed *sticking* extra protons into the nucleus of carbon. As physicists gained insight into nuclear physics, they proposed that the 'carbon cycle' was the essential source of energy in the Sun. We now know that the carbon cycle, while effective in stars more massive than the Sun, is barely operative in the Sun. The main source of energy in the Sun comes from the proton – proton cycle.

Not only did Hans Bethe in 1939 discovered the carbon cycle but he also computed approximately at what rate this process would produce energy. He found that the rate of energy production depends strongly on temperature, because with increase in temperature the number of collisions between atomic nuclei greatly goes up. He estimated the temperature of the solar core to be about 40 million degrees. At that time this was considered as just the right order of magnitude to make the carbon cycle proceed at a rate roughly sufficient to account for the observed solar luminosity.

Immediately after this success, however, a new difficulty arose due to improved computations of energy production rates. It was found that the carbon cycle resulted in a much higher solar luminosity than was observed. Scientists started wondering where mistake was that led to this discrepancy. The reason was found in Eddington's assumption that hydrogen and helium could be neglected. He used Russell's earlier solar elemental abundance determinations from surface spectroscopic data. A short computation showed that by introducing a reasonable percentage of helium into the solar interior the result was a lowering of the computed temperatures, which in turn appreciably reduced the carbon cycle. In fact, to make the carbon cycle operate exactly at the right rate, and also to satisfy the mass-luminosity relation for the Sun, it was realized early in the 1950s that theoretical solar models should have approximately sixty five percent (65%) hydrogen, thirty percent (30%) helium, and about five percent (5%) of heavier elements. With this composition the temperature at the center of the Sun worked out to 19 million degrees Kelvin, with a central pressure of the order of 10^{12} atmosphere. With improved solar models now available, using helioseismological techniques, this figure of central temperature has been further reduced.

With better determinations of the chemical composition and better theoretical methods used in building solar models, it was recognized that the carbon cycle in the Sun contributes only about 1% of its energy. The carbon cycle is important in hotter stars, but in the Sun the most efficient nuclear reaction is the proton – proton cycle.

3.1.2.1 *Proton – Proton (p-p) Chain*

It is now well established that at the center of the Sun, where the temperature is about 15 million degrees Kelvin, the proton – proton chain dominates. The first step in the p-p chain is the formation of deuterium ${}_1H^2$ (heavy hydrogen), a positron e^+, and a neutrino ν, from the fusing of two protons. The neutrino escapes from the Sun directly, but the positron collides with an electron and they annihilate each other, with a release of energy:-

$$ {}_1H^1 + {}_1H^1 \rightarrow {}_1H^2 + e^+ + \nu. \qquad (3.6) $$

Next the deuterium nucleus fuses with another ordinary hydrogen nucleus within about 10 seconds to form a helium isotope with 2 protons and emit a gamma ray,

$$ {}_1H^2 + {}_1H^1 \rightarrow {}_2He^3 + \gamma. \quad (3.7)$$

Finally, two of three helium isotopes fuse to make one nucleus of ordinary helium plus 2 nuclei of ordinary hydrogen,

$$ {}_2He^3 + {}_3He^3 \rightarrow {}_2He^4 + 2{}_1H^1. \quad (3.8)$$

In the p-p cycle, 6 hydrogen nuclei, one at a time, fuse and wind up to form one helium plus two hydrogen nuclei. There is a net transformation of four hydrogen nuclei into one helium nuclei, but the 6 protons contain more mass than the final single helium nuclei plus 2 protons, a positron and a neutrino. If we compare the mass of these 6 protons at the beginning of the p-p cycle with the masses of the reaction products in percentage terms, about 0.7% of the mass is lost. This *mass loss* appears as radiation or energy release. In other words, every kilogram of hydrogen processed in this manner, looses 7 gram of mass, which is completely converted into pure energy. According to Einstein's famous equation, $\mathbf{E} = \mathbf{mc^2}$, where **E** is the energy released by converting completely the mass **m** into energy, and **c** is the velocity of light (3×10^{10} cm/sec). The small difference in mass that disappears in the p-p process is completely converted into 19.78 MeV of energy. This assumes that the mass (m) can be converted to energy (E) with complete efficiency. From the p-p chain, it is easy to estimate that burning of only about 5% of the Sun's hydrogen supply in the core into helium, would be sufficient for its vast energy output during its life time of nearly 5 billion years. Although about 85% of the Sun's energy is thought to be produced by the p-p chain, there is an alternate possibility that in the Sun the helium–3 particles, once formed, can bump (about 5% chance) into helium– 4 and make beryllium–7, with the release of a gamma ray:-

$$ {}_2H^3 + {}_2H^4 \rightarrow {}_4Be^7 + \gamma. \quad (3.9)$$

This can then branch out in one of the two ways that either the beryllium-7 catches an electron, forming Li^7 and a neutrino ν,

$$_4Be^7 + e^- \rightarrow {}_3Li^7 + \nu, \tag{3.10}$$

$$_3Li^7 + {}_1H^1 \rightarrow {}_2He^4, \tag{3.11}$$

or it catches a proton and makes boron–8, which being unstable, immediately decays to beryllium–8 and two helium particles:

$$\begin{array}{llll} _4Be^7 + {}_1H^1 \rightarrow & _5B^8 + \gamma & & \\ & \downarrow & & \\ & _5B^8 \rightarrow & {}_4Be^8 + e^+ + \nu & \\ & & \downarrow & \\ & & {}_4Be^8 + 2\, {}_2He^4. & \end{array} \tag{3.12}$$

Out of the three alternate mechanisms, the principal source for energy generation in the Sun is due to fusion of 4 hydrogen nuclei and producing two helium nuclei. Although the two alternate branches in which $_2H^3$ nuclei reacts with $_2H^4$ to produce beryllium and lithium isotopes play relatively unimportant role in view of the energy generation, but these reactions do produce neutrinos in the high energy range. These can be detected at the Earth, and provide a handle for understanding the nuclear reactions. In subsequent Section 3.1.4, the solar neutrino process will be discussed. In Table 3.1 is given a summary of p-p chain reactions and the release of energy in its various stages.

The nuclear energy generation in the Sun and stars depends on the central temperature and pressure. We can estimate the energy generated at the center from the Sun's surface luminosity. Every second, the Sun consumes about 650 million metric tons of hydrogen, which is converted into helium. When our Sun came into existence a little over 70% of its mass was hydrogen. At present the central hydrogen is depleting at a rate of 5 million tons per second through thermonuclear reactions. In spite of this enormous depletion of fuel (hydrogen) from the solar core, our Sun will last for another 4.5 billion years. Thus our Sun can be said to be middle aged.

Table 3.1 Summary of p-p chain reaction and energy release in each step is given in the right column in MeV, together with the neutrino energy in parenthesis.

$4\,{}_1H^1 \rightarrow {}_2He^4 + 2e^+ + 2\nu$	Q = 26.73 MeV
${}_1H^1 + {}_1H^1 \rightarrow {}_1H^2 + e^+ + \nu$	1.442 (1.442)
${}_1H^2 + {}_1H^1 \rightarrow {}_2He^3 + \gamma$	5.493
${}_2He^3 + {}_3He^3 \rightarrow {}_2He^4 + 2\,{}_1H^1$	12.859
or	
${}_2He^3 + {}_1H^1 \rightarrow {}_2He^4 + e^+ + \nu$	19.795
or	
${}_2H^3 + {}_2H^4 \rightarrow {}_4Be^7 + \gamma$	1.587 (9.625)
${}_4Be^7 + e^- \rightarrow {}_3Li^7 + \nu$	0.862
${}_3Li^7 + {}_1H^1 \rightarrow {}_2He^4$	17.347
or	0.135
${}_4Be^7 + {}_1H^1 \rightarrow {}_5B^8 + \gamma$	
${}_5B^8 \rightarrow {}_4Be^8 + e^+ + \nu$	15.079 (14.02)
${}_4Be^8 + 2\,{}_2He^4$	2.995

3.1.3 *Energy Transport and Solar Model*

How does the heat flux or energy transport take place in the Sun from the solar interior to the surface? Heat flows only where there is a temperature difference, only from hotter to cooler regions. As we know that energy flows from the core of the Sun to the surface, there must be a temperature difference, decreasing steadily from the core to reach a minimum at the surface, the photosphere. But how does this decrease in temperature occur? This depends on the particular mechanism by which the heat or energy is transported. Heat is transferred by three processes: *conduction, convection and radiation*. Conduction occurs only in solids and is irrelevant for stars and Sun. Through convection energy is transferred by the bulk transport of heated matter. By radiation processes, it is directly transported at the speed of electromagnetic waves, like X-rays, light and radio waves etc.

Let us consider that heat flows by convection as hot material rises to overlaying cooler layers. In the case of the Sun, through the first 70% of the solar radius heat is transported by radiation. As we saw in the earlier section, the photons liberated in the core in thermonuclear process, were high energy gamma rays. These bump their way along making countless collisions with electrons and atomic nuclei. This bumping process gradually increases the number of the photons and lowers their energy as they diffuse outwards, first to X-rays and the extreme ultraviolet, then to ultraviolet, and finally to the visible light. In fact it takes an estimated 10,000 years for the energy generated in the solar core to diffuse out to the surface.

As the photons approach the last one-third of their journey from the center an important change occurs, the gas properties become such that convection sets in. What has actually happened is that, starting from the core, the temperature decreases steadily until one encounters atomic nuclei which are cool enough and therefore traveling slowly enough, to be able to hang on to electrons. Thus ions are formed. These ions have a drastic effect on the propagation of photons, which are transformed by now into ultraviolet rays from the more energetic gamma rays. Ultraviolet photons can be readily absorbed by ions. Thus the bulk of the solar material is in the right condition to absorb radiation, instead of kicking around from nucleus to nucleus. In other words, the solar material at this distance becomes completely opaque and the energy gets trapped in this layer. The gas responds to this input 'dumping' of energy by bubbling furiously. The whole situation becomes locally unstable, leading to turbulent convection as the opaque gas, which is blocking the radiation, is forced to rise upwards through the Sun to cooler layers. This is convection. Convection is a highly efficient way of moving energy. This region of the solar interior extending from about 0.7 $R_\odot$ to almost the solar surface is called the convection zone. In Figure 3.2 is shown a schematic diagram of the solar interior, indicating where energy is transferred by radiation in the inner regions, and by convection in outer region through giant convection cells.

Within the hot gas of a star, light is emitted. The hotter the gas, the more light. Consider two cubic elements one on top of the other. Let the lower one be hotter than the top one. Then the lower one will send more radiation upwards than the top one sends downwards. Hence there will be a net flow of radiation upwards. If there is a steady temperature decrease layer by layer, from the center towards the surface, then through

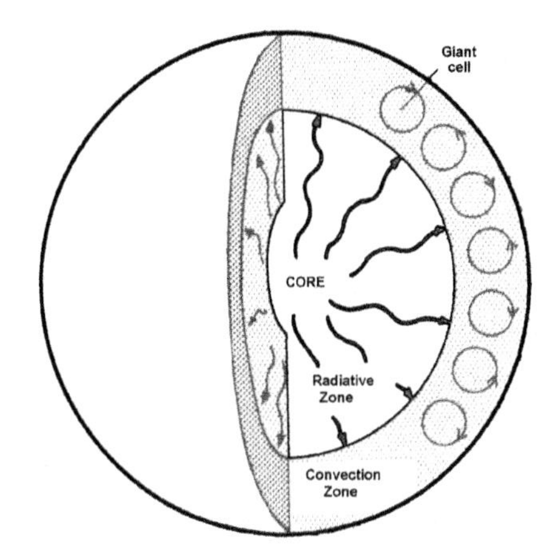

Figure 3.2 Schematic diagram showing the core, the energy generation region, radiative, and convection zones.

every layer there will be heat flowing outwards in the form of radiation. The strength of the radiation will depend on the temperature and the opacity of the gases which also absorb radiation. The greater the temperature per kilometer, the larger will be the energy flow, but the more opaque the material; the smaller will be the flow.

Then what is the conclusion of these thermodynamic considerations in the building of our solar model so far as the transport of energy is concerned? Starting from the core, we must decrease the temperature layer by layer, and choose this decrease taking into account the opacity of the material. And the flow of energy through every layer must equal the observed luminosity of the Sun.

Eddington was the first to build a 'Model Sun' based on the two conditions we just described. He built his solar model out of elements mixed in the proportion to what Russell had found in 1929, from eye estimates of line strengths in the solar spectrum. He also assumed that the lightest elements, hydrogen and helium, were present in the Sun in negligible amounts. He found that the solar gases must be highly concentrated toward its center and the central temperature of the Sun should be 40 million degrees Kelvin. Eddington made one further important discovery. If you want to build a star model such as the Sun with a given mass, radius and chemical composition, then this star must have certain luminosity. If it has a higher luminosity you would need a very steep increase of temperature inwards, ending up with so high a central temperature that the star would blow up. On the other hand, too small a luminosity would lead to such a low temperature in the interior that thermonuclear reactions would not take place. The model star has to satisfy conditions depending very much on the initial mass of the star; the larger the mass the higher the luminosity. This relation is called the *mass-luminosity relation*, and it was tested by Eddington on the Sun and a few other stars for which the mass and luminosity were known. It was found that stars approximately follow this relation.

After this first success, a difficulty appeared when Eddington computed what the luminosity of Sun should be according to its mass. He found a value ten times larger than the observed luminosity. For the computation of the solar luminosity, he needed exact values of the opacity of the gases, and these values were not very reliably known then. However by 1932, physicists had improved their computation of stellar opacity and there was little doubt on its accuracy. Therefore, something had to be changed in Eddington's original model of the Sun to make the computed luminosity come out equal to the observed one. It was discovered simultaneously by Eddington in England and Bengt Strömgren in America that the trouble lay in Eddington's assumption that hydrogen and helium were negligible. By making in their theoretical Sun, thirty five percent of hydrogen and the remaining sixty five percent Russell's mixture of heavier elements, they came out with a computed luminosity just equal to the actual luminosity of the Sun. The introduction of hydrogen changed the general appearance of the theoretical solar model very little, but it decreased the central temperature from 40 to 20 million degrees Kelvin. The central temperature further decreased with improved abundance and opacity measures to about 15 million degree K.

Leo Goldberg and his colleagues in 1960 carried out a detailed study of solar elemental abundance, based on curve of growth and spectrum synthesis techniques. They came out with the striking result that the Sun consists mainly of hydrogen, with a mixture of about 8% helium, and a much smaller concentration of heavier elements. This brings us up to date. In Table 3.2 is given the physical conditions in the solar interior obtained by averaging a number of recent solar models. It gives the run of temperature, density, mass, luminosity and pressure with distance from the center.

THE STANDARD MODEL OF PHYSICAL CONDITIONS IN THE SOLAR INTERIOR

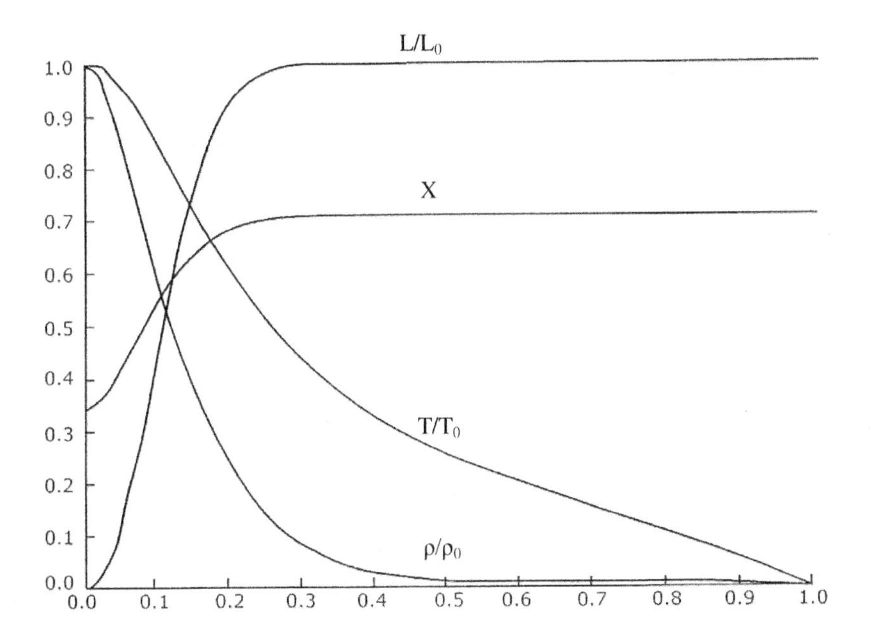

Figure 3.3 Plot of the physical conditions in the solar interior.

In Figure 3.3 are plotted the computed values of these parameters obtained from a model by Bahcall and Ulrich (1988) assuming a homogeneous chemical composition with fraction of hydrogen as $X = 0.71$, fraction of helium as $Y= 0.27$, and fraction of other elements as $Z = 0.02$, and considering that Sun's present age as 4.6 billion years. In this model the central temperature attains a value of about 15 millions degrees Kelvin and decreases to about 1.9 million degrees at the bottom of the convection zone. However, these values have been further slightly modified from recent helioseismology results.

Table 3.2 Conditions in Solar Interior. The data tabulated are averaged and smoothed from a number of solar models (from Allen's Astrophysical Physical Quantities– Cox's edition 4th 1999)

Central values:

Temperature	$T_c = 15 \times 10^6$ °K
Density	$\rho_c = 160$ g cm^{-3}
Pressure	$P_c = 3.4 \times 10^{17}$ dynes cm^{-2}
Composition	$X_c = 0.38$,

where T = temperature, ρ = density, P = pressure, M_r = mass within radius r, L_r = Luminosity or energy generated within radius r. $R_\odot$, $M_\odot$, $L_\odot$ = radius, mass, energy generation of whole Sun.

r		T	ρ	M_r	L_r	log P
$R_\odot$	10^3 km	10^6 °K	g cm^{-3}	$M_\odot$	$L_\odot$	in dyn cm^{-2}
0.00	0	15.5	160	0.000	0.00	17.53
0.04	28	15.0	141	0.008	0.08	17.46
0.1	70	13.0	89	0.07	0.42	17.20
0.2	139	9.5	41	0.35	0.94	16.72
0.3	209	6.7	13.3	0.64	0.998	16.08
0.4	278	4.8	3.6	0.85	1.00	15.37
0.5	348	3.4	1.00	0.94	1.000	14.67
0.6	418	2.2	0.35	0.982	1.000	14.01
0.7	487	1.2	0.08	0.994	1.000	13.08
0.8	557	0.7	0.018	0.999	1.000	12.18
0.9	627	0.31	0.0020	1.000	1.000	10.94
0.95	661	0.16	0.0^34	1.000	1.000	9.82
0.99	689	0.052	0.0^45	1.000	1.000	8.32
0.995	692.5	0.031	0.0^42	1.000	1.000	7.68
0.999	695.3	0.014	0.0^61	1.000	1.000	6.15
1.000	696.0	0.006	0.0	1.000	1.000	-

3.1.4 *The Neutrino Behavior*

3.1.4.1 *Neutrino Flux*

In the last section we saw that the energy generation in the Sun's core by the p-p cycle is through thermonuclear reactions by conversion of 4 hydrogen nuclei into one helium nuclei plus two hydrogen nuclei, and with the emission of two particles called neutrinos. The name neutrino means "little neutral one". This particle of sub-atomic nature is most remarkable, as it has no electrical charge, no mass, or it is negligibly small, and it travels with the velocity of light. Neutrinos are just bundles of energy, but quite different in character from photons, which are perceived as electromagnetic energy. Neutrinos have amazingly high penetrating power because they practically do not interact with matter once they are created. In the Sun $2 x 10^{38}$ neutrinos are produced every second. The question is - what happens to them? As neutrinos hardly interact with matter, they simply stream out from the Sun unhindered into space. An interesting situation occurs in the Sun in that electromagnetic radiation takes thousands of years to get out, because the energy bounces around interacting with the solar material, whereas the neutrino zips away from the solar core, almost unimpeded reaching the Earth in 8 minutes at the velocity of light. At the distance of Earth, the neutrino flux can be approximately given as:-

$$F(\nu) = 2L_o x 4\pi (AU)^2 / 25MeV = 3.5x10^{12} \text{ neutrino/cm}^2\text{/sec}, \quad (3.13)$$

where $L_\odot$ is solar luminosity, and AU is the astronomical unit (mean Sun-Earth distance). This large flux of neutrinos is passing though our bodies without our awareness.

3.1.4.2 *Detection of Solar Neutrino*

If a detector for solar neutrino could be made, scientists would be able to "see" inside the Sun, and study today's condition of the solar core as it is important to compare the predicted neutrino flux obtained from theoretical solar models with the real observed situation. An American physical chemist Raymond Davis first devised such a neutrino detector in the early sixties. He proceeded with the idea that there are really two classes of neutrino, neutrinos and anti-neutrinos, the later being the mathematical opposite of the former. When a particle collides with its anti particle they both vanish with a release of energy. Davis wanted to distinguish between neutrinos and anti-neutrinos, and to do this he proposed to use the atoms of chlorine-37 to make a "neutrino-catcher". The reasoning behind this was that if an atom of chlorine-37 could grab a neutrino of the right energy, it would be transformed into an atom of argon-37, and an electron is kicked out in the process. Anti-neutrinos can never have this effect on chlorine-37. Davis first attempted this task near a nuclear reactor, where large flux of neutrinos was available, according to the reaction:-

$$Cl^{37} + \nu_0 = Ar^{37} + e^- . \tag{3.14}$$

In his pioneering 1955 experiment Davis used 15,000 liters of perchloroethylene, because this fluid is rich in chlorine-37. After about 35 days, argon-37 was detected. He had detected neutrinos from the nuclear reactor and also placed an upper limit on the accepted neutrino flux from the Sun. To actually measure solar neutrinos, Davis (1968) built much more sensitive "Solar Neutrino Telescope". This was a tank holding 450,000 liters of highly purified perchloroethylene. The chlorine rich atoms were expected to capture the solar neutrinos and change into argon-37 atoms. As cosmic rays can also produce neutrinos in the Earth's atmosphere, the solar neutrino detector must be shielded from cosmic rays. Therefore it was located in the 1½-km deep Homestake gold mine in South Dakota, USA.

Davis's chlorine-37 detector could not detect *every* kind of neutrino from the Sun, particularly not the ones from the simple proton-proton chain because of their very low energy. The neutrinos produced through the decay of boron-8 (^8B), as shown in the earlier section Equation (3.12), features the third of the three possible branches of the p-p cycle. These have the right energy, about 14.06 MeV, to kick the chlorine-37 and form argon-37. This is the only solar neutrino that can be detected at present with this method. Bahcall (1965) calculated the resulting flux of neutrinos from ^8B decay as $2.5x10^7$/cm/sec or 40 Solar Neutrino Units (SNU) (1 SNU = 10^{-36} neutrino absorptions/sec/target atom). It is known that neutrinos formed through this process occur only in 0.05% of the p-p chain and depend very strongly on the temperature. Hence, the measurement of such neutrino flux gives us a way of estimating also the temperature of the deepest layers in the Sun.

3.1.4.3 *The Case of Missing Neutrinos*

From his chlorine experiment, Davis found a rate of 2.56±0.23 SNU. There was a running debate between Davis's observed neutrino flux and the theoretician's values derived from Standard Solar Model (SSM). Every time Davis would improve the sensitivity of his detector, the theorists would bring down the expected neutrino flux obtained from SSM. Finally a value of 7.6 ± 1.2 SNU was arrived at, which was still about three times higher than the observed neutrino flux. It was well established that the Davis measurement could not be in error by more than 10%. Although the neutrinos produced through ^8B decay represented only $5x10^{-5}$ of solar neutrinos, it was thought that perhaps the calculated probability of their production might be in error. One would like to measure the low energy p-p neutrinos directly, which could be done by measuring the capture of neutrinos by Gallium to produce Germanium. About 15 tons of Gallium was required for this experiment. There are 3 such radio chemical experiments, namely GALLEX, SAGE and GNU that use Gallium and are capable of detecting the low energy p-p neutrinos at relatively low threshold of 0.233 MeV. All these experiments, reported by Anselmann, *et al.* (1999), measured solar neutrino counting rates of 74.7 ± 5 SNU, while the SSM (Standard Solar Models) predicts a neutrino capture rate for the Gallium experiments as 128 ± 8 SNU, again a large deficient in measured neutrino flux. A number of implausible explanations for the low flux had been presented, for example:-

1. That the Sun's interior is completely mixed, so that the central temperature is still that of the young Sun, and therefore the neutrino flux is low. The mixed model gives about 1.5 SNU.

2. That the Sun, occasionally undergoes contraction and during those periods, it does not use fully its nuclear fuel.
3. That the abundance of heavy elements in the solar interior is much less than what SSM assumes, hence the interior opacity and the central temperature is lower and thus producing less ^{8}B.
4. That the SSM need to be modified.
5. That there is some problem with the fundamental particle – neutrino physics, or
6. That the neutrinos which are detected on the Earth, might have transformed to a different "flavor" or variety, than those which started from the solar core.

3.1.4.4 *Kamiokande and Sudbury Neutrino Observatory Results*

Another effort to measure intermediate and high energy ^{8}B neutrinos released through the third branch of the p-p chain was made by Japanese scientists (Fukunday, *et al.* (1996)). This experiment consists of a 680-ton water tank, located at about 1 km underground in Kamiokande mine, and the charged particles are detected by measuring the Cherenkov light, through *elastic scattering* with the hydrogen nuclei. This and later Super Kamiokande experiments (Fukunday, *et al.* (1999)) are sensitive only to intermediate and high energy ^{8}B neutrinos. The measured flux from Super Kamiokande experiment was found again to be deficient by about 50%, over the total flux predicted by the SSM.

None of the measurements of neutrino flux by chlorine, water and Gallium experiments were consistent with each other, and all results showed deficit neutrino flux, as compared to the theoretically predicted values. Recently, another experiment is being carried out at the Sudbury Neutrino Observatory (SNO) in Canada which uses 1000 tons of heavy water (the deuterium isotope) located at a depth of 2 km underground in an Ontario nickel mine. This experiment detects intermediate and high energy solar neutrinos. In both heavy water and ordinary water experiments, the neutrinos can elastically scatter electrons to produce Cherenkov light radiation, but such electron scattering can be generated by any of the three neutrino "flavors" or types, namely *electron, muon,* and *tau* neutrinos. The SNO is capable of measuring the ^{8}B neutrinos, through the following three reactions:-

The "charged–current" (CC), in which a solar electron neutrino ν_e hits a deuteron producing the reaction

$$\nu_e + d = e + p + p. \tag{a}$$

This is called the "charged-current" interaction, because it involves the virtual exchange of the charged intermediate vector boson ($w^{\pm}$). The result of CC events was that the ν_e flux was roughly one third of the ^8B neutrino, as predicted by SSM. To see what becomes of the missing ⅔ of the predicted flux, the following second possible reaction was examined

$$\nu + d = \nu + p + n. \tag{b}$$

This quasi-elastic breakup of a deuteron into a proton and neutron is called a "neutral-current" (NC) reaction, because it occurs by a neutral intermediate vector boson (Z^0). This reaction (b) can be produced with equal probability by a neutrino of any of the three flavors. It serves as a direct measure of the total flux of solar neutrinos, disregarding any flavor. The NC reaction yield corresponds to the total flux of ^8B neutrinos, regardless of 3 flavors, of $(5.09 \pm 0.62)\ \mathrm{x}10^6\ \mathrm{cm}^{-2}\ \mathrm{s}^{-1}$. The Standard Solar Model flux prediction for ν_e produced by ^8B decays in the solar core is $(5.05 \pm 0.91)\ \mathrm{x}10^6\ \mathrm{cm}^{-2}\ \mathrm{s}^{-1}$. But Sudbury Neutrino Observatory's measurements of the flux of ^8B electron neutrinos, based on 2000 CC events observed was only $(1.76 \pm 0.11)\ \mathrm{x}10^6\ \mathrm{cm}^{-2}\ \mathrm{s}^{-1}$. This strongly suggests that solar neutrinos change *flavor* en-route from the Sun. Thus the neutrino flavor seems to *oscillate*. This argument becomes even stronger when one considers the 3rd type of reaction, namely the elastic scattering of solar neutrinos with electrons

$$\nu + e = \nu + e. \tag{c}$$

If an incident neutrino is a $\nu_{e,}$ both Z and W exchange and contribute to this reaction. For neutrino flavors only the smaller Z-exchange term can contribute to the scattering of electrons. The calculations show that the total rate for reaction (c) is proportional to the incident flux ν_e plus about 15% of the combined incident flux of muon (ν_μ) and tau (ν_τ) neutrinos. Although the reaction (b) provides the most direct measure of the total incident flux, regardless of the flavor, one can also measure the total flux by combining data from reactions (a) and (c). And this was how SNO scientists were able to estimate the total neutrino flux.

3.1.4.5 *Solution of the Solar Neutrino Puzzle*

From the latest experimental results obtained from the SNO and Super Kamiokande, as reported by Ahmad *et al.* (2002), it is now convincingly shown that the Sun's output of electron neutrinos flux is just about what astrophysicists predicted from the Standard Solar Models. About ⅔rd of the highest energy electron neutrinos emerging from the solar core are transformed either to muon ν_μ or tau ν_τ neutrinos by the time they arrive at Earth in changed guises (flavors) to interact with deuterons or electrons in heavy water.

There is a good evidence that there are only three varieties of neutrinos, namely the electron, muon and tau in nature.

Looking at the historical perspective of the neutrino puzzle, it may be pointed out that when Raymond Davis and the theoretician John Bahcall, first reported in 1968 the shortfall of observed solar neutrinos, as compared to the predicted values obtained from SSM, then and even until recently, many scientists believed that the problem lay with the solar models. Now it's quite clear that the inadequacy was in our understanding of the neutrino physics and that the theoretical models of solar internal structure were correct. The new technique of helioseismology is providing a complementary tool to probe the solar interior, which will be discussed in Chapter 9.

3.2 The Solar Constant - Solar Irradiance

The Solar Constant, now generally known as Solar Irradiance was originally defined as the total radiation falling on a 1 cm^2 area in one minute (note that the unit of time is minute) placed normal to the sunbeam, outside the Earth's atmosphere, and at the mean distance to the Earth. This datum is of fundamental importance to infer:-

1. The heat balance of the terrestrial surface,
2. The condition in the solar atmosphere, and

3. The interior of the Sun, since the radiation leaving the Sun should be equal to the radiation produced inside.

At the Astrophysical Observatory of the Smithsonian Institution in Washington, Charles Abbot and his colleagues made extensive measurements of the Solar Constant, during the early part of the last century. Their basic instrument was the pyrheliometer, which measured the total amount of radiation per minute per square centimeter. Although such measurements had fair degree of accuracy, but it was difficult to precisely eliminate the extinction by the terrestrial atmosphere. Normally this was done by observing the Sun at different altitudes and extrapolating to zero air mass. The extrapolation is reliable only if it is made for each wavelength and combined with spectro-bolometric records. The extrapolation to zero air mass gave a value of the solar constant as 1.90 cal cm^{-2} min^{-1} or about 130 milli watts per cm^2. In spite of very careful observations made by Abbot, the accuracy of the ground-based measurements was about 1.5%.

When observations from space became possible, several instruments were flown to measure the solar irradiance at a much higher accuracy. One such instrument is the Active Cavity Radiometer Irradiance Monitor-I (ACRIM-I), which was flown on Solar Maximum Mission (SMM) during 1980-1989. Since then several space missions such as HF, ACRIM-II and VIRGO have obtained precise measurements. In Figure 3.4 is shown a complete plot of the Total Solar Irradiance (TSI), which replaced the term Solar Constant. These measures were made from 1979 through 2002. This plot shows short-term variations in the solar irradiance of about 0.2%. The dips in the Sun's overall intensity correlate with the appearance of sunspots crossing the solar disk, indicate that at least temporarily sunspots do prevent energy escaping from the Sun. The ACRIM instruments showed a slight but steady decline of the solar irradiance between 1980 and 1985 and seems to flatten around 1986 and again increase with the ascending phase of solar cycle. From these measurements it seems that the so-called solar 'constant' is not in fact constant, as earlier scientists thought. Even slight changes in solar energy output would profoundly affect our Earth's climate. The longer term variations correlate with the solar activity cycle. The Sun is brighter at solar maximum by about 0.1%. This increase arises mainly from

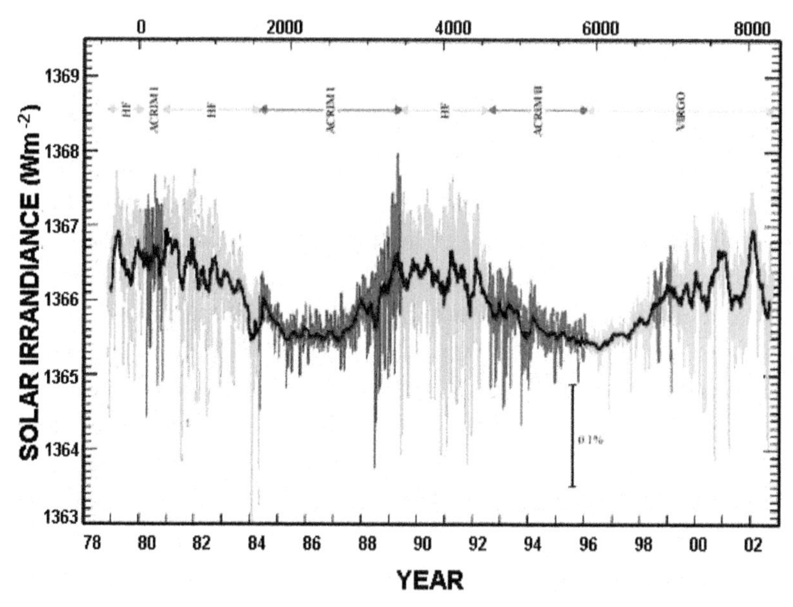

Figure 3.4 Plot showing solar irradiance measurements obtained by various spacecraft experiment during 1979 to 2002. Sharp peaks indicate variations due to appearance of sunspots, large variation on the order of 0.1% correlate with the solar cycle variation.

faculae and the so-called network structures. They are magnetic in origin. It would be of interest to investigate if the TSI varied during the 70-year period of Maunder Minimum from 1645 to 1715 AD. In fact climatic changes were indeed noticed as a Mini-Ice age during that period, apparently affecting the global temperature.

3.3 Limb Darkening

Besides the sharp boundary, as mentioned above, one also notices less brightness near the limb of the solar disk as compared to the central portion. This phenomenon is known as '*limb darkening*'. When we look directly at the photosphere at the center of the solar disk, we look deep to the hottest part of the photosphere. But when we examine the limb, the relatively darker portion of the solar disk, we receive radiation coming from the higher layers of the solar photosphere, Figure 3.5. Lower intensity means less temperature, indicating that the higher layers of the photosphere are cooler as compared to the deeper layers. Limb darkening intensity measurements provide a means to probe the temperature and density structure of the solar photosphere. In other words, this is a way to determine a photospheric '*solar model*'. On an average the brightness at the disk center corresponds to a temperature of about 6390°K, characteristic of deep photosphere, while the brightness near the limb corresponds to about 5000°K, in upper layers of the photosphere. The effective or the mean solar photospheric temperature, averaged over all

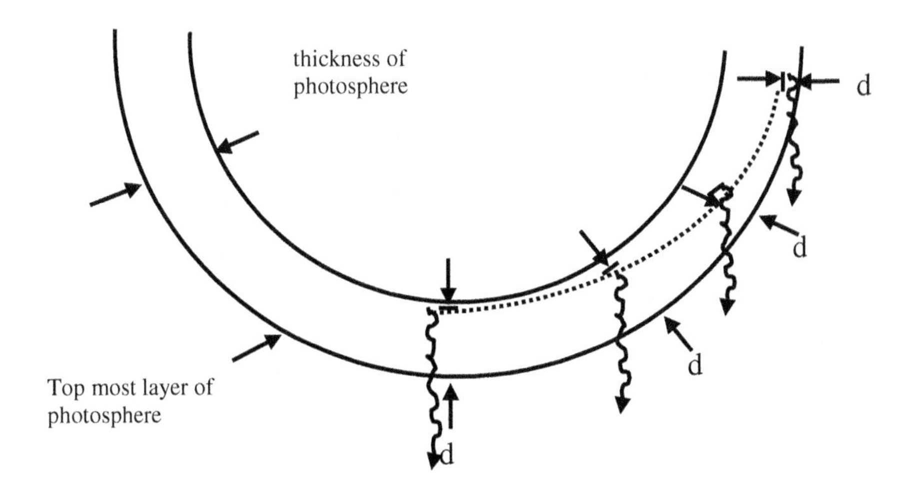

Figure 3.5 Schematic showing the effect of limb darkening. The depth 'd' to which we *see* in the photosphere decreases when our line of sight moves from the center to the limb. The broken arc represents the loci of all the points from which we receive the radiations. As we look towards the solar limb we receive radiation from higher and higher photospheric layers, which are darker thus must be cooler.

the locations is about 5800°K. The easiest procedure to measure the limb darkening profile is to stop the telescope drive and allow the solar image to drift across a light detector. The Earth's rotation will then do the job of scanning the solar disk in a very uniform way. Limb darkening depends strongly on the wavelength, and is much pronounced in the violet than in the red region of the spectrum, as shown in Figure 3.6.

In the far infrared at 9 microns, the opacity of the solar material is so high that we receive radiation only from the outermost layers of the photosphere at both the limb and disk center. The limb darkening curve in far infrared is almost flat. Limb darkening is caused by the photospheric temperature gradient and it disappears in the far infrared. In the ultraviolet, around 0.15 micron, it would correspond to the temperature minimum region, where $dT/dh \approx 0$, and again it disappears.

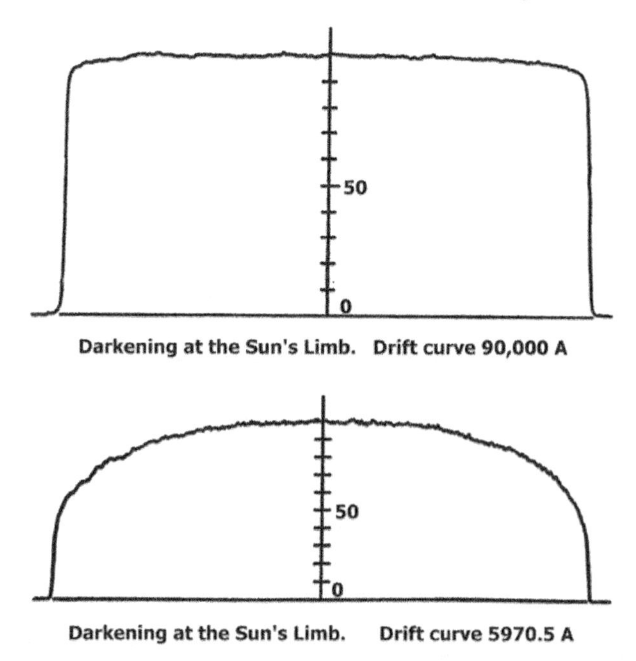

Figure 3.6 Limb darkening profiles at 90,000 Å (top curve) showing an almost flat intensity distribution, indicating that infrared radiation comes from the upper most layers of the photosphere. In visible light, at 5970.5 Å (lower curve), there is strong limb darkening.

Determination of the wavelength, where the disk intensity profile changes from limb darkening to limb brightness can provide a means to locate the height of the photospheric temperature minimum layer. Evidence of a limb brightness of about 25% in the last 30-arcsec ($0.03R_{\odot}$) from the limb has been observed at 6-cm (Furst *et al.* 1979), (Marsh *et al.* 1981), and about 10% within 20-arc sec from the limb at 1-mm (Horne *et al.* 1981). In practice it is difficult to precisely observe the profile near the limb, because diffraction leads to the low spatial resolution at long wavelengths. Broadly these results are consistent with the location of the temperature minimum obtained from photospheric models.

3.3.1 *Limb Polarization*

Lyot in 1948 (Lyot 1948) detected traces of linear polarization near the extreme limb of the Sun. The component perpendicular to the radius was slightly stronger than the one parallel to it. The polarization in an area of diameter 1 arc minutes at 2 arc minutes from the limb was found to be between 2×10^{-5} and 39×10^{-5}. This small amount of polarization rises from anisotropic scattering in the highest photospheric layers. Since the radiation from the Sun is mainly directed radially the tangential vibrations are dominant. An exact solution of the observed polarization, based on pure electron–scattering atmosphere, was derived by Chandrasekhar (1950). The effect of a superposed continuous absorption coefficient reduces the predicted polarization by a large factor and leaves only a small amount that has been observed (Code, 1950, Pecker, 1950).

3.4 Solar Rotation

Soon after the invention of the telescope in 1610 by Lippershey and his associates in Holland, Galileo in Italy began observations of the Sun in white light. He noted a westward drift of sunspots with a period of about one lunar month. Around the same time in 1611, Scheiner also began systematic observations of the Sun and he published the first observational determination of the Sun's rotation period and the inclination of solar rotational axis to the elliptic. He noted that sunspots on either side of the equator at higher latitudes had longer periods of revolution than those near the equator. Modern observations substantiate these early findings of Scheiner that the Sun does not rotate like a solid body. It exhibits differential rotation from pole to equator such that, at the Sun's equator it rotates once in about 27 days and at its polar regions it takes 31 days.

Soon after the introduction of spectroscopy in astronomy, and using the Doppler Effect discovered in 1872, the earliest spectroscopic measurements of solar rotation rates were made in the nineteenth century. Note that the apparent rate is called *synodic* and is the rotation rate relative to the Earth, while the *sidereal* rate is the true rate of rotation of the Sun on its axis relative to the stars. The two differ due to the Earth's orbital velocity of $0^{\circ}.9865$ per day. The synodic period is slightly longer because the Sun rotates a little further to face the Earth's new position each day, as both Sun's axial rotation and Earth's orbital motion being in the same direction.

Interest in the determination of solar rotation continued with improved solar spectrographs. The spectroscopic method has an advantage over sunspots because it can be applied over the entire surface and is not confined to the sunspot zone. Fairly consistent results on rotation rate were obtained by Adams and Lasby (1911), St. John (1932) at the Mount Wilson observatory and by J.S. Plaskett (1915) and H.H. Plaskett (1916) at the Oxford University, and Stoney (1932) in England. However, all these measurements had systematic errors because of the scattered light and other factors not taken into account. The early technique was to measure the rotation by exposing spectra near the east and west limbs at several latitudes and then subtract the Doppler shifts obtained using one or several spectral lines. Substantial errors in earlier determinations of solar rotation were detected due to various obliterating effects like; the influence of granulation and super granulation motions, which have amplitude ranging from 0.2 to 0.5 km/sec, the steep variation of the 'limb red-shift' (Adams 1959) near the limb, and also from small errors in position on the solar disk.

During the early sixties modern photoelectric-spectroscopic techniques to measure the solar rotation became available. Using the solar magnetograph in its velocity mode at Mount Wilson observatory, Howard embarked on very systematic photoelectric measurements of the solar rotation. Howard and Harvey (1970) published detailed solar rotation velocities measures. These observations gave higher accuracy due to high sensitivity and the large number of measurements that were made over the whole disk. They also took account of the geometrical corrections, limb red-shift, supergranulation motions, and the orbital velocity of the Earth. At the John Wilcox Solar Observatory at Stanford, rotational velocity observations were also carried out using essentially similar spectroscopic photoelectric magnetographic techniques as at Mount Wilson (Scherrer et al. 1980). A slight difference of about 1% has been detected between the Mt. Wilson and Stanford's rotational velocity measures. This may be due to the difference in the epoch of observations.

The sidereal rotation period, Ω is normally given in form of a polynomial:-

$$\Omega = A - B\sin^2\theta - C\cos^4\theta \text{ deg/day,} \tag{3.15}$$

where A, B, C are constants and θ is the heliographic latitude. Howard and Harvey gave values of constants A, B and C as 14.19, 1.70 and 2.36 respectively from their spectroscopic determinations.

3.4.1 *Solar Rotation from Sunspot Tracers*

As mentioned earlier, the first indication of solar rotation came from sunspots used as tracers by Galileo, Scheiner, Hevelius and Carrington. In 1874 the Greenwich Observatory began systematic synoptic observations of sunspots on white light full disk photoheliograms and continued this until 1976. Similar white light observations were started in 1902 at the Kodaikanal Observatory in India and in 1977 at the Heliophysical Observatory in Debrecen, Hungary.

Howard *et al.* (1984), from the Mount Wilson photoheliogram archives, obtained rotation rates using individual sunspots for the period 1921-82. Newton and Nunn (1951) determined the rotation rate from 136 single-spot recurrent groups using the Greenwich data set for the period 1934-44. The solar sidereal rotation rate obtained by Newton & Nunn is given by the following relation:-

$$\Omega = 14.38 - 2.96\sin^2\theta \text{ deg/day.} \tag{3.16}$$

In Figure 3.7 is shown the variation of solar rotation rate with latitude over various time intervals obtained by several authors using different

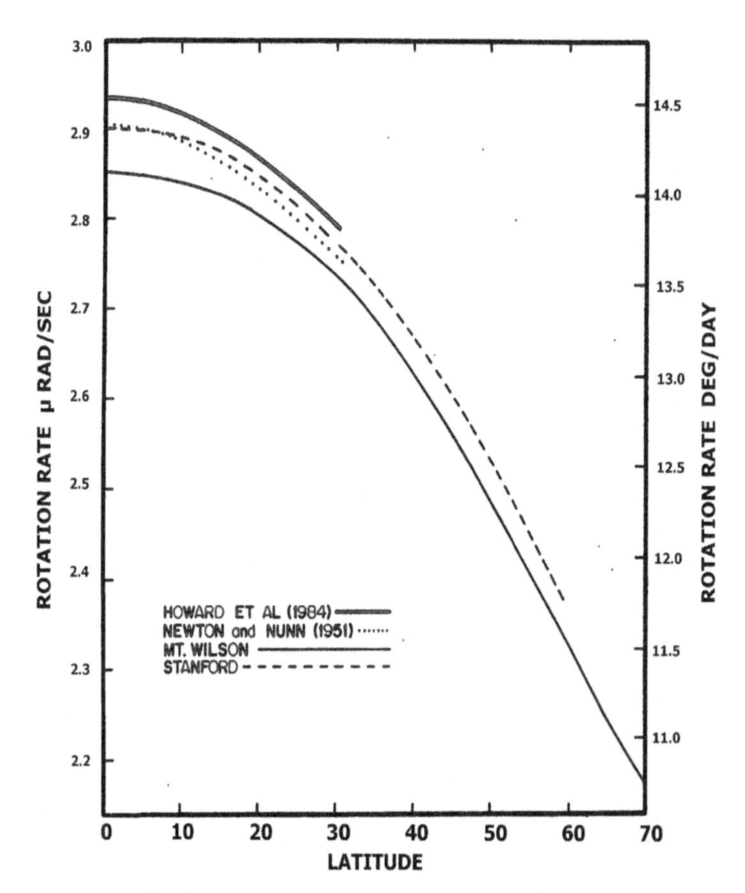

Figure 3.7 Latitude dependence of solar rotation rate obtained using spectroscopic data and sunspots as tracers, by various authors and at different epochs. Thick continuous curve refers to Howard *el al.*, using Mount Wilson's sunspot data for the period 1921-82. Dotted curve refers to Newton and Nunn's results based on 136 single spot recurring groups during 1934-44. The thin continuous curve refers to Mount Wilson's Doppler data for the interval 1967-82 and the dashed curve is from spectroscopic determination by the Stanford group during 1976-79.

methods. These measurements show some systematic differences. The Mount Wilson values are higher by about 1% as compared to the Greenwich values. This difference could be due to different epochs of the two measurements, or may be selection effects in the choice of sunspots and spot groups. From a very extensive study of 96,283 spots, extending over a period of 62 years (1921-82), Howard *et al.* (1984) showed that small spots of area less than 5 millionths of the solar disk rotate faster by about 0.8 % as compared to large spots of 15 millionths, see Figure 3.8.

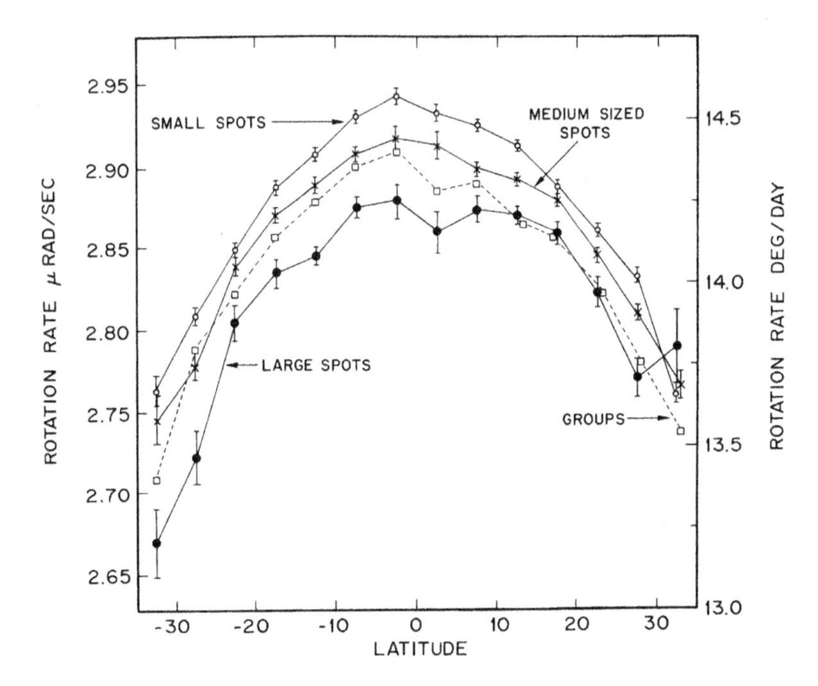

Figure 3.8 Sidereal rotation rate using sunspots of various sizes and sunspot groups. Small spots are less than millionths of the solar disk, medium spots are from 5 to 15 millionths, and large spots are greater than 15 millionths. From these plots it is easily seen that small spots having lower magnetic strength rotate faster in all latitude belts, while large spots with very strong fields rotate the slowest.

Comparing the rotation rates obtained by sunspots and those obtained by Doppler spectroscopic method, which essentially gives the velocity of the solar plasma, it will be noticed from Figure 3.7 that there is a marked difference between the two. The rotation rates obtained using sunspots as tracers near the equator show about 3% higher rate compared to the spectroscopic rates. But around ±30° latitude the difference between the two decreases. From these results, it appears that the spots and spot groups 'plough' through the solar photosphere (Gilman & Foukal 1979). It is as if a boat is peddling with its own energy faster than the flowing river. The question is; how is this additional motion is imparted to sunspots?

Howard *et al.* (1984) have also shown that there is a significant difference between the rotation rates obtained from individual sunspots, sunspot groups and large spots. As mentioned earlier the spots of different sizes rotate at different speeds – the largest spots rotate slowest. The large sunspots and sunspot groups, having high magnetic field strengths, rotate slower by about 0.8 % than the average small size spots. It seems that large spots and groups having strong magnetic field do not seem to 'plough' through the solar plasma as fast as small low magnetic field sunspots.

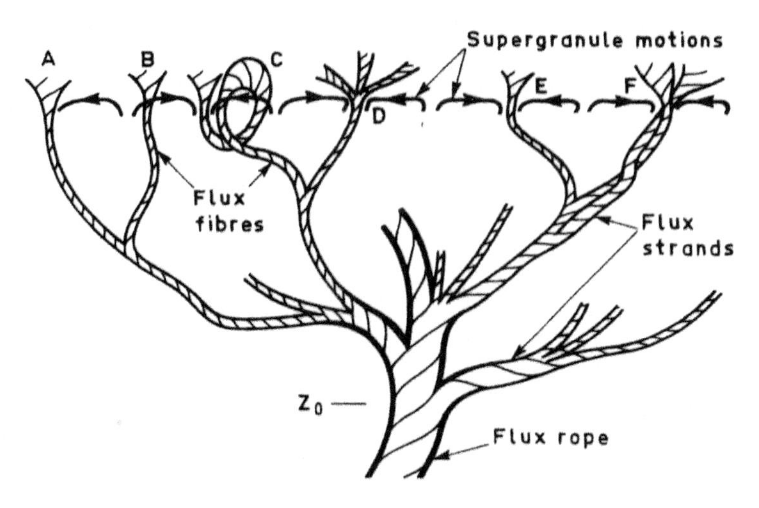

Figure 3.9 Schematic drawing of flux tubes inside the Sun. Drawing after Piddington, 1979.

A scenario of magnetic flux ropes inside the Sun was presented by Piddington and is shown in Figure 3.9. This schematic depicts that sunspots appearing on the solar surface are manifestation of ropes like magnetic flux tubes and fibers, similar to "tree trunks" and "tree branches". In the upper convection zone, which opens up at the solar surface, they appear as sunspots of varying sizes, as shown in Figure 3.9. Bigger sunspots would correspond to larger flux ropes and are 'anchored' in deeper layers, while the smaller spots correspond to smaller flux tubes anchored in the uppermost layers of the convection zone, just below the photosphere. The observed difference in the rotation rates, corresponding to spots of different sizes, may be due to being 'anchored' at different depths in the Sun. The difference between the observed rotation rates obtained for large spots and those from small spots may also be due to some kind of 'magnetic braking'. Large flux tubes, as compared to thin flux tubes, have greater magnetic field strengths.

3.4.2 *Variation of Rotation Rate with Solar Cycle*

From the earliest times sunspots served as tracers for determination of the solar rotation. Eddy, Trotter and Gilman re-constructed the pattern of solar rotation obtained by examining the original drawings of sunspots made by Scheiner and by Johannes Hevelius in their books. Scheiner's book entitled 'Hosaursina' was published in 1630 AD, just before the start of the Maunder minimum. Hevelius's book entitled '*Selenographia*' was published in 1647 AD and gave sunspot position data just at the beginning of Maunder minimum in 1643. In each of these two books daily drawings of the Sun and sunspots are shown for nearly continuous period of about 2 years. From these drawings, Eddy (1974) derived the solar rotation rates shown in Figure 3.10. The trend of the solar rotation obtained from these ancient records seems to tally fairly well with the modern determinations. It is clearly seen from Figure 3.10 that around the beginning of the Maunder minimum - 1642-1644 AD, when the sunspots number started declining, the solar rotation rate was systematically higher as compared to the earlier period before the minimum around 1625-1626 AD. The *equatorial velocity* appeared to increase three fold during the minimum period, as compared to higher latitudes (± 20°). In other words the Sun's equatorial region speeded up, completing one full day rotation (about 0.5 deg. per day) faster at the start of the Maunder minimum in 1642-1644 AD than it had before the minimum in 1625-1626 AD. If this is true, two important questions arise;

how and why the Sun's differential rotation changed during the Maunder minimum? And does it happen during all sunspot minimum periods?

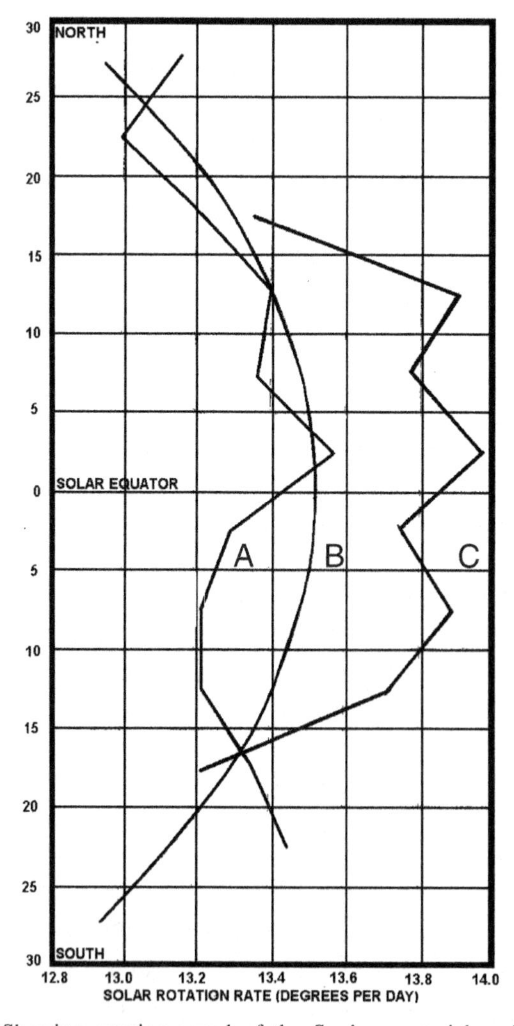

Figure 3.10 Showing rotation speed of the Sun's equatorial region. Curve A, shows Scheiner's observations that the Sun's rotation rate in 1620 was (before Maunder minimum) much the same as it is today. Curve B shows the modern rate of rotation at different latitudes plotted in degrees per day. Curve C shows Helevius's observations in 1640, during Maunder minimum, indicating that the Sun's equatorial rotation rate, compared with the polar rate had increased by almost three times.

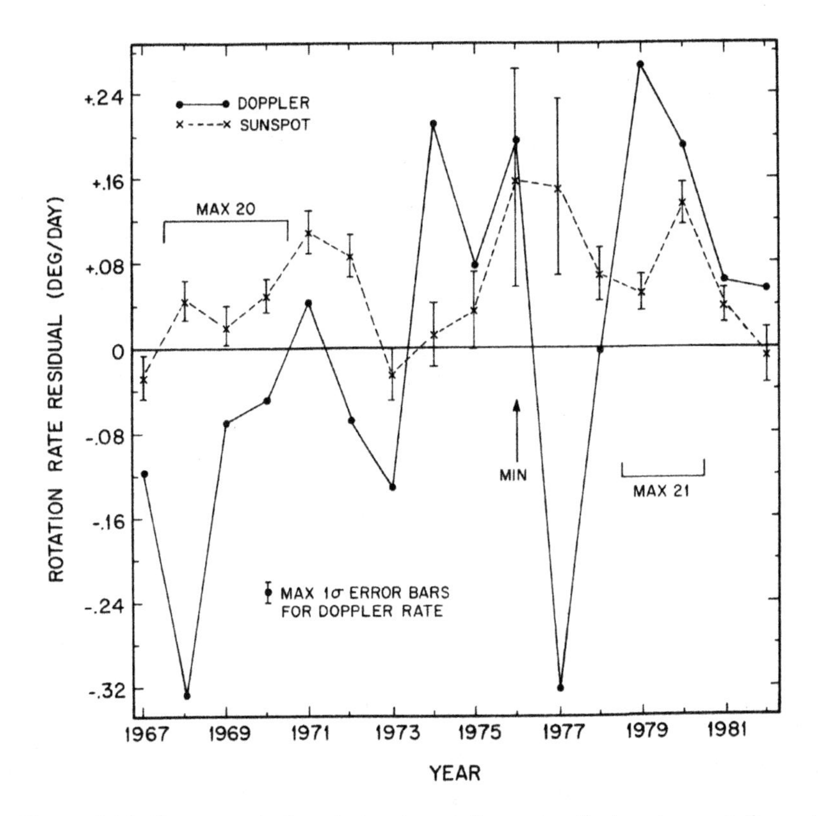

Figure 3.11 Shows marked variation in rotation rate, displayed as rotation rate residual, obtained using Doppler measurements shown by continuous line, and that obtained from sunspot positions by dash line, during 1967 to 1982 period, covering about 1½ solar cycle.

From a study of Mount Wilson's Doppler and sunspot data for the period 1967-1982, Howard *et al.*, (1984) have shown that the rotation rate varies with solar cycle. It is noticed that there is a marked variation of about 1% in the rotation rate during a solar cycle as shown in Figure 3.11. From this Figure it will be noticed that the amplitude of the variation is greater by a factor of about two in the photosphere, obtained by Doppler measures, as compared to that obtained by sunspots as tracers. Tuominen and Virtanen (1987) have also confirmed from Greenwich sunspot measurements that the rotation rate changes with solar cycle.

Hathaway and Wilson (1990) have obtained a correlation between the rotation rate and sunspot number. Recent advances in helioseismology have also broughtout this fact, that the solar rotation rate is correlated with solar activity cycle. From the GONG (Global Oscillation Network Group) and MDI(Michelson Doppler Imager) data, Jain, Tripathi and Bhatnagar (2001), and also other authors, have shown that the equatorial rotation rates show variations with the activity cycle, as seen in Figure 3.12.

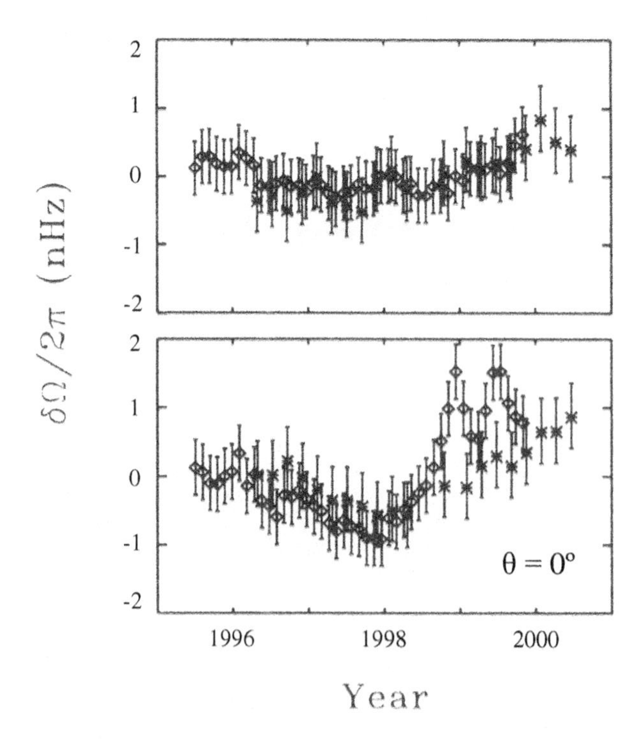

Figure 3.12 Shows the residuals of rotation rates at equator and 45° latitudes, obtained from GONG data, indicating variation of rotation with solar cycle during 1995 to 2000.5 period. Vertical lines indicate the rsm errors.

An increase by almost 1-2 %, is seen in the residual rotation rate during solar cycle 22-23, in the equatorial region ($\theta = 0^o$) while at higher latitude range ($\theta = 45^o$) the effect is marginal. At even higher latitudes of

85°, in the solar interior ($0.98R_0$, $0.95R_0$, $0.90R_0$, $0.80R_0$), Basu and Antia (2001) have shown (Figure 3.13) from the GONG and MDI data, that the rotational rate residuals decrease with increasing solar activity. The rotation rate in the outer layers of the polar region seems to decrease with time, during the rising phase of the cycle 23 (1995–2000). This decrease extends to a depth of at least $0.1R_0$, but beyond this depth, at all latitudes the variation in rotation rate with time is not detectable. How the solar rotation is affected by solar activity is an interesting problem in solar physics, and at present no adequate explanation is available.

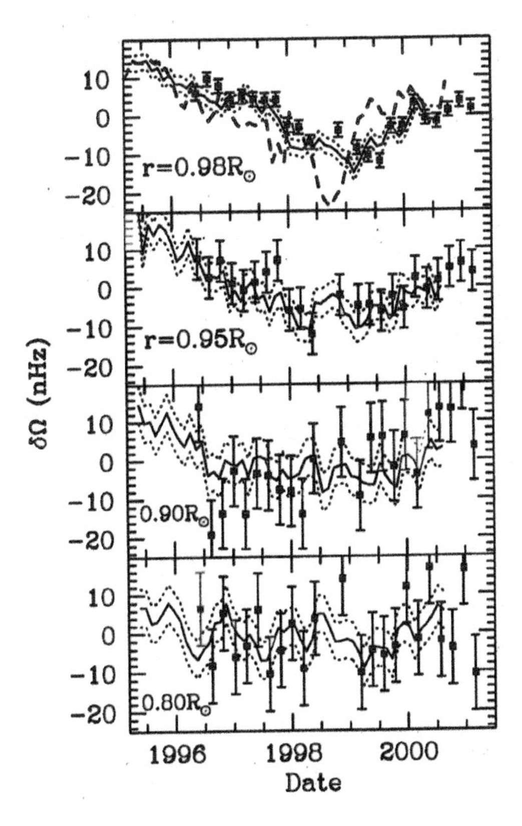

Figure 3.13 Shows the rotational rate residuals at 85° latitude, plotted as a function of time at a few radii ($0.98R_0$, $0.95R_0$, $0.90R_0$, $0.80R_0$) in the solar interior, obtained from the GONG and MDI data. These curves distinctly show a variation in rotation rates with the solar cycle in the interior of the Sun up to at least $0.1R_0$, from the solar surface beyond this, the effect is not discernable.

3.4.3 *Rotation of Photospheric Magnetic Field*

Using the Mt. Wilson full disk magnetic field data, Wilcox and Howard (1970) examined the rotation of *non-spot* magnetic fields on the Sun and found that the solar magnetic field rotates at a rate similar to large recurrent sunspots. This was in agreement with Newton and Nunn's (1951) findings at the equator, but at higher latitudes the magnetic field rotates faster than sunspots. Snodgrass (1983) also examined the rotation rate obtained from the Mount Wilson magnetic field data and found a close agreement with the sunspot rotation rates obtained by Newton and Nunn near the equator, but at higher latitudes zones he found good agreement with Mount Wilson's Doppler measurements. The following expression gives the angular rotational velocity of magnetic fields with latitude:-

$$\Omega = 14.37 - 2.30\mathrm{Sin}^2\theta - 1.62\mathrm{Sin}^4\theta \quad \mathrm{deg/day}. \tag{3.17}$$

where θ is the latitude. Further, the slope of differential rotation for the magnetic data is intermediate between sunspots and that for the spectroscopic Doppler measurements.

Ye and Livingston (1998) studied rotation right at the solar poles. For this they used the CO lines at 4.6 microns, and made measurements at 5 arc sec away from the limb around the poles. Taking advantage of the seasonal tilt of the B_0 angle, they actually observed over the pole so that the direction of rotation reversed. The advantages in this IR work were that: no supergranule disturbances, low scattered light, and no limb effect. There had been predictions of a possible spin up from models. None was found. Over the limited time the observations were made (3 months) there seemed to be an occasional cessation of rotation, but this may be a merely a statistical fluctuation.

3.4.4 *Rotation in the Solar Interior*

Using the new technique of helioseismology, high precision values of solar rotation rates from the surface to almost the core has been experimentally determined. In principle the rotation in the solar interior

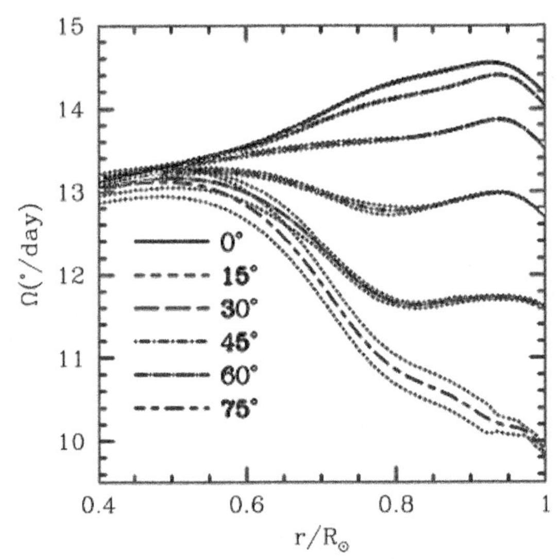

Figure 3.14 Showing rotation rate at various latitudes (0°, 15°, 30°, 45°, 60 and 75°) as a function of radial distance from the Sun's center. It is seen that the differential rotation at the surface continues through the convection zone but near the base of the convection zone, located at 0.65 R_0 there is a sharp transition from differential rotation to a solid body rotation

is measured from the frequency splitting of the global pressure p-mode oscillations (Rhode *et al.* 1979).

Tremendous progress has been made during the last two decades to determine the solar rotation rate from several global oscillation networks such as IRIS (International Research on the Interior of the Sun), BISON (Birmingham Solar Oscillation Network), GONG (Global Oscillation Network Group), TONG (Taiwan Oscillation Network Group), and space based Stanford University's SOI (Solar Oscillation Instrument)-MDI (Michelson Doppler Imager) experiment on the SOHO (Solar Orbiting Heliospheric Observatory) spacecraft. The latest picture of the variation of rotation rates with depth from almost the solar surface to $0.4R_0$ at 0°, 30°, 45°, 60°, 75° latitude zones is shown in Figure 3.14 obtained from the GONG data.

Around the sunspot zone of 0° - 30° up to about 0.75 R_0 the rotation rate varies between $14^{\circ}.6$ and $13^{\circ}.6$ per day and is higher as compared to deeper layers up to the base of the convention zone around $0.65R_o$. At higher latitudes (60°-75°) a sudden increase in the rotation rate is observed around 0.7–0.8 R_o as compared to higher layers at these latitudes. From Figure 3.14, it is clearly seen that the observed differential rotation at the surface continues through the convention zone but near the base of the convention zone, located at a radial distance of $0.65R_o$ there is a sharp transition from the differential rotation to a *solid body* rotation.

In Figure 3.15 is shown the rotation rate in the form 'iso-tach' contours of constant rotation velocity with radial distance from the Sun's center.

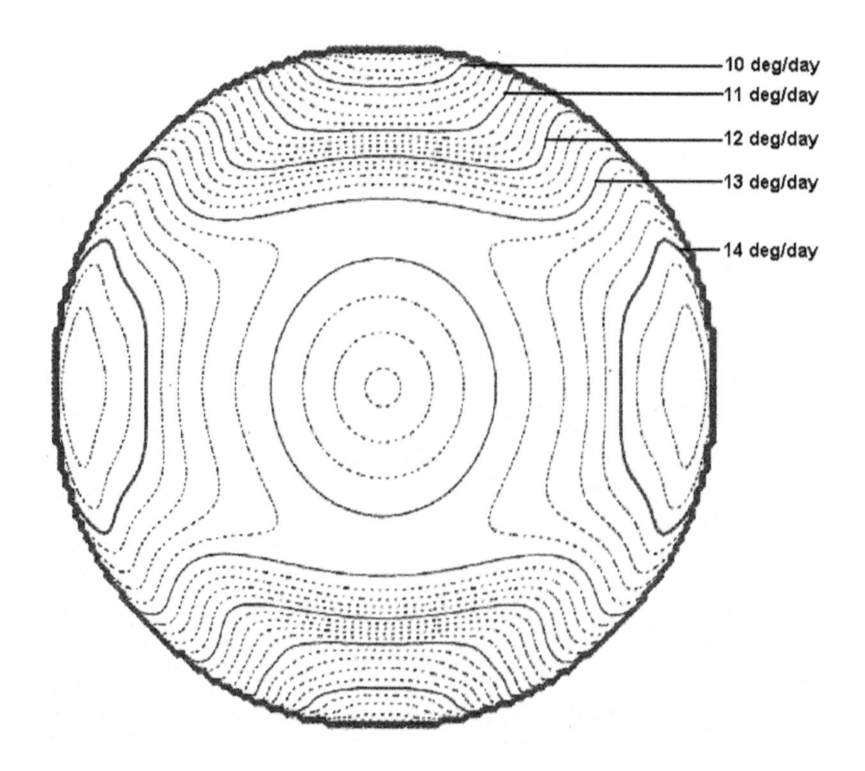

Figure 3.15 Shows the rotation rate in the form of 'Iso-tech' contours, at constant rotation rates with radial distance from the Sun's center at various latitude zones.

The rotation rate derived from helioseismology techniques at the surface is found in good agreement with spectroscopic Doppler measures, but is somewhat lower than the rotation rate obtained from the sunspots. The rotation rate derived from sunspots as tracers; correspond to the rate closer to that found at a depth of 0.03 R_0 below the surface. The determination of solar rotation in *very* deep solar interior obtained from the helioseismology techniques is rather uncertain, as only very low degree p-modes penetrate in the core and the determination of the rotational splitting is not reliable. The best measurements of low degree splitting are obtained from the full disk integrated sunlight experiments by BISON and IRIS network observations. From the results of these experiments, Tomcezyk *et al.* (1995) have shown that the solar core is rotating slower than the equatorial rotation rate at the surface. Recently in 2003, Simon Kras, using the GONG data, has shown that the solar core rotates at about the same rate as the surface. The earlier claims made by Dicke in the seventies, that solar core is rotating two to nine times faster than the surface is now discredited.

3.5 Fast and Slow Streams-the Torsional Oscillations

From the large amount of full disk Doppler observations obtained during 1967-1982 at Mount Wilson observatory, Howard and LaBonte (1980), LaBonte & Howard (1982a, b) discovered that the Sun undergoes low-amplitude '*torsional velocity oscillations*' about the solar rotational axis. These torsional velocity oscillations were found as residual motions after removing the mean rotational velocity in particular latitude belts. The main feature of the torsional oscillations is seen as streams moving faster than the average rotational speed, which is considered positive, or westward, and the slower or eastward motion for particular latitudes. The amplitude is about 5 m/s. These torsional oscillations consist as shear waves that start at very high latitudes, perhaps quite close to the poles (although it is very hard to observe there), and move equator ward taking nearly 22 years, with a latitude drift speed of about 2 ms^{-1}. This means that in each hemisphere, at any time, there are four zones of alternating streams with slightly different speeds. Solar activity tends to be centered

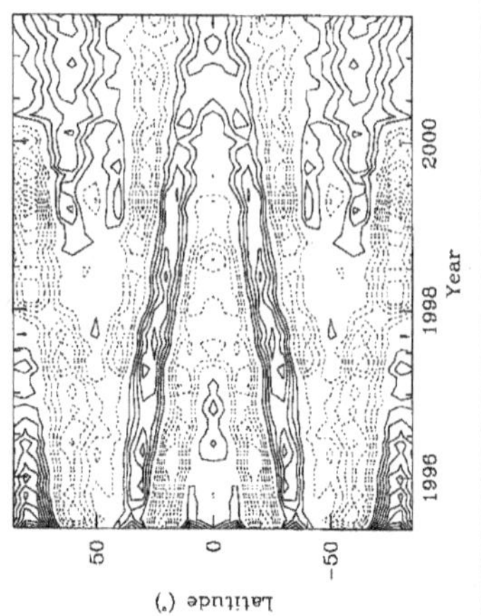

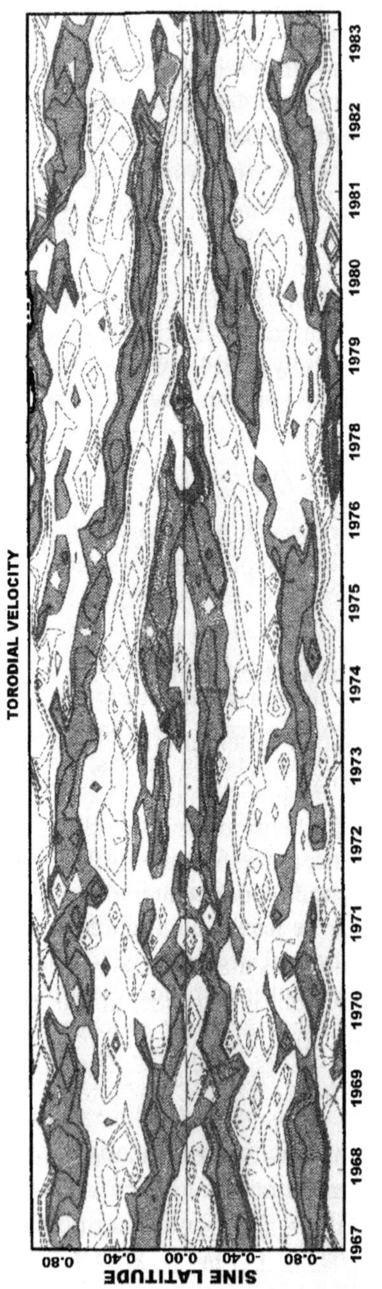

Figure 3.16 Left panel showing the fast (continuous lines) and slow (dotted lines) streams, the so-called 'Torsional oscillations' on the solar surface during 1967 to 1982 (Howard & LaBonte, 1980). Top panel shows torsional oscillations from the GONG data during 1995.5 to 2000, inside the Sun, contours indicate the rotational rate residuals, indicating torsional oscillations at r = $0.98R_0$ (Basu and Antia, 2001).

on one of the shear zones, as it moves equator-ward, through the active latitudes in each hemisphere. In Figure 3.16 (a) is shown the torsional oscillations, as bands of fast and slow moving streams on the Sun seen from 1967 to 1982 and in Figure 3.16 (b) from GONG-helioseismic data obtained for the period 1995 to 2000 AD, referred to a depth of 0.98 R_o.

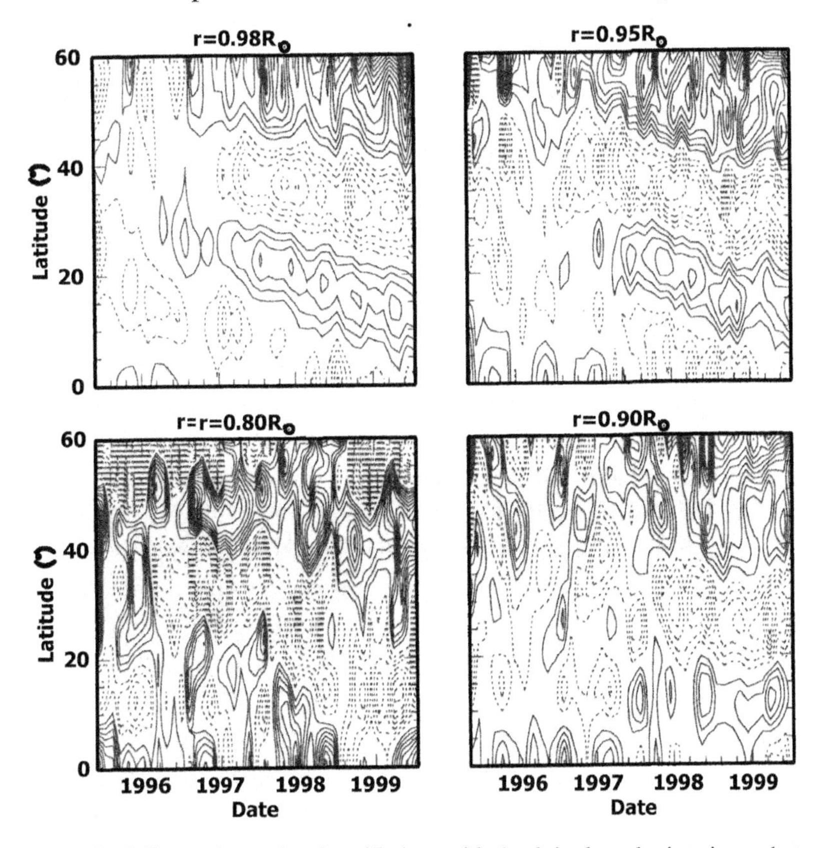

Figure 3.17 Shows the torsional oscillations with depth in the solar interior and at various latitudes (15°, 30°, 45° and 60°). From these diagrams it appears that the fast and the slow streams are inclined or slanting in the solar interior. At higher latitudes ~ 60° and in deeper layers streams are not discernable, due to noise in the data.

From helioseismological observations, similar fast and slow streams have been also observed to continue inside the Sun. Helioseismology has shown that these fast and slow streams persist even down to a depth of at least $0.1R_0$, thereafter the streams are **not** discernable, due to increase in noise level in the data. In Figure 3.17 is shown the variation of torsional oscillation with depth in the solar interior obtained by Antia *et al.* (2000). It appears that the Sun is easily excited in torsional oscillation modes of low-amplitude. So far there is no theoretical explanation for the excitation of these torsional oscillations nor for the connection found with solar activity.

3.6 Rotation of the Chromosphere and Corona

Chromospheric features like plages, faculae and network patterns observed in CaII K display solar rotation. Many authors, namely Maunder and Maunder (1905), Hale (1908), Fox (1921), Schröter and Wohl (1975), Schröter, Soltau, Wohl, and Vazquez (1978) have determined the rotation rate in the chromosphere. Generally the results obtained by these authors agree best with the rotation rates for small spots rather than the slower rates of large spots. Rotation rates obtained using the polar faculae as tracers, show the slowest rates of 10° per day as obtained by Müller (1954) and by Waldmeier (1955). Livingston (1969) reported that the features in the chromosphere agree with the spectroscopic results obtained using the H-α line, and that the chromosphere rotates faster by about 3%. Differential rotation in the chromosphere appears less than the photosphere. Several authors such as Belvedere *et al.* (1978), El-Raey & Scherrer (1972), Antonucci & Svalgaard (1974), and Belvedere *et al.*, (1977) reported that the rotation of chromospheric features depends on their lifetime. Smaller, younger and compact faculae rotate faster as compared to larger, older and more diffuse faculae. This may be because of the fact that smaller and more compact faculae are more rigidly 'anchored' to fast-rotating, sub-surface magnetic flux tubes. As the chromospheric features age and become diffuse, the subsurface flux connection becomes weaker and thus the older faculae rotate more closely to the surface plasma rate.

Coronal rotation rate has been measured by several methods. Hansen *et al.* (1969) have used K-corona observations for coronal rotation measurement at many latitudes. Due to instrumental problems, the resulting rotation rates are somewhat uncertain, but it is clear that the rotation at the equator is in fair agreement with the sunspot results and shows less variation with latitude at higher latitudes than as seen in the chromosphere. The coronal green line (Fe XIV, 5363 Å) has been also used to measure the rotation rate at higher latitudes by Waldmeier (1950) and by Loops and Billings (1962). These results also indicate a faster rate as compared to the sunspots, again suggesting a much lower differential rotation rate in the corona.

For determining the rotation of the transition region, Simon and Noyes (1972) used the bright emission points in the Lyman continuum and the Mg X line at λ625 Å from the Skylab EUV spectroheliograms. They found that the rotation rates were similar to the underlying plages, in spite of the great difference in height. Dupree and Henze (1972), using the same Lyman continuum data and number of EUV lines from Skylab observations, found significantly slower rotation rate ($\omega = 2.72 - 0.604 \sin^2 \theta$ μrads^{-1}) but with a steeper latitude dependence and no height gradient. Measurements of daily motions obtained from the solar coronal features on a spectroheliogram, should give the true rotational rates. Using the Skylab X-ray data, Gulub and Vaiana (1978) found that the shortest lived X-ray emission features rotate at about the photospheric rate, while the larger long lived features rotate faster by 5%. This could be understood if the small X-ray features are influenced by photospheric plasma, while large X-ray features are associated with sunspot groups and active regions.

Coronal holes as observed from the Skylab and *Yohkoh* spacecrafts have been used to determine the rotation rate of the outer corona. Also the HeI 10830Å maps from the National Solar Observatory (NSO) have been used to determine the rotation. According to Stepanian (1994) coronal holes participate in differential rotation as a rule. Only some which span more than 20 degrees in latitude have rigid rotation rates. These are determined solely by the active regions and as such they seem to rotate more rigidly than the underlying photosphere. There is a general agreement, both from direct observations and from magnetic extrapolation models, that the corona becomes more rigid with height. Estimates of the absolute rotation rate are rather subject to systematic errors since both short-term and long-term evolution of coronal structures affect accurate period determination.

3.6.1 *Coronal Rotation from LASCO Observations*

Several attempts have been made to determine the rotation rate of the inner and outer corona from the white light observations of coronagraphs and even during total solar eclipses. However, with the new coronagraphs on board the SOHO spacecraft, called *LASCO* (**L**arge **A**ngle **S**pectrometric **Co**ronagraph), it is now possible to observe continuously the white light corona from $1.1R_o$ to $30R_o$. To cover this large sky area, three coronagraphs are used. These are called C1, C2 and C3. LASCO's C1 instrument has a Fabry-Perot filter which allows FeXIV emission line imaging of the inner corona between 1.1 R_o and 3 R_o. C2 observes from 2.5 R_o to 6.0 R_o, C3 from 8.0 R_o to 30.0 R_o. From a detailed study of the data from LASCO, made by Lewis *et al.* (1999) for over a year during May 1996 to May 1997, they concluded that rotation of the corona displays a radially rigid rotation of 27.5 days synodic period from 2.5 R_o to > 15 R_o. Beyond this the period determination becomes difficult. The emission line corona within 2 R_o was found to rotate at a slightly faster rate.

Chapter 4

The Quiet Sun

4.0 Introduction

The Sun is like as onion. You peel one layer after another and see to different depths, different structures and different temperatures. The deepest layers from which the visible and near infrared radiation can escape is only about 400-500 km below the so called *surface* of the Sun. This is the photosphere or the sphere of light. On the photosphere one can see fine mottled structure- the granulation; near the limb one can also see bright patches called faculae; and there are the small to sometimes large dark regions famous as sunspots. All these features undergo temporal variations ranging from a few minutes to several years. For example, the normal life time of a granule is between 5 and 8 minutes, faculae may last for a few weeks, while sunspots last from a few hours to several months and show a periodic cyclic variation in their numbers of 11 years.

Just above the visible surface of the Sun lies the chromosphere- the sphere of color, extending to some 10,000 km in height and consisting of a tenuous gas having a density much less than photosphere. In this layer also one can see interesting features and phenomena through appropriate narrow band optical filters. Some of these phenomena exhibit rather slow variations while other show violent activity. Beyond the chromosphere lies the extended and extremely tenuous layer or better to call it an outer atmosphere of ionized gas; the corona. In the corona there are features which display both slow variations and violent changes. In this Chapter we shall discuss the quiet aspects of these layers of our Sun. The more violent and active phenomena will be explored in Chapter 5.

4.1 The Quiet Photosphere

4.1.1 *Granulation*

4.1.1.1 *Early Visual Observations*

Visually observing the Sun through his speculum mirror telescope of 3-meter focal length, Sir William Herschel (1738-1822) was perhaps the first person to record the unevenness or "mottled" structure of the solar disk. What Herschel saw was probably a large scale pattern of brightness fluctuations, which appears when the disk is viewed with inadequate resolving power or under poor 'seeing' conditions. A realistic drawing of the solar granulation (even by today's standard) was made by Father Secchi (1818-78) with a small telescope and is shown in his book *'Le*

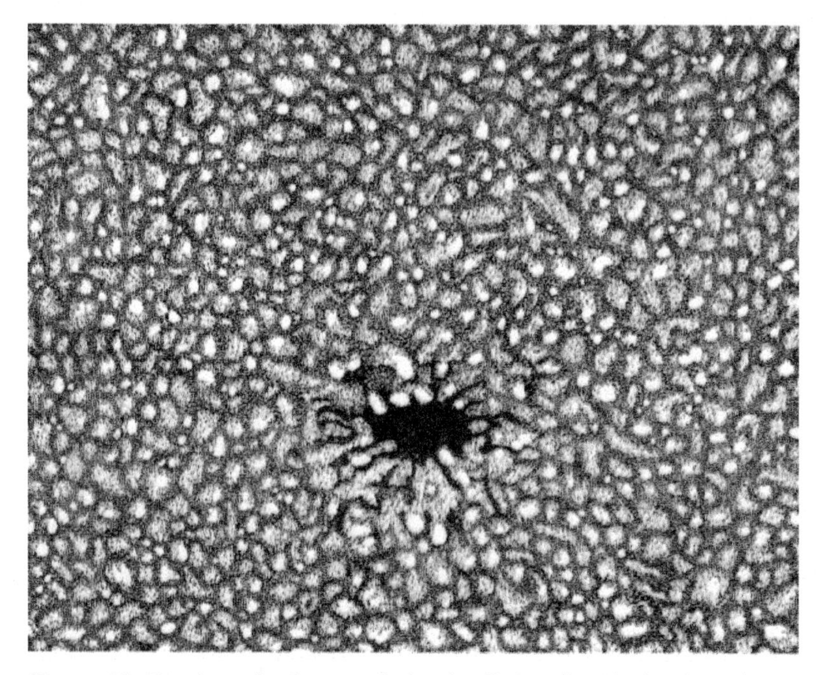

Figure 4.1 Drawing of solar granulation by Father Secchi showing well defined bright and dark features and a small pore.

Soleil' (1875), see Figure 4.1. He described the solar surface as being covered by small bright grain-like features, separated by dark lanes. Secchi estimated the size of these grains as about 0.3 arc sec, whereas the modern average value is about 1.3 arc sec.

Another early keen observer was S.P. Langley (1834-1906). He used the 33-cm Alleghney observatory's refractor in USA, and found the average diameter of grains between 1 to 2 arc sec. Among the early visual solar observers was also the English astronomer Sir William Huggins (1824 –1910) who preferred to call the grains as *granules*. It is indeed remarkable that the early solar observers like Secchi, Langley, and Huggins could visually see such fine structures on the Sun, using the then available best quality telescopes.

4.1.1.2 *Early Photographic Observations*

The French astronomer Pierre Jules Janssen (1824-1907) took the first photographs of the solar granulation using a 13.5 cm aperture refractor and enlarging the solar image to 30-cm diameter. One photograph taken on 1st April 1894 clearly shows the well-defined pattern of bright granules with diameters between 1–2 arc sec, separated by narrow dark lanes. Janssen's photographic solar observations extended over a period of some 20 years. They were published in 1896 in a volume which contains reproductions of 12 of his original photographs. Several early observers like Father Stanislas Chevalier (1852-1930) at Zō-Se observatory in China, and Hermann Strebel (1868-1943) in Germany at Munich Observatory also made important contributions to granulation studies. In 1933 Strebel noted irregular polygonal shape of solar granulation, which was *re-discovered* and confirmed only in 1957 by Schwarzschild and Danielson's balloon-borne 30-cm telescope Stratoscope-I project. It is curious that early expert solar observers, like George Hale and John Evershed, did not show much interest in granulation observations.

Interest in granulation was revived with the work of Richardson and Schwarzschild in 1950 through their spectral study of granulation. At Mount Wilson they tried to look for a correlation between the granulation brightness fluctuations and Doppler shifts of lines in the solar spectrum. If the granulation is a manifestation of convection one would expect bright granules to be blue shifted with respect to the dark lanes. Their results showed only a weak correlation between the brightness and velocity fluctuations, which to some extent caused doubt on the convection theory of granulation. To further investigate the nature of

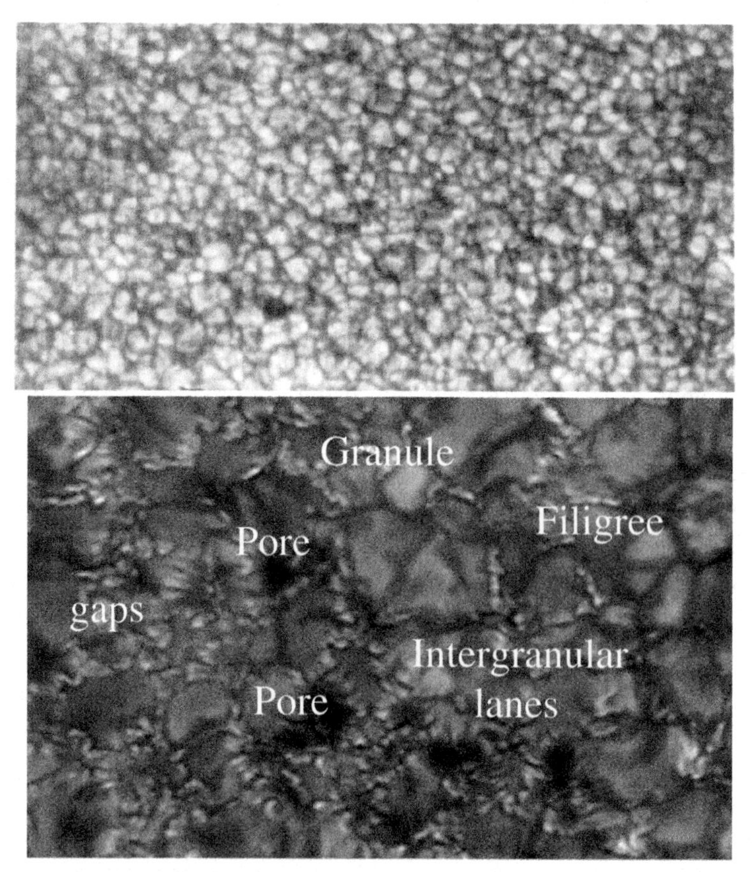

Figure 4.2 Top panel - Photograph of solar granulation taken from a balloon–borne 30 cm aperture Stratoscope telescope in 1957 from a height of 24 km. Bottom panel– Modern recording taken at the 1-m New Swedish Solar Telescope with the aid of adaptive optics and phase diversity techniques.

granulation, Schwarzschild decided to obtain the highest possible resolution photographs, which could then be obtained only from above the disturbing Earth's atmosphere. Under his direction, at Princeton University, a 30-cm balloon borne automatic telescope was built and flown to photograph the Sun in white light. It attained a height of 24 km in the Stratosphere and was known as 'Project Stratoscope'. A granulation picture taken by project Stratoscope is shown in Figure 4.2, and compared to modern high resolution picture taken from ground.

4.1.1.3 *Granules as Convection Cells*

In 1930 Albrecht Unsöld, the distinguished German astrophysicist, laid the foundation of modern convection theory to explain the origin of the solar granulation. He showed that due to the increase in hydrogen ionization with depth, a zone of convective instability must exist just below the photospheric layers. An elementary volume of gas moving upwards through the hydrogen ionization zone is heated by release of ionization energy. The buoyancy of the element is thus increased and it continues its journey upwards. In this way convection cells are generated which are manifested on the surface as granulation. Following this theory, H. Siedentopf in 1933 suggested that the granules are in fact bubbles of hot gas pushing their way upwards through cooler descending material. Now it is well established that granules are due to convection, the bright granules are hot rising material, while surrounding cooler material descends along the dark intergranular lanes.

4.1.1.4 *Shape of Granules*

From photographs as shown in Figure 4.2, it will be seen that the granules have very irregular shapes, and generally are polygonal with their cell boundaries elongated and common. The best pictures show that granules have almost parallel straight boundaries and that dark lanes separating them are generally of uniform width. But lower resolution pictures give them a roundish appearance due to the lowing of contrast. In summary, granulation can be described as irregular cellular patterns of bright elements, separated by narrow dark lanes. This pattern provides one of the chief reasons for considering granulation as due to convection. On very high resolution pictures such as taken by the new 1-m

Swedish vacuum telescope shown in Figure 4.2, it will be noticed that tops of the majority of granules show intensity variations across the granules, and the inter-granulation dark lanes almost vanish. This indicates that the earlier estimates of the width of the inter-granular lanes were wrong due to inadequate resolution and contrast in the pictures.

4.1.1.5 *Granule Size, Brightness and Contrast*

At first sight it might seem that the average granule diameter would be a convenient parameter to measure and use. However, precise measurement of the diameter of individual granules is difficult due to photometric inaccuracy and the need for corrections for the instrumental profile of the telescope and seeing. The true measurement of the granule diameter can be inferred only if the influence of these factors is precisely known. Recent application of adaptive optics, coupled with high quality vacuum telescopes has radically changed our concept about the granule size or 'diameter'. Visual inspection of high resolution pictures taken from the NSST (New Swedish Solar Telescope) and other high resolution telescopes, show spectrum of granule sizes range from a fraction of second of arc to about 3 arc sec (2300 km). There is a large dispersion in granule size, the smallest granule that can be distinguished may have an apparent diameter as small as 0.3 arc sec (~210 km), a figure comparable to the resolving power of many telescopes. However a mean diameter of granules obtained by several authors is about 1.3 arc sec (970 km) close to the strong magnetic field regions like sunspots. A systematic shrinkage of the mean diameter has been reported by some authors (Schröter, 1962). A more objective determination of the granule diameter Dg can be obtained through granule area A_g, as proposed by Karpinsky (1980) and is given by:-

$$D_g = 2\left(\frac{A_g}{\pi}\right)^{\frac{1}{2}}.$$

The measurement of granule areas used here to determine the granule diameter can also yield the total area A_g covered by granules on the Sun.

If A_T is the total area sampled, the fractional area of granules is given by:-

$$\phi_g = \frac{A_g}{A_T}.$$

Estimates of ϕ_g obtained by several authors (Rösch 1959, Pravdjuk *et al.* 1974, Keil 1977, Wittmann & Mehltretter 1977, Namba and Diemel 1969) gave an average value of 0.49, implying that there is little or no difference between the fraction of the solar surface covered by granules and that covered by the intergranular lanes.

To overcome the difficulty of accurately measuring the diameter of granules, Bray and Longhead (1959, 1977) defined it as the average distance between centers of adjacent granules. This parameter is independent of the photographic contrast and instrumental profile, provided that individual granules are resolved. The mean cell size is a better criterion for theoretical consideration of convection. Direct measurement of the distance between granules yields an average separation of 1.9 arc sec. For a characteristic cell size of ~ 1.8 arc sec, the mean surface density of granules obtained by several authors in a 10x10 arc sec square area on the Sun is 31.5, implying a mean cell area of 1.67×10^6 km^2. This leads to the total number of granules on the solar surface as 3.7×10^6.

The latest granulation pictures taken with the Dutch Open Telescope (DOT), Kiepenheuer Vacuum Telescope and NSST, have revolutionized our earlier ideas about granule brightness and contrast. Individual granules show a considerable diversity in their brightness. Precise determination of granule brightness and contrast is difficult. No measurements of brightness and contrast are yet available from these high resolution observations. However, from earlier measurement by Bahng and Schwarzschild (1962) from the Stratoscope pictures, they found the rms temperature fluctuation in the granules (determined from the brightness fluctuation) to be about ± 100°K. They also found a weak correlation between the brightness and the size of granules, implying that greater radiative cooling time to be expected for larger granules.

4.1.1.6 *Evolution and Life-times of Granules*

Cinematography in solar astronomy was first introduced by Bernard Lyot (1897–1952) at Pic-du-Midi Observatory in early 1940's and by R.R. McMath (1891-1962) at McMath-Hulbert Observatory in 1933. They were mainly interested in prominences and other chromospheric phenomena. It was soon realized that 'white light' cinematography could be used for recording changes in photospheric granulation. This technique was fully exploited by J. Rösch at Pic-du-Midi, and later by Bray and Loughhead at CSIRO in Australia. D.E Blackwell, D.W. Dewhirst and A. Dollfus (1959) used a manned balloon borne telescope for making time lapse observations of the solar granulation, from a height of 6.7 km above the ground. Blackwell and co-worker's results showed that for better resolution observations should have been made from much greater heights, if the residual effects of atmospheric seeing was to be completely eliminated.

This effort prompted Martin Schwarzschild in 1957, to launch an unmanned balloon-borne 30-cm aperture telescope, which could take high resolution solar white light photographs from a height of 24 km. Very high resolution white light observations were obtained over long period in time, which provided good data on the evolution of granules and their life time. Following the launch of project Stratoscope I & II in 1957 and 1959, Spektrostratoskop was flown by German scientists in 1975, to a height of 28 km with a 32-cm aperture telescope and a spectrograph. They obtained high resolution photospheric and spectrographic observations for more than 6 hours.

Several authors have worked on the difficult observational problem of evolution of granules. This question is of great importance for understanding the related hydrodynamical processes. Time dependent theoretical models of the granulation have been proposed and the process through which granules evolve, and how they vary in size, shape and brightness etc., throws light on the physics of non-linear convection believed to occur in the Sun.

From the results of several observers [Bray and Loughhead (1958), Rösch (1962), Mehltretter (1978), LaBonte, Simon & Dunn (1975), Kawaguchi (1980)] the common picture that emerges about the evolution

of granulation is as follows:-
1. That there is a greater predominance for granules to increase in size than to decrease, but increase and decrease in brightness seem to occur with equal frequency,
2. It was also observed that after a granule is formed, its diameter begins to increase in general, until it reaches about 2 arc sec, then it breaks up into several small granules, which fade and vanish. The fragmentation begins by formation of a dark central spot, which gives granule a ring like appearance. The dark spot then develops into dark radial lanes that fragment the granule. Some authors have described this phenomenon as an 'exploding' granule. Sometimes, the dark spots evolve into a dark 'notch', connecting to the intergranular boundary, and then into a new lane fragmenting the granule in the short time of a minute of so,
3. It is noticed that small granules tend to fade away, but almost all large granules fragment or occasionally merge,
4. All authors agree that granules are born, almost without exception, from previous granular fragments. Occasionally a granule appears to develop as a faint patch of brightening in the intergranular region, but most granules represent resurgence from an earlier manifestation.

Several authors have studied granulation photographs taken over long duration. They report that the mean life time of photospheric granules range from 6 to 16 minutes. It is also noticed that their life times depend on the size of granules.

Now with the availability of extremely high resolution granulation pictures of solar granulation obtained over long period in time with large aperture vacuum telescopes and using adaptive optics techniques, it will be possible to study and obtain much better understanding of the evolution of granules and their life time.

4.1.1.7 *Center-limb Visibility of Granulation*

The visibility of granules from center-to-limb on the solar disk provides a qualitative measure of the height to which the granules extend into the upper photosphere. This question is of importance, as we need to know the structure of the inhomogeneous photosphere in higher layers of the photosphere where the Fraunhofer lines originate. In the past several attempts have been made by Rösch (1959), Edmonds (1960), Bray and Loughhead (1958) and Müller (1977), these authors examined the question of disappearance of granulation near the solar limb. From their observations it was concluded that large granules seem to disappear between the heliocentric angles of 80°-84°, corresponding to a distance of 5-10 arc sec from the limb.

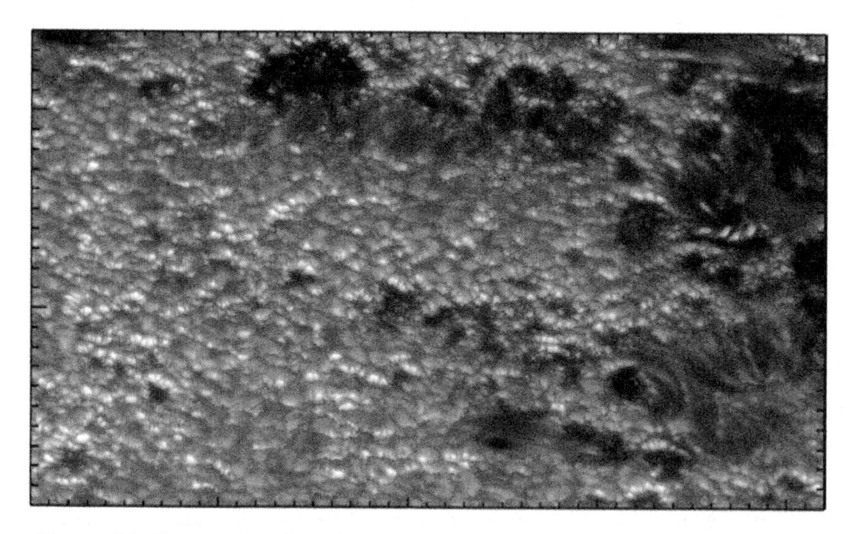

Figure 4.3 Photograph taken through the 1-meter New Swedish Solar Vacuum telescope in white light near the solar limb on 24 June, 2002, showing effect of granules elevated over the general surface.

If granules are visible up to 5 arc sec from the limb, it implies that the convective pattern can be followed up to $\mu = \cos 84°$. Thus granulations are seen to far lesser optical depth above the photosphere than $\tau = 1$. On filtergrams taken in the FeI 6569.2 Å line, Bray and Loughhead have reported traces of inhomogeneous pattern observed even within the last 2 arc sec from the limb. Extremely high resolution pictures recently taken near the solar limb, through the 1-m New Swedish Solar Vacuum telescope at La Palma show a *3-dimensional* kind of view of the solar granulation, as seen Figure 4.3. From this photograph it is interpreted that the granules are perhaps elevated features over the solar 'surface' and extend up to 100 to 400 km in height.

4.1.1.8 *Granule Velocity and Brightness Variation*

As stated earlier, granulation is a manifestation of convection, which is caused in a fluid, when it is heated from below. Hot columns of gas rise and cool columns descend downwards. This is a standard picture of solar granulation. The upward and downward velocities of the solar material can be directly measured by Doppler shift of Fraunhofer lines. One

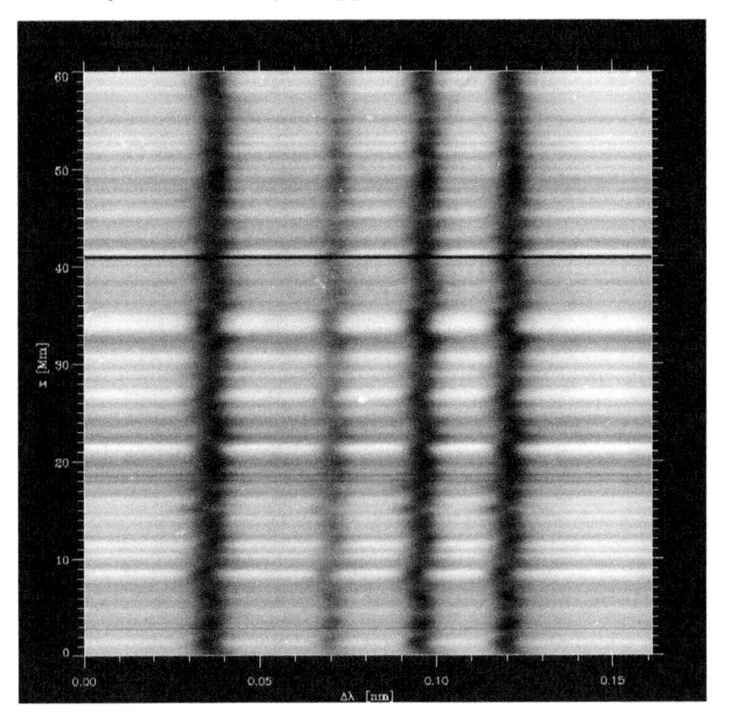

Figure 4.4 Showing a high resolution spectrum of solar granulation, bright and dark streaks are due to granule brightness variation and zigzag pattern in absorption lines is due to vertical granular motion. It will be noticed that the line shifts in darker regions generally appear towards the longer wavelength side (right side), implying a downward motion, away from the observer. Spectrum taken from German Vacuum Telescope

must be careful to note that the Doppler shifts measured using line shifts refer to higher layers of the photosphere, while the granulation brightness refer to the continuum, formed deeper by about 100-200 kms.

Using the Mount Wilson observatory's 150-foot tower telescope and its Littrow spectrograph, Richardson in 1949 was perhaps the first to take a solar spectrum showing prominent bright and dark streaks running parallel to the dispersion and showing a zigzag pattern in Fraunhofer lines. The bright and dark streaks observed in the continuum were interpreted as due to granulation and the zigzag pattern in lines, as due to the granule motion. With improved telescopes, gratings and spectrographs, higher spatial and spectral resolution became available and later several authors have obtained, what is now called as "wiggly line spectra". In Figure 4.4 is shown a high-resolution granulation spectrogram obtained by Nesis and colleagues at the German telescope on Tenerife in the Canaries.

The strong correlation found between the brightness and velocity fluctuations, puts the convective origin of the granulation beyond doubt. The spectrographic observations of granulation and associated velocities display only the one–dimensional sampling of granulation pattern and do not show the peak velocity of individual granule or the surrounding intergranular lanes. To determine individual granule velocities and brightness, a better technique is to use tunable narrow band filters or spectroheliographic technique which yields 2-dimensional and Doppler solar images. Beckers (1968) at Sac Peak Observatory, placed a Wollastron prism behind a narrow band Zeiss 0.25 Å passband filter to produce two simultaneous images of 0.25 Å pass band separated by 0.12 Å, from the line center at 6569.6 Å line of Fe I. Beckers found that the brightness variation in the red wing were much greater than the variation in the blue wing pictures. Based on this, Beckers estimated that the velocity difference between the two components would be about 6 km/s. In a subsequent paper, Beckers and Morrison (1970) analyzed the filtergrams in much more detail and found a maximum outflow speed of 250 m/s to occur at a distance of 450 km, from the center of the average or composite granule. Bray, Loughhead and Tappere (1976) further refined the measurements using the 30-cm refractor of the CSIRO Solar Observatory, Australia, and derived a velocity difference of 1.8 km/s between the granule and intergranular lanes. However there is a large variation in the granulation velocity field and precise determination is rather difficult.

4.1.1.9 *Granulation and Magnetic Fields*

The possibility of detecting magnetic fields in granulation had attracted the attention of many workers, mainly Steskenko (1960), Semel (1962), Livingston (1968), Howard and Bhatnagar (1969), these authors placed an upper limit of the field strength ranging from 2 to 50 gauss. Interest in small scale magnetic flux tubes grew with Stenflo's idea that solar magnetic fields are confined in fine bundles of flux tubes of strong magnetic field of the order of thousand gauss. Title and co-workers at Lockheed, using the Swedish Solar Vacuum telescope at La Palma, obtained very high resolution magnetic field maps, showing magnetic features of less than 0.5 arc sec. These observations indicate field strength of nearly 1000 gauss or more.

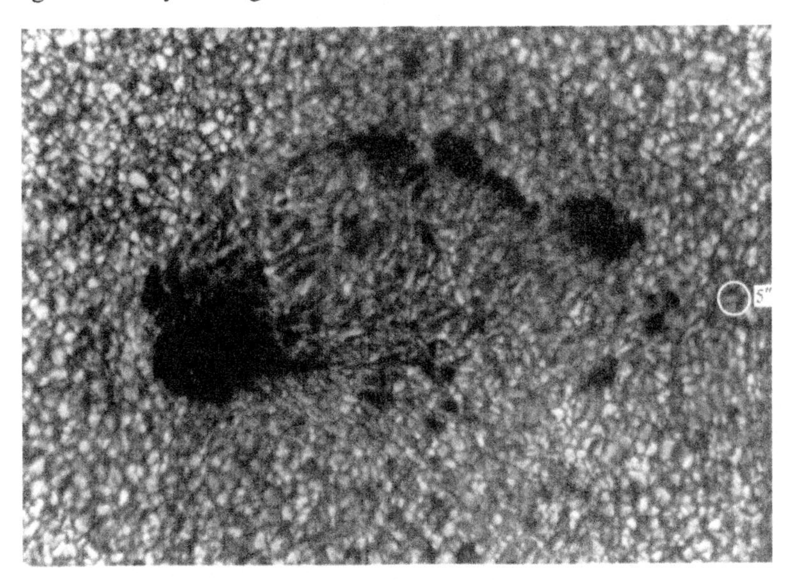

Figure 4.5 Photograph taken with the 76-cm Dunn vacuum tower telescope at Sac Peak Observatory, displaying disturbed granulation pattern near a developing active region. The dark elongated structures seen between the sunspots and pores are due to emerging magnetic flux.

In an emerging flux active region, it has been observed that normal granules show abnormal elongated structures, between sunspots and pores, as shown in Figure 4.5 taken with the 76-cm vacuum tower telescope of the Sacramento Peak observatory. Such abnormality observed in granulation pattern in an active region indicates that the magnetic field or the flux tubes emerging from below the solar surface influence and stretch the overlying granulation pattern.

The question; whether the field strength is higher in the inter-granulation region as compared to bright granules, was partly answered by Howard & Bhatnagar (1969) through their analysis of a high resolution "wiggly-line" spectrum. They placed an upper limit of about 50 gauss field higher in the dark lanes. On filtergrams taken in the far wings of the H-α line, and using the Sac Peak's vacuum telescope, Dunn and Zirker (1973) observed very fine (<0.3 arc sec) bright structure within the dark intergranular lanes and called them as "filigree". A possible relationship between "filigree" and the photospheric magnetic elements has been considered by several authors (Dunn & Zirker, 1973, Mehltretter, 1974, Simon & Zirker, 1974). It seems that the observed brightness in the far wings of H-α line may be due to the enhanced magnetic energy, leaking through the intergranular lanes which excite the overlying material.

4.1.2 *Supergranulation*

As we have seen in the preceding section granulation displays vertical motion. From the center of granules, material moves upward with characteristic rms velocities of 1 km/s and drains down through the intergranular region. In the course of spectroscopic determination of solar rotational velocity, Hart (1954, 1956) discovered at the Oxford Solar Observatory, the existence of *another* kind of large-size horizontal motion on the Sun. She found indication of large scale individual motions occurring on an irregular scale ranging from 25,000 to 85,000 km that persisted for several hours with velocities of about 0.5 km/s.

The real breakthrough in detection of true large-scale horizontal motion came from the 2-dimensional Doppler spectroheliograms obtained by Leighton in 1960. At the Mount Wilson 60-foot tower, using a new spectroheliographic technique, Leighton and his colleagues (Leighton *et al.*, 1962) obtained a detailed picture of the structure of the horizontal velocity field. A classic photograph of the Doppler spectroheliogram is show in Figure 4.6, where large-scale cells displaying horizontal motions are seen away from the disk centre. This is a typical full disk Doppler photograph of the Sun taken in CaI 6103Å. The dark and bright regions refer to line-of-sight velocities directed away

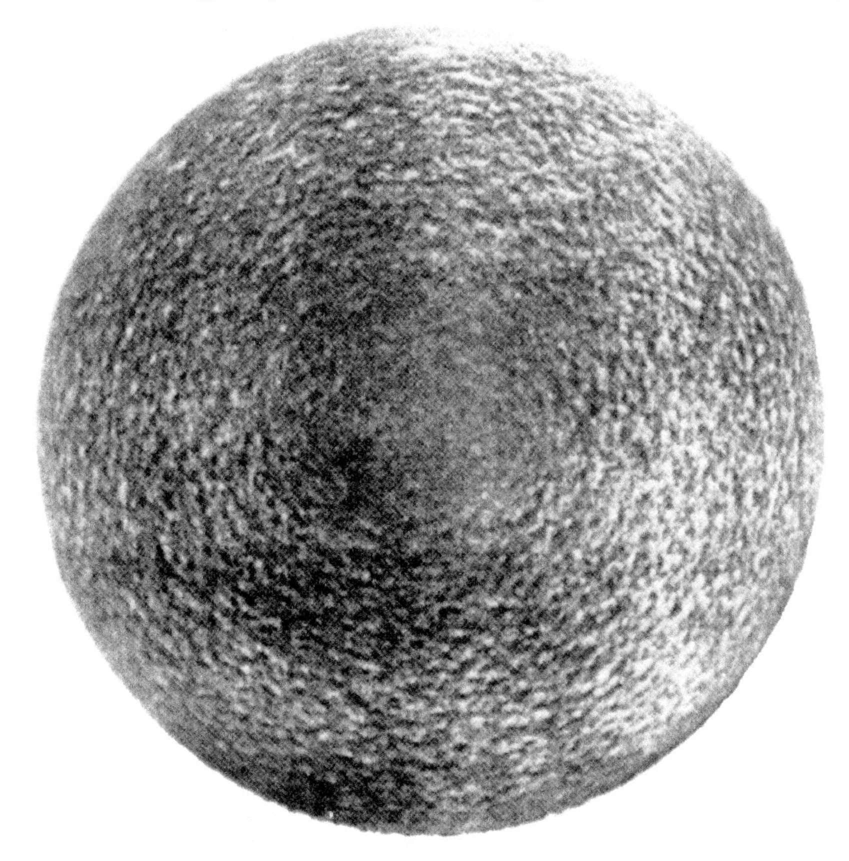

Figure 4.6 Full disk Dopplergram obtained by Leighton showing large-scale solar velocity field known as Supergranulation cells. Near the disk center the cells are not discernable. Away from center they are seen as dark and bright elongated cells. The dark areas indicate receding motion.

and towards from the observer. In this picture the effect due to the solar rotation and granulation motion has been largely removed. From this Dopplergram it is evident that the velocity field is made up of numerous velocity 'cells' thousands of kilometer in diameter.

Near the disk center the 'cells' are hardly visible, but towards the limb in all directions the cells show evidence of geometrical foreshortening. There is no difference in the appearance of cells between the equatorial and polar regions. This observation of the velocity field in the cells is interpreted as implying that the material motion in each cell is horizontal and directed outwards from the cell center towards the cell boundary, as displayed in a schematic shown in Figure 4.7. From the observed similarity between the photospheric granulation and these large velocity cells, Leighton and colleagues called them as '*Supergranulation*', because their spatial and temporal scales were much larger than the granulation. From the flow pattern in the cells, it was concluded that these supergranulations are manifestation of large-scale convection currents in the solar atmosphere.

Most of the characteristics of supergranulation are described in the following:-

1. *Mean cell size*: At a given time, there are approximately 2500 cells on the visible hemisphere. From this, one may easily derive an estimate of the mean cell size or the average distance between the centers of the adjacent cells, which comes to $\approx$ 35,000 km. From an autocorrelation analysis carried out by Simon & Leighton (1964), it was shown that there is a fairly large spread in the cell diameters, ranging from 20,000 to 54,000 Km.

2. *Life time*: Hart (1956) and Leighton *et al.* (1962) have shown that the life time of individual cells is at least several hours. A more precise life time determination is difficult to make, because near the disk center, the cells loose their identity. Some auto-correlation studies have indicated life times on the order of a day.

3. *Correlation with Ca II brightness network*: Simon and Leighton (1964) found one-to-one correlation in position of supergranulation and the bright network seen on CaII K-line spectroheliograms. From K-line spectroheliograms, Simon & Leighton derived a life time of supergranulation of about 20 hours. They also found that the K-line network boundaries represent regions of stronger magnetic field.

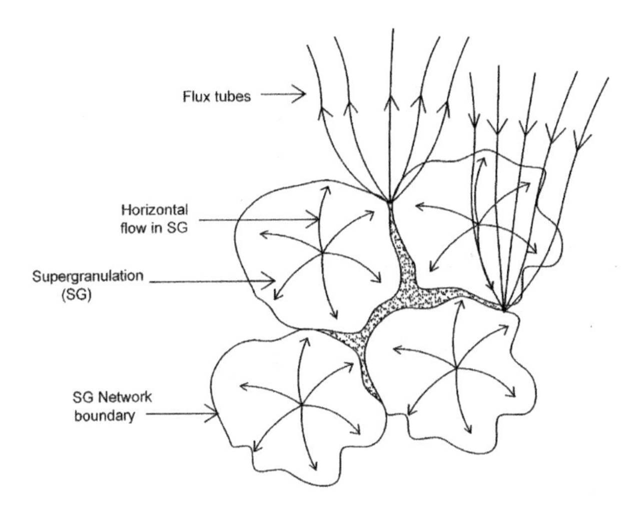

Figure 4.7(a) Schematic drawing of supergranulation pattern showing velocity vectors directed towards the cell boundaries from the center.

4. *Magnitude of horizontal velocities*: The average peak horizontal velocity within an individual cell was found to be about 0.3–0.5 km/sec by several authors (Hart, 1956, Evans & Michard, 1962b, Simon & Leighton, 1964). The velocity decreases with height in the atmosphere.

5. *Vertical velocities*: As interpreted by Leighton and colleagues that the supergranulation is a manifestation of large scale convection, then one should expect to find evidence of vertical motions both at the cell center and the boundary of individual cells. In fact, near the disk center, Simon & Leighton (1964) detected very faint pattern of rising velocities at the center of each cell and descending velocities on the order of 0.1–0.2 km/s at the cell at boundary.

6. *Supergranulation & magnetic field*: From the observed close correlation between the CaII K-line spectroheliograms and the supergranulation, it is well established that the boundaries of the supergranulation tend to contain magnetic flux. Frazier (1970) had shown that the down flows at the vertices, where several cells meet, are

much more prominent than the rest of the cell boundaries. At these points the magnetic field appears enhanced. This may be because over the supergranulation cells, the lines of force are 'swept' by the horizontal velocities in the cell to the cell boundaries, and thus the magnetic flux concentrates there. In Figures 4.7 is shown supergranulation motions on magnetic lines of force in horizontal and vertical planes.

7. Supergranulation is **not** discernable on a white light photograph of the Sun (except near the limb where it is seen as faculae).

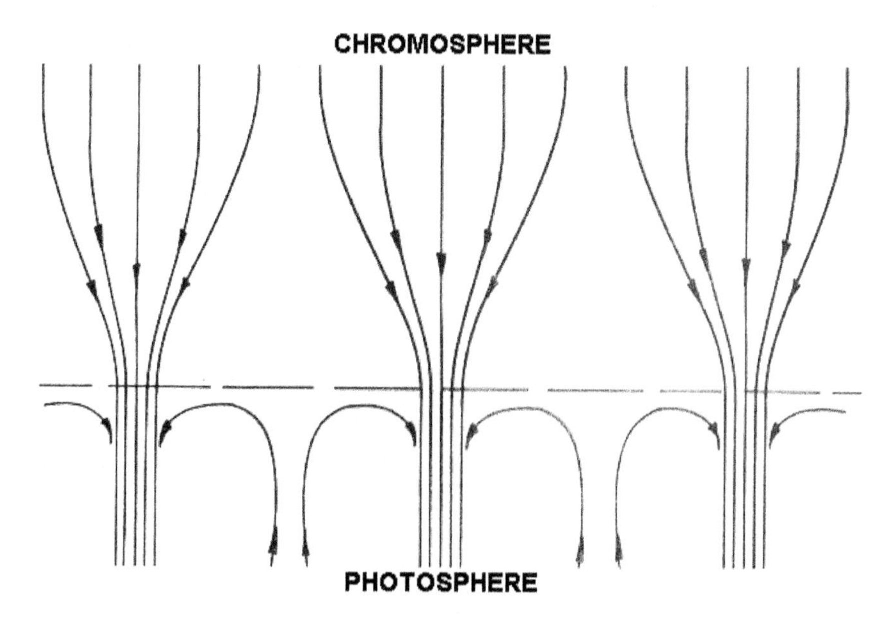

Figure 4.7(b) Schematic picture of the supergranulation horizontal motion on a weak magnetic field. The lines of force are swept by the motions in the cells towards the cell boundaries.

Beckers (1968) and several other workers tried to detect supergranulation by averaging out small granulation fluctuations, and found that the mean photospheric brightness is slightly brighter at the boundaries than at the center. If these results are true, then the correlation between brightness and velocity would be **contrary** to that expected of convective phenomenon. Worden (1975) made observations in the infrared line at 1.64 μm, where the solar opacity is minimum which refer to the deepest photospheric layers; and found that the centers of the supergranulations are slightly brighter than the boundaries.

4.1.3 *Mesogranulation*

November, *et al.* (1981), detected a cellular velocity pattern in the upper photosphere having a scale of 5000-10,000 km (7-14 arc sec). It has a life time of at least 2 hours with an rms velocity of about 60m/s, comparable to the vertical velocities in the supergranulation, but much smaller than the fine granulation velocities. As these features seem to have larger scale, it is called as 'mesogranulation'. It is not yet clear if there is a relation between granules and mesogranules. Perhaps mesogranules are groups of individual granules displaying a common velocity pattern. However, more observational work is required before one can fully understand the mesogranulation phenomenon. In Table 4.1 is given a summary of characteristic property of granules, supergranulation and mesogranules.

Table 4.1 Properties of granules, supergranules and mesogranules.

Feature	Diameter	Mean cell size	Life time	Velocity Km/sec
Granules	1″-2″	1″.9	~10-16 minutes	~ 1 (rms, vertical) ~ 2 (rms, horizontal)
Supergranulation	30″-70″	48″	≥ 1 day	0.3-0.5 (horizontal)
Mesogranulation	7″-14″	10″	~ 2 hours	0.06 (rms vertical)

4.2 The Quiet Chromosphere

4.2.1 *Introduction*

In this Section we shall discuss some of the interesting properties of the quiet chromosphere. From the bottom to the top of this layer the temperature rises steeply from about 5000 K to nearly 100,000 degrees in just a few thousand kilometers, indicating that the source of heating can **not** be thermal but some non-thermal processes must be playing a major role in heating the chromosphere. The chromospheric surface features are best seen in the strong Fraunhofer lines of ionized Calcium, Magnesium, and Hydrogen. The chromosphere is also the seat of most interesting phenomena on the Sun, such as spicules, prominences, CaII and Hα-network, emerging active regions, solar flares etc. Chromosphere also presents us with one of the most challenging problems in astrophysics, that is; **what heats the solar chromosphere?**

4.2.1.1 *Early Observations of the Chromosphere*

The two Englishman; Solar Physicist Sir J.N. Lockyer (1836-1920) and Chemist E. Frankland (1825-1899) first proposed the name '*chromosphere*' or the '*coloured sphere*'. This term was derived from the distinct red or pink 'arcs' seen surrounding the eclipsed Sun, during the 1851 total solar eclipse. We know now that the red hue is due to the emission from the brightest red Hα line of hydrogen in the chromospheric spectrum. Perhaps the earliest serious study of the chromosphere was undertaken by Father Angelo Secchi (1818-1878) and by a British amateur astronomer Warren de la Rue (1815-1889) during the total eclipse of 1860. At this eclipse the chromosphere together with the corona and prominences (*protuberance*) were observed. Several eclipse observers reported seeing little 'flames' or 'jets' projecting upwards and outwards from the chromospheric boundary, which were distinctly different from the large prominences seen earlier. Father Secchi in 1877 visually recorded the shape, size and structure of these fine 'flames' or spicules as we now know. He estimated the width of these features to be only 100 or 200 km, however the modern value is around 800 km.

In Figure 4.8(a) is shown Father Secchi's drawing of the chromospheric spicules in 1877 and in Figure 4.8(b) is shown modern high resolution photographs taken by Dunn from the Sacramento Peak observatory. Notice the close resemblance between Father Secchi's drawings and the modern day pictures of spicules. The close resemblance between the two is the testimony of Secchi's skill of observations and perhaps the quality of his instruments.

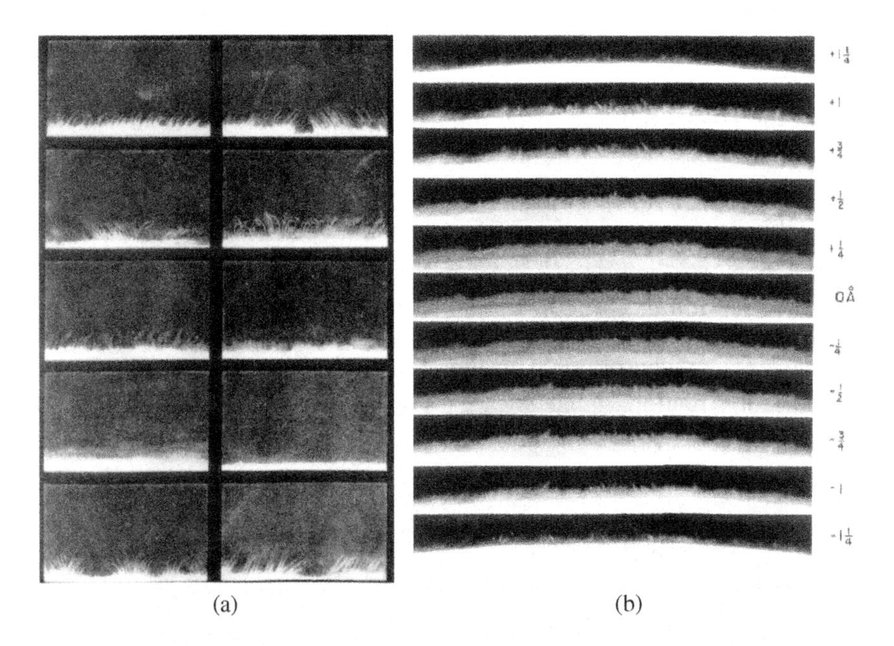

Figure 4.8 (a) Drawings of spicules by Father Secchi as given in his book-*Le Soleil* in 1877. (b) Photograph of the solar limb but at various distances from the Hα line center, taken by Dunn with the 40-cm chromospheric telescope at the Sacramento Peak Observatory. Note the close resemblance between Secchi's drawings and the modern pictures.

4.2.1.2 *Early Spectroscopic Observations*

In mid nineteenth century number of astronomers got interested to apply the new technique of spectroscopy to study the Sun and to examine the nature of the prominences seen during the total solar eclipses, for this they had to await for the 18 August, 1868 total eclipse. The French astronomer, Pierre Jules Janssen (1824-1907) was first to use a spectroscope at this eclipse and he was greatly impressed by the brightness of the prominences and wondered if they could be seen also without an eclipse. To his amazement and delight he could see the prominences next morning through his spectroscope, by widening the slit. Sir Norman Lockyer also had a similar idea to observe the chromosphere and prominences in the bright lines of Hα and Hβ, but he could try this only at the 20 October 1868 eclipse in India. At this eclipse he discovered Helium in the flash spectrum. By rotating the spectrograph around the solar limb and using a tangential slit, Lockyer demonstrated that the chromosphere surrounds the photosphere. Following Janssen's and Lockyer's discovery, it was no longer necessary to wait for a total solar eclipse to observe the prominences and the chromosphere.

4.2.1.3 *Observations of the Flash Spectrum*

The next major advancement in the study of solar chromosphere came from the observations made by C.A.Young, during the total eclipse of 1870. At this eclipse, Young positioned the spectrograph slit tangential,

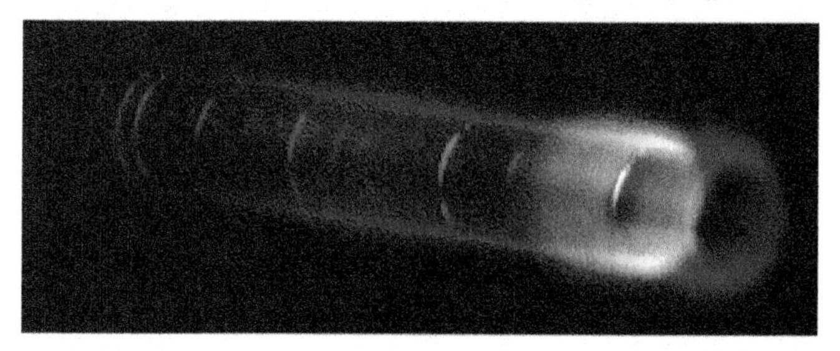

Figure 4.9 Flash spectrum taken from Hyderabad, India, at the February 16th 1980 total eclipse.

instead of radial at a point near the solar limb, where the last ray of the sunlight disappeared just before the totality and suddenly like a 'flash' he saw numerous emission lines appeared instead of the absorptions lines. Young could see this phenomenon only for a few seconds, since the advancing Moon rapidly covered the lower layers of the chromosphere, where the emission lines were brightest. Young believed that the bright lines seen in the '*flash spectrum*' corresponded one-to-one in position with the 'dark' Fraunhofer lines of the photosphere, but reversed in intensity. Due to this phenomenon the lower chromosphere was earlier called as the '*reversing layer*'. The first flash spectrum was photographed at the 1883 eclipse. During the 1898 total solar eclipse and using a slit-less prismatic camera of only 2.5 inch aperture, John Evershed (1864-1956) succeeded in photographing 313 emission lines extending from 3340 to 6000Å, including 28 lines of Balmer series and the Balmer continuum. Since the early observations made by Evershed and other workers, extensive study of the solar chromosphere has been made from the flash spectrum observations. Since then number of workers have obtained high resolution flash spectrum, for example the High Altitude Observatory's expedition obtained very fine slit-less spectrum extending from 3,500 to about 8700Å at the 1952 total solar eclipse at Khartoum, Sudan. Such chromospheric limb spectrum provide us the most direct information about the chromospheric chemical composition, temperature and pressure, and is the only unambiguous source of information on the height dependent properties of the chromosphere.

To achieve finer height resolution in the chromosphere, the Udaipur Solar Observatory's group took a color movie of the flash spectrum during the February 16, 1980 total solar eclipse from Hyderabad, India, at a fast rate of 24 frames per second through a 50-mm objective prism–lens arrangement. In Figure 4.9 is shown one color frame of the movie showing some of the brightest chromospheric lines (Bhatnagar *et al.* 1981).

Under exceptionally clear sky condition at the Mount Wilson observatory, Hale and Adams in 1909 obtained chromospheric emission spectrum *without* an eclipse. In recent times also a chromospheric emission line spectral atlas has been produced by Keith Pierce at the National Solar Observatory, Kitt Peak, USA. It is noticed from the flash spectrum that the metallic lines confine to lower heights, while the hydrogen, ionized calcium and magnesium lines extend to greater heights.

4.2.2 *Chromospheric Heating and it's Spectrum*

As early as 1860 or even earlier, it was recognized that the entire Sun is a gaseous body including the photosphere. The chromosphere and the prominences are also gaseous in nature as shown by their spectra, corresponding to laboratory emission spectrum of incandescent gas.

Following Kirchhoff's law and comparing the laboratory hydrogen spectra with the chromospheric spectra, Lockyer in 1869 concluded that the gas pressure in the chromosphere is far smaller than that of the Earth's atmosphere. He erroneously assumed that above the photosphere, the temperature of the solar gases continuously decreases and that the height attained in the solar atmosphere by any particular element depends on its atomic weight. Now we know that the distribution of emission lines with height is not a true indicator of the actual stratification of the elements in the chromosphere, but depends on following two conditions:-

1. Intrinsic strength of the transition responsible for emission of the line, and
2. Condition for excitation and ionization of atoms, which in turn depends on the temperature and the gas and electron pressures.

Due to lack of understanding of the atomic spectra and the laws governing the excitation of atoms, the early workers failed to correctly interpret the chromospheric spectra.

A real break through in the understanding of the chromospheric flash spectra came in 1920, with the publication of the well known ionization equation applied to the solar chromosphere by the Indian physicist, M.N. Saha (1893-1956). The Saha ionization equation explicitly indicates that in addition to temperature, the gas pressure exercises a controlling effect on the degree of ionization of a gas. The emission lines of ionized metals observed in the flash spectrum, compared with the Fraunhofer absorption spectrum, provides the correct explanation in terms of lower pressure and higher temperature in the chromosphere as compared to the photosphere. However, a major difficulty still remained that many emission lines such as the hydrogen Balmer lines, H & K lines of CaII and of HeI & He II lines were observed to extend to heights greater than 10,000 km in the chromosphere. How these great heights could be accounted for by an atmosphere in *hydrostatic equilibrium* at a temperature of 5000 to 6000 K. The variation of temperature, pressure and density in the outer chromospheric layers of the Sun is enormous. Therefore it is necessary to specify to what height each 'layer' (photosphere, chromosphere and corona) extents in the solar atmosphere. A quantity called the '*scale height*' is generally used to describe the pressure and density structure in the solar atmosphere, and is defined as the rate at which the density or pressure changes by a factor of 1/e (e = 2.718) with height.

In an isothermal atmosphere, the density distribution is given by the following expression:-

$$P = \rho_o \exp(-\beta * h), \tag{4.1}$$

where $1/\beta^*$ is the density scale height, given by $1/\beta = kT/mg$, and k the Boltzmann's constant, T is the temperature, g the acceleration due to gravity and m the atomic mass.

Early workers in 1920's assumed that the major constituent of the chromospheric atmosphere was hydrogen at a temperature of T=5000 K, at this temperature the density scale height, $1/\beta^*$ turns out to only 150 km. From number of flash spectrum lines photographed at various eclipses, the calculated density scale height was compared with the observed emission scale height. The observed scale heights for hydrogen and ionized calcium ions were much larger than the density scale height calculated for an isothermal hydrogen atmosphere, hence it was concluded that there should be some other mechanisms at play supporting the extended chromosphere.

4.2.2.1 *Heating by Turbulent Motion*

W.H. McCrea in 1929 proposed that turbulent motions through certain volume elements, might supply the extra energy to support the chromosphere. However he did not identify any known chromospheric features, but proposed that the features have a Maxwellian velocity distribution with the mean square value - U^2. This would result in an increase in the gas pressure from $p = N_H KT$ to an effective gas pressure $p^* = N_H KT^*$, where $KT^* = KT + 1/3 m_H U^2$, where m_H is the mass of hydrogen atom, N_H number density of neutral hydrogen atoms, assumed as the only major constituent. Based on the observed width of emission profiles of H and K lines of CaII lines, McCrea took U as 18.4 km/sec and temperature T = 5000K. With these figures he obtained an effective temperature T* of 18,400K and the density scale height of ~ 560 km. This figure was four times larger than the earlier estimates, where no turbulence motion was invoked. However, this value was in good agreement with the observed emission scale height.

McCrea's theory is of great importance as it introduced the theory hydrodynamic mechanism for the first time to explain the chromosphere's anomalous extension in height. But during 1930-40, it became evident that the constant temperature of the chromosphere between 5000 and 6000 K, as assumed by early workers, could no longer be accepted in the face of the mounting evidence of very high temperature of the corona of million degrees. This led number of investigators to look for other possible *non-radiative* mechanism for heating of the chromosphere.

4.2.2.2 *Heating by Wave Motion*

Biermann (1946, 1948) was the first to propose that the heating of the chromosphere might be due to the dissipation of the energy of upward propagating shock waves. His basic idea was that; above the photospheric granulation (τ_{5000}=1) level as seen in white light, a convective stable region extends up to the lower chromosphere. The bottom of this layer is continually pushed by the rising granules. This granular motion gives rise to generation of acoustic waves, which then propagate upwards, carrying with them some portion of the original kinetic energy of the granules. As these acoustic waves proceed upwards, their velocity amplitude increases rapidly due to marked decrease in the gas density with height in the chromosphere. In a short distance the amplitude of these waves becomes comparable to the sound speed; thereafter the acoustic waves develop into shocks. Subsequent dissipation of these shocks at higher levels was considered responsible for heating of the chromosphere. Some authors (Thomas 1948) attributed the heating of the chromosphere to a system of supersonic jets and identified them with spicules.

Biermann (1946) and Schwarzschild (1948) had proposed that heating of the chromosphere might be also due to dissipation of train of acoustic waves generated by photospheric granulation, which may generate shock waves and dissipate their energy to heat the chromosphere and the corona. To understand the energy dissipation through this mechanism, it is necessary to know whether the kinetic energy transported by the granules is sufficient to make good to the energy loss from the chromosphere and the corona. To estimate the energy loss, let us assume that the kinetic energy carried by a single rising granule is given by:-

$$E_{granule} = \tfrac{1}{2}\,\rho A d v^2, \tag{4.2}$$

where $\rho \approx 3 \times 10^{-7}$ gm cm^{-3}, is the mean density at the top of convection zone, $A = 2.8\times10^{16}$ cm^2, is the average surface area of a granule, $d \approx 3.5\times10^{7}$ cm, is the average thickness of convective zone, and $v = 1\times10^{5}$ cm sec^{-1} is the mean upward velocity of a granule. If $N = 2.2\times10^{6}$ is the total number of granules on the Sun and $t = 600$ seconds, their average life time. It turns out that the kinetic energy brought upto the top of the convection zone, over the surface of the Sun per unit time is:-

$$L_{granules} = NE_{granule}/t = 5.4\times10^{30} \text{ ergs sec}^{-1}. \tag{4.3}$$

On the other hand, the net flow of energy per second from the chromosphere and corona, including the solar wind is estimated to be:-

$$L_W = 3.9\times10^{29} \text{ergs sec}^{-1}. \tag{4.4}$$

The net energy loss from the chromosphere and corona is an order of magnitude less than the kinetic energy transported upward by the granules.

Of course, one would not expect that the kinetic energy of the all the granules is transformed into acoustic wave energy or that the wave generation region to be confined only to the stable layers, overlying the convection zone. Some scientists believe that the waves are excited in the convective zone itself through turbulent motion. Besides the acoustic waves, there is also a possibility that internal gravity waves may be generated in the stable layers above the convective zone. And some amount of heating of the chromosphere and corona can be accounted for through dissipation of gravity waves by thermal convection. To find an answer to this crucial question; which particular type of wave, that is acoustic or the gravity is responsible for chromospheric and coronal heating? Several workers have examined this question. From all the available evidence, it is evident that acoustic waves are mainly responsible for heating of the chromosphere, as there is no obstacle to the upward propagation of the acoustic waves, provided that the periods are less than about 195 seconds. On the other hand, the gravity waves cannot propagate in the upper photosphere, due to their longer period ~ 210 second or more, and are heavily damped in the very low chromosphere compared to the acoustic waves.

4.2.2.3 *Heating by Magnetic Field*

Alfvén (1947, 1950) suggested that the magneto-hydrodynamic (MHD), waves may contribute to heating of the chromosphere and the corona. He proposed that the motion of granules, in a magnetic field of about 25 gauss can give rise to such waves traveling up into the chromosphere and the corona. It is known that in the absence of gravity, three types of waves are possible, either the Alfvèn wave (is a wave motion occurring in magnetized plasma in which the magnetic field oscillates transverse to the direction of propagation, without a change in magnetic field strength), or fast and slow mode magneto-acoustic waves. It was shown by Osterbrock (1961) that both Alfvèn and slow-mode magneto-acoustic waves are heavily damped in the upper photosphere and low chromosphere due to the Ohmic dissipation, unless the magnetic flux exceeds 50 gauss or so. In the case of supergranulation networks, which are seen as CaII enhanced brightness regions, the field strength is less than this value, hence only the fast-mode waves could develop into shocks due to decrease in the density. These shocks could propagate upwards, which may be responsible for heating of the chromosphere above the supergranulation boundaries. Recently, several workers including Parker, Ulmschneider, Hasan, and others have been working on the idea of invoking magnetic flux tubes responsible for heating of the chromosphere and the corona. It seems that the acoustic waves, heat the low and middle chromosphere while the Magneto-hydro-dynamic (MHD) waves heat the magnetic regions up in higher chromospheric levels. Ulmschneider (2003) suggested that the magnetic regions become more dominant contributor to heating, due to faster rotation in the chromosphere at higher levels, however it may not be that the chromosphere rotates faster at higher heights. From several investigations it seems that the highest layers of the chromosphere need additional non-wave heating mechanisms. Some scientists have suggested that the required energy may come from *micro-flares*, which arise due to reconnection of magnetic lines of force in the chromosphere.

4.2.2.4 *Heating by 5-minute Oscillations*

The 5-minute solar oscillations discovered by Leighton and colleagues in 1960 were also considered as a possible source for heating the chromosphere. The 5-minute oscillations are mainly vertical oscillations occurring in the upper photosphere and the lower chromosphere. The question whether they can generate acoustic waves and shocks responsible for heating the chromosphere and corona has been examined. It has been shown that in the upper photosphere, the oscillation with periods of about 300 seconds have mainly 'damped' wave characteristics. Thus these cannot generate acoustic wave motion responsible for heating the chromosphere.

4.2.3 ***Quiet Chromospheric Structure***

4.2.3.1 *Chromosphere on the Disk*

Earlier the chromosphere had been observed only at the time of solar eclipse, and that too only at the solar limb, but now using narrow passband filters, spectroheliograph or spectrohelioscope centered on strong chromospheric lines like Hα and H and K lines of CaII lines, the chromosphere could be seen even without an eclipse, both at the limb and on the disk. Very fine observations of chromosphere on the limb and disk are now available, which display variety of interesting phenomena, such as; spicules, bright and dark mottles, network structure, plages, filaments, active regions, flares etc. The disk observations indicate that the chromosphere is a highly non-uniform region of the solar atmosphere, as seen in Figure 4.10, showing spectroheliogram taken in K-line.

By simply adjusting the 'tuning' of the filter or the Spectroheliograph's second slit, one can see different levels of the solar

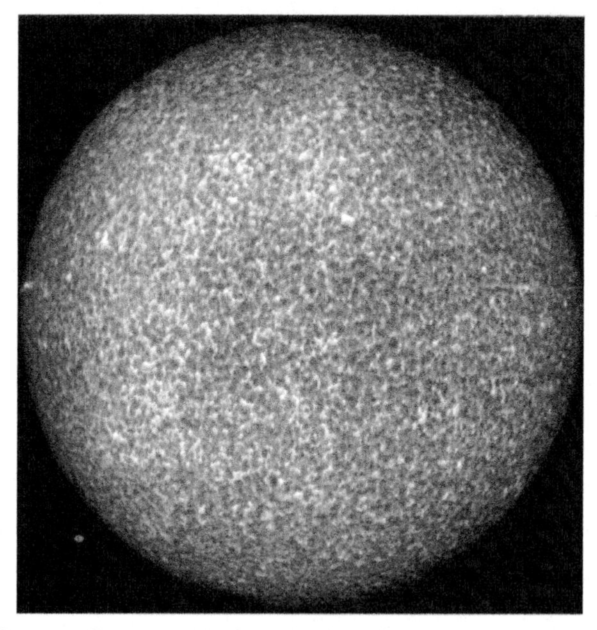

Figure 4.10 Spectroheliogram taken from the Kodaikanal observatory in K-line of CaII, displaying chromospheric network structure.

chromosphere, because different parts of the line profile produce images at different levels in the solar atmosphere. For example, a filtergram taken at the Hα line center refers to the top most layer of the chromosphere, while pictures taken in the wings of a line, produce images successively of the mid-chromosphere (~1500 km above the photosphere) down to the photosphere. Similarly pictures taken in the CaII, K_2 or H_2 emission peaks and K_3 or H_3 core show features in the upper and mid-chromosphere, while images in the K_1 or H_1 wings refer to the temperature minimum region. The fine chromospheric features show rapid changes due to mass motion and consequently their emission and absorption lines are Doppler shifted. The chromosphere has a very complex morphology and is far from being homogeneous and uniform. The most conspicuous features seen on Hα and K-line chromospheric spectroheliograms or filtergrams are the bright patches called the *plages*, elongated dark structures called the *filaments*, and fine dark lines known as *fibrils*, besides these features the entire chromosphere appears to be made up of bright network pattern, called the *chromospheric network* with cell sizes of nearly 30,000-35,000 km.

Filtergrams taken in the wings of Hα line show coarse bright and dark mottled structures, which mark the boundaries of network, as shown in Figure 4.11.

4.2.3.2 *Chromosphere at the Limb – Spicules*

Warren de la Rue first noticed *Flames* or spicules as small jets emanating from the solar limb during the total solar eclipse of 1860 and later several observers also reported these features. These limb observations generated considerable interest in the mid-1900s.

Lead by R. Roberts of the Harvard Observatory in early 1950s, Dunn (1960) made extensive high resolution observations of spicules and took pictures across the Hα line, ranging from Hα +1.25 to Hα-1.15, using the 40-cm aperture chromospheric telescope of the Sacramento Peak Observatory and the Zeiss 0.25 Å pass band filter. In Figure 4.8 is shown a sequence of the limb photographs taken by Dunn at these wavelengths. Spicules seem to be most visible on the pictures taken around +3/4Å and 1Å away from Hα. At some places around the limb they appear in 'clusters'. From the micro-photometry of large number of spicules, Dunn deduced the widths of spicules in the range of 300-1100 km (0.4 -1.5 arc sec). The emission from the spicules decreases gradually with height, consequently they do not have sharply defined upper boundary. The observed variation in height could be due to the exposure time and also due to the intrinsic height variation. From Dunn's observations, Beckers (1968) deduced an estimate of the scale height of 2200 km (~3 arc sec), being the distance over which the spicule intensity drops by a factor of e. The height of spicules is conventionally measured from the photospheric limb, down to which they appear to extend when observed in the far wings of Hα line (> ± 1.0Å). Very few workers have given explicit value of average height of spicules, which range between 6500 and 9500 km. In some cases spicules have been reported extending to even 15,000 km as reported by Lippincott (1957). From a statistical investigation, Athay (1959) noted a decrease in the apparent height of spicules from the poles to the equator. Generally the available spicule heights refer to the projected height in the sky plane, which is not the true height above the solar surface. The corrections due to the curvature of the limb and the location of spicules on the Sun have to be accounted to obtain the true height of spicules.

4.2.3.3 *Spicules on the Disk*

We have seen in earlier section that spicules on the limb appear as small hairy-jet like structures, protruding out from the Sun. But when viewed on the disk in filtergrams taken at say 0.5–1.0 Å away from the

Figure 4.11 Showing dark mottled structure photographed at Hα +0.5Å note the spiky appearance of mottles suggesting that these features correspond to spicules seen on the disk.

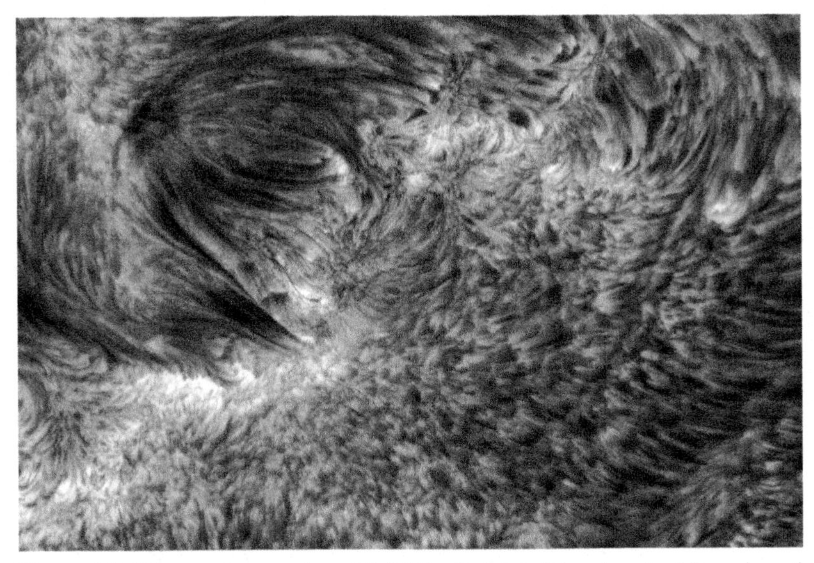

Figure 4.12 Picture taken on June 16, 2003, slightly off-band in the blue wing of Hα line, of spicules on the disks from NSST showing fine spicules as hairy structures on the upper left hand corner of the picture.

Hα line center in the wings, the general appearance of the chromosphere turns into a *network* of dark mottled structure, as shown in Figure 4.11, taken at Hα+0.5Å. There had been considerable debate on the true nature of these dark mottles, but now it is well resolved and agreed by the solar community that these dark mottles appearing in clusters on the disk pictures, are in fact spicules seen projected against the bright background. Their average width range from 1 to 3 arc sec and several times longer, and are arranged in network of nearly 30,000 km in diameter, as shown in Figure 4.11. At the end of dark mottles, a very tinny sub-arc second size roundish bright mottle or dot has been occasionally observed.

The appearance of spicules or the dark mottles drastically change, as one compares the pictures take in Hα line center and in the wings. This is due to two reasons:-

1. In the wings of lines we see deeper in the solar atmosphere, thus to a different level, and

2. Due to the motion in spicules which manifests as Doppler shift.

Time lapse movies of 'disk spicules' taken in *off-band* Hα line show continuous motion both in position and size indicating a dynamic nature of spicules. Filtergrams taken in other chromospheric lines, such as in CaII lines also show dark mottle network structure corresponding to spicules on the disk. Recently extremely high resolution images in Hα of spicules on the disk has been obtained on June 16, 2003 from the New Swedish Solar Telescope showing fine hairy structure of fraction of arc sec (0.4 arc sec) width and several second of arc in length, as shown in Figure 4.12.

4.2.3.4 *Evolution of Spicules*

Spicule appears initially as a small bright lump or *mound* in the low chromosphere. It then rapidly elongates upwards and increases in brightness, attaining its maximum height within a minute or two of its first appearance. Subsequently, it may either fade away or descend back to the low chromosphere with the same initial or a different speed and may shrink to a small *mound*. Some spicules may fall back taking a different path also. The average life time of spicules range between 2 and 6 minutes.

Several authors have made number of measurements of the upward and downward velocities of spicules from time lapse observations. From these measurements it has been noticed that the upward motion is usually regular and continuous. The spicules seem to rise with constant velocity, stop abruptly at a maximum height and then descent either smoothly or in jerks. Lippincott (1957) had reported upward velocities range between 5 to 60 km/s while the downward velocities range from 0 to 70 km/s. An average value between 25 km/s for both upward and downward spicules is an accepted value. Spectroscopic observations of spicules provide the line-of-sight velocity component due to the Doppler shift of the Hα line. While comparing the velocities obtained by the two methods (i.e., velocity obtained in sky plane and by Doppler measures), one should bear in mind that the direct time lapse observations give velocities in the sky plane, but the Doppler measures provide the component of the velocity along the line-of-sight. The average line-of-sight velocity obtained by several authors range between 2.5 and 16 km/s, taking into account the possible tilt of spicules to the solar surface and the curvature of the solar surface, the two values seem to be in agreement.

The emission spectrum of spicules show a 'tilt' with respect to the dispersion in the horizontal direction, this suggests that the two 'sides' of a spicule may have different velocities and direction, implying a *rotational motion* in individual spicules. This idea was suggested by Beckers (1968) and by Pasachoff *et al*., (1968). It is proposed that the material in spicule rises in a *helix* and fall back along a similar path and that the spicules may follow magnetic lines of force. However, no direct magnetic field measurements in spicules have been made to-date, but the direct evidence that the spicules in the polar regions of the Sun are aligned parallel to the coronal rays, suggests that the local magnetic field seems to have influence on the orientation of spicules and coronal rays.

4.2.4 *Quiet Chromospheric Model*

The quiet chromosphere has a very complex morphology, it is far from being homogeneous and uniform, it display's conspicuous non- uniform structures such as spicules, mottles, cells and networks. Under these circumstances it is difficult to construct a unique single *chromospheric model,* giving the distribution of temperature, density, pressure and other parameters with height. To account for the observed non-homogeneity due to chromospheric network, several authors have worked out two-component chromospheric models {Thomas and Athay (1961), Hiei (1963), Kanno (1966), Beckers (1968)}. Here we shall describe one of the two-component models proposed by Vernazza *et al*., (1981) which is based on Extreme Ultra Violet (EUV) *Skylab* observations of the continuum emission between 400 and 1400 Å, and assuming hydrostatic equilibrium. The spatial resolution of these observations was 5 arc sec, which was sufficient to distinguish between the cells and the network boundaries. Although this is a one–dimensional model, as each of the two components is assumed to have a horizontal extent compared to the depth of the layer. This assumption is reasonably valid in cells whose horizontal dimension of about 30,000 km far exceeds the height of the low chromosphere, but in network it may *not* be valid as here the horizontal and vertical scales are of comparable dimensions.

In Figure 4.13 are shown the chromospheric temperature profiles with height in cells and in networks.

The temperature profiles of both these structures show a rapid temperature rise from the minimum values around 4000 K at 500 km above $\tau_{0.5} = 1$, to a broad temperature plateau from 6,000 to 7,000 K extending over almost 1500 km. This is followed by a very steep rise in temperature to 20,000 K and higher within a few hundred kilometers in height. This chromospheric model is based mainly on the observations in the EUV and visible spectrum, particularly CaII, He I, He II lines, but do not take into account the highly dynamical nature of spicules.

Using this model Vernazza *et al.*, (1981) have worked out the heights of formation of the various chromospheric radiations and lines. In Figure 4.13 (b) is shown a schematic diagram giving the approximate regions of formation of the various radiations.

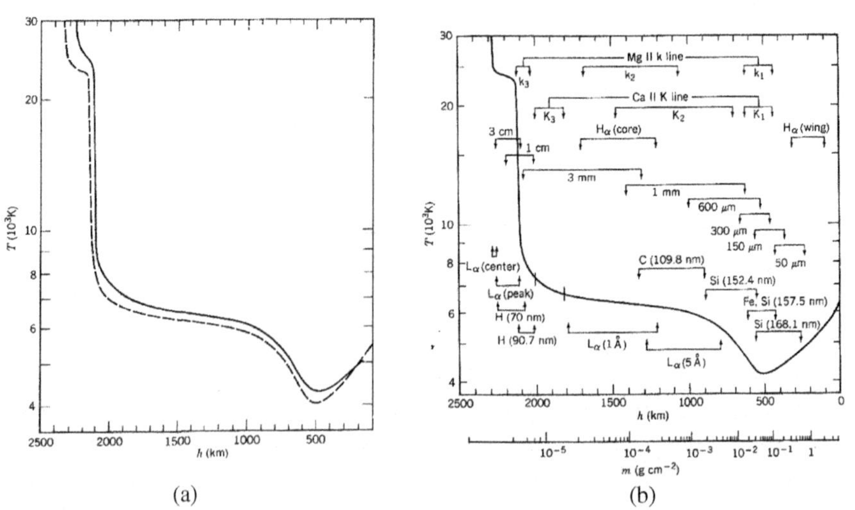

Figure 4.13 (a) Chromospheric model showing the run of temperature versus height for the cells shown by dotted line and in network by solid line. (b) Diagram showing approximate regions, in height and in mass column density of formation of the various chromospheric radiations, ranging from Lyman α to Hα wings.

4.3 Transition Region

A sharp increase of temperature from about 10^4K at the top of the chromosphere to 10^6 K in the corona occurs in a very narrow region of few hundred kilometers of the solar atmosphere, as shown in Figure 4.21, this region is called the *transition* region. This region is best observed in the strong EUV resonance lines of heavy ions of Fe XV, Mg IX, Ne VII, O VI and OIV.

Observations made using different lines which arise at different temperatures can be used to *see* various layers of the solar atmosphere. For example the singly ionized helium, HeII line at 304 Å arises at about 60,000 K, which is formed in the lower part of the transition region near the chromosphere, while the images taken in Fe IX, Fe XII and Fe XV lines by TRACE are formed respectively at 1.0, 1.5 and 2.0 million degrees Kelvin. These radiations refer to the upper transition region and the corona. In Figure 4.14(a) is shown a full disk image taken in the

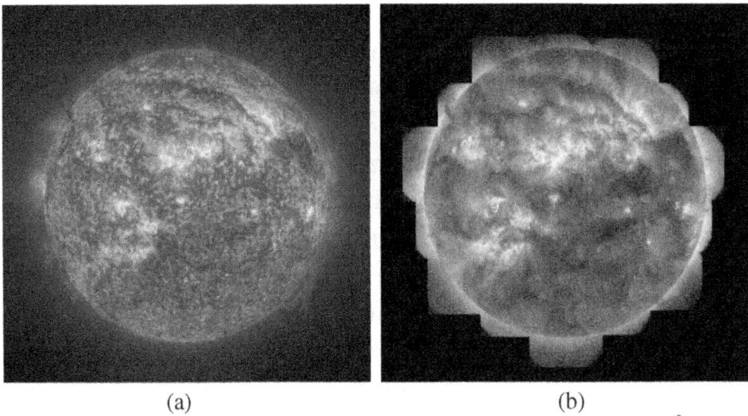

(a) (b)

Figure 4.14(a) Picture of taken from SOHO in the He II line at 304 Å on 14 September 1997. (b) Picture taken from TRACE in Fe IX/X line at 17.1 nm, note the difference in the surface details in the two pictures, the first refers to the lower transition region, while the other to higher region.

HeII line from SOHO, while in Figure 4.14(b) a solar image taken in Fe IX line from TRACE spacecraft. Both these images show brightness over active region and dark filament channel. There is slight difference in appearance in surface details, for example HeII image shows mottle structure while TRACE image does not show.

4.4 The Quiet Corona

4.4.1 *Introduction*

Were it not for the remarkable coincidence that the apparent angular diameter of the Sun and the Moon happen to be nearly the same and that the Moon's orbit and the ecliptic lie in almost the same plane in the sky, we would have never known the existence of one of the most spectacular views of any celestial phenomenon – the *crown* around the Sun – called the solar corona. Historical reports of solar eclipses are found dated back to the 14th century BC, but unambiguous description is not available except for a possible Babylonian eclipse in 1063 BC, described by Fotheringham (1920) that, "on the seventh year the day was turned into night and fire in the midst of the heaven". From this sentence, "fire in the midst of the heaven", it appears that, this refers to the appearance of extended coronal streamers. Another reference to the sighting of the corona is found in *Plutarch*, published by the Harvard Press, Cambridge, Mass. 1957, that, "an eclipse occurred a few years earlier showed a kind of light visible around the rim which keeps the shadow from being profound and absolute". Although this sentence does not perfectly describe the extended structure of the corona, but may indicate that some sort of coronal brightness was visible then. The rock paleograph, as shown in Figure 1.6 of a total solar eclipse of July 11, 1097 AD, made by Native American Indians clearly shows that extended corona was indeed observed by ancient civilization. During a total solar eclipse, the *dark Sun* appears surrounded by extremely faint and much extended atmosphere called the corona. The maximum intensity of the corona, near the edge of the solar disk is almost a million times (10^{-6}) less than the visible disk of the Sun, and it rapidly decreases to billion times (10^{-9}) within a one solar diameter from the solar limb. Due to the scattered light in our Earth's atmosphere, which is several times brighter than the corona, we are unable to see the faint extended corona without an eclipse. During a total eclipse, the Moon besides blocking direct vision of the solar disk, it also subtends a cone of darkness over the Earth's atmosphere, which lowers the sky brightness by a factor of nearly 4 orders of magnitude, which is quite sufficient to reveal the beauty of the corona. With the use of special telescope called *coronagraph* and from high altitude locations, where one gets very clear and dust free atmosphere, it has become possible to see the inner corona near the limb even without an eclipse.

4.4.2 *Coronal Components, Brightness and Structure*

Broadly speaking there are three main components of the solar corona, essentially identified by the mechanism they are produced. These components are:-

1. K- (*Kontinuierlich*) corona,
2. F- (*Fraunhofer*) corona, and
3. E- (*Emission*) corona.

Recently a fourth, T-*thermal* emission component has been also proposed. The emission mechanisms of their formation are quite different and are described in the following. The K-corona displays a continuous emission spectrum and is strongly polarized. It arises from the photospheric light scattered by electrons in the corona. Due to the very high coronal temperature of the order of a million degrees or more and extremely low density, the electrons travel in the corona at very high speeds, therefore the Fraunhofer spectrum lines, which are scattered by the electrons, are highly broadened and appear washed out. During a total eclipse, the K- and the F-coronae appear as white light halos around the Sun's disk. The K-corona shows high degree of linear polarization, while the F-corona has no polarization. If the Sun was a point source and the scattered light occurred normal to the line of sight, then we would expect a 100% linear polarization. However, due to the complicated geometry of the coronal structures, the observed polarization of K-corona is less than 100%. The linear polarization of the K-corona varies from about 18% near the Sun to nearly 66% at about 3 to 4 solar radii. The degree of polarization varies with distance from the Sun, and also depends on the particular coronal feature and on the epoch of the observation.

The F-corona arises due to the scattering of the Fraunhofer spectrum by the dust particles in the interplanetary atmosphere. In fact the F-corona is not a solar phenomenon at all; it could be considered an extension of the zodiacal light. The F-corona displays even dark absorption Fraunhofer lines due to scattering of sunlight by the dust particles, but no polarization.

The E-corona represents the actual emission component, arising due to the extremely high temperature of the coronal gas of the order of million degrees and low density of the order of 10^{10} cm^{-3}. Until 1938, it was a challenging problem in astrophysics to identify these emission lines in the corona. The major breakthrough was made by Grotrian and by Edlén in 1939, who identified some of the E-coronal emission lines as arising from forbidden lines of highly ionized ions of Fe, Ca, Si, Ar, Ni. Although the total integrated brightness of the E-corona is relatively

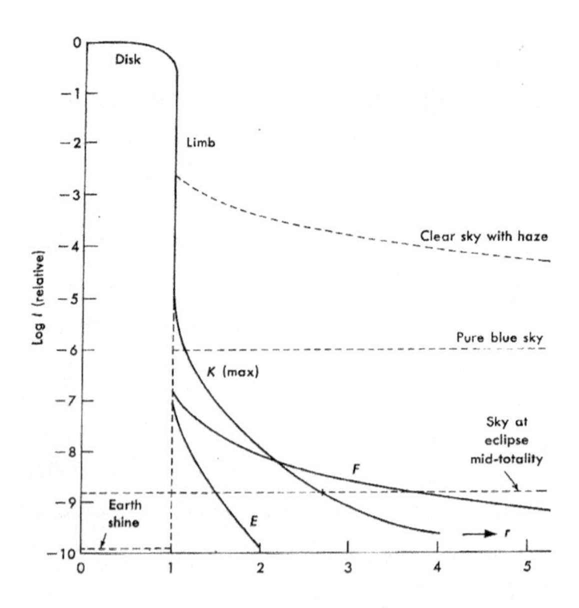

Figure 4.15 Plots of relative intensities of various components of coronal light and sky brightness verses distance from solar limb. The blue sky brightness can be million times less than the Sun's intensity, while at totality it may drop to about 1000 million times Sun's brightness, hence the faint K & E-coronae become visible.

small as compared to the K- and F- coronae, but the strong emission from the E-corona which is confined to narrow bands makes it possible to see the E-coronal features during the eclipse, and also outside the eclipse, through narrow band filters centered on the emission lines.

A fourth component–T (thermal) -corona has been recently proposed, which is caused by thermal emission in Infrared region by the interplanetary dust; the same dust particles that produce the F-corona. Since T-corona is due to the thermal emission, it is detected only by Infrared sensors, which adds to the continuum in all infrared measures.

In Figure 4.15 is shown the brightness variation of the three coronal components, as a function of the radial distance from the edge of the Sun, and the contribution due to the sky. A typical ground level sky brightness is of the order of $10^{-5}I_o$ of the Sun's intensity. Under exceptionally clear blue sky condition available from high mountains, the sky brightness could decrease by an order of magnitude, but during the total solar eclipse, it could reach even as low as to 10^{-9} I_o . Thus the K- and the E- coronae, whose intensities are of the order of $10^{-6}I_o$ or less become visible. The intensity of the K-corona steadily decreases with distance from the Sun.

Air borne observations from high flying aircrafts have been used to take pictures of extended corona up to 20 solar radii during 1980 solar eclipse by Keller *et al.*, and a description of five airborne eclipse expeditions to determine the temperature and density structure is given by Keller (1982). During the 1995 total solar eclipse, an attempt was made by Bhatnagar *et al.*, (1996) to photograph the corona from a height of 25 km, from the cockpit of MiG-25 plane of the Indian Air Force, flying at a speed of 2 mach. The idea was to take pictures of the corona from the highest point attainable from an airplane, as at such heights the atmospheric scattered light is further reduced and the corona may be seen to very large distances. The high flying pilots of MiG-25 tell us that from 25-km height the sky looks dark gray, and during the totality the corona appeared extended very far out to more than 16-18 solar radii.

4.4.3 *Coronal Structure*

The solar corona is highly structured region of the Sun, consisting of complex fine filamentary coronal loops as seen in Figure 4.16 and outwardly directed streamers, plumes, helmets-shape features, extending far out from the Sun (see Figures 4.17).

Figure 4.16 Coronal loops over the eastern limb of the Sun was taken by TRACE in FeIX/X line at 171Å pass band on November 6, 1999, at 02:30 UT, this line is formed at one million degree temperature, in the lower corona and the transition region. Shows numerous thin magnetized loops extending to hundreds of thousands of kilometers above the solar surface. These detailed images suggest that most of the heating occurs low in the corona near the base of the loops, as they emerge and return to the photosphere. From the emission measures, it is found that heating is not uniform along the entire length of the loop.

As the corona is optically *thin* at visible wavelengths, therefore the optical radiations are **not** *absorbed* while traversing the corona over the solar disk, as in the case of the chromosphere and photosphere, hence in the visible light the coronal features **are not** seen on the solar disk. To see coronal features on the disk one has to make use of soft X-rays or and EUV radiations. When viewed at the limb many coronal features, like streamers and helmets add together and one gets an integrated effect of the all the features lying along the line of sight.

The solar magnetic field plays an important role in defining the shape and structure of the solar corona. As seen in Figure 4.17, the shape of coronal streamers, helmets, plumes etc., seems to be *controlled* by the photospheric magnetic fields. The shape and the intensity of the corona also vary with solar activity cycle, as shown in Figure 7.6.

In Figure 4.17 is shown solar eclipse photographs taken during the solar maximum and minimum periods in 1980 and 1994 respectively. Near the minimum the corona appears symmetrically elongated about the equator, displaying long extended streamers on either side of the equatorial region. This is because the active regions during the minimum period are generally located at low latitudes around the solar equator and that the distribution of the coronal material is correlated with surface magnetic field. During high active phase of activity, the corona appears much more symmetrical, because the active regions appear at higher latitudes also and that the magnetic lines of force form close loops over active regions along which the coronal plasma is confined. The helmet-shape streamer seen on the lower bottom right in Figure 4.17(b) consists of bright coronal arches, appears roughly in the sky plane. Occasionally at the base of helmet streamers a prominence may also be seen. The helmet arches demark the magnetic field configuration in the corona. The shape of the corona also varies with distance from the limb.

Near the polar regions, the coronal intensity is generally much less as compared to the equatorial region, particularly during the minimum period. Around the poles, short high latitude streamers called *polar plumes* are seen. These polar plumes deviate considerably from the radial direction and tend to bend towards lower latitudes, apparently following the magnetic lines of force of a global *'bar'* magnetic field. From a detailed analysis of polar plumes, observed on two large scale eclipse plates taken during 21 January, 1898 and 21 September, 1922, Bhatnagar and Rahim (1970) found that maximum frequency of plumes occurs in an annular zone around 10 degrees from the poles and is least over the poles. From these observations they concluded that the length of a *hypothetical bar magnet* is $0.5R_o$, located at centre of the Sun. The frequency and orientation of the polar plumes also vary with the solar cycle. What these observations and measurements mean is not clear, but

(a) (b)

Figure 4.17 (a) Showing photographs of solar corona taken on 16 February 1980 during maximum and (b) taken on 03 November 1994 during the minimum period of solar activity. Note the symmetrical shape of the corona and streamers all over the Sun seen during maximum and extended streamers seen near the equatorial region, during the minimum period.

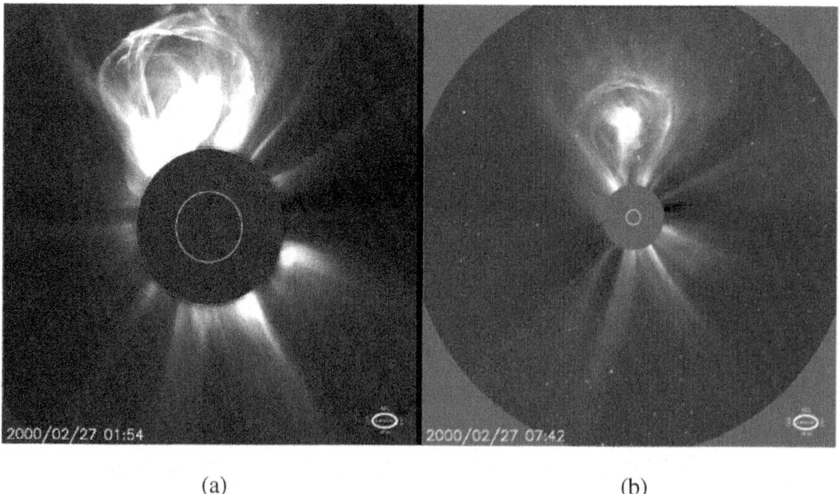

(a) (b)

Figure 4.18 (a) Coronal Mass Ejection (CME) observed on 27 February 2000 with C2 coronagraph and (b) picture taken with C3 coronagraph with LASCO instrument on SOHO spacecraft. CME is observed upto 30 solar radii from the Sun. The inner circle represents the location and diameter of the Sun.

it is definite that the global magnetic field of the order of 5-10 gauss in the polar regions seems to control the configuration of polar plumes.

Since 1996, it has become possible to observe the white-light solar corona up to 30 solar radii from space with LASCO's (Large Angle

Spectrometric COronagraph) 3 coronagraphs on board the SOHO spacecraft. In Figure 4.18 is shown pictures of a huge Coronal Mass Ejection (CME) taken from LASCO's C2 and C3 coronagraphs on February 27, 2000 extending to more than 30 solar radii. LASCO has been taking extremely important CME data for the last 8-9 years, this has greatly helped to understand the interactions of high speed solar plasma and geomagnetic field.

4.4.4 *Observations in Short Wavelengths & Coronal Hole*

The solar corona on the disk is best seen in broad band soft X-ray and EUV images. With rocket and space–borne instruments on board *SkyLab,* soft X-ray pictures of the solar corona were first obtained. But now from the Japanese satellite–YOHKOH (since 1991 until 2001), and from TRACE satellite and other X-ray and EUV space-borne telescopes, it has become possible to obtain X-ray and EUV images of the corona over the disk, on a regular basis. In X-rays and EUV radiations the corona is optically thin, but as the photosphere is cool and does not emit short wavelengths, therefore these radiations (EUV & X-ray) from the solar corona appear as bright emission, against a dark background. Early images taken from the rocket–borne X-ray telescope on July 7, 1970 during the total eclipse and those from *SkyLab,* provided a wealth of information on the solar corona. The first and foremost discovery was that of *coronal holes* - areas of very low or even zero X-ray emission. On X-ray images of the Sun, coronal holes appear as dark regions which is due to sharp decrease in density and the corresponding decrease in emission. These are known as coronal holes because they appear as void in the X-ray coronal images. Perhaps the first indication of coronal holes came from M. Waldmeier's observations, who noticed them from limb observations of the coronal green line as early as in 1957, and called them as Löcher (in German meaning holes). Observations from NASA's Orbiting Solar Observatory (OSO) in 1960 and in early 1970, also indicated presence of areas of low intensity in extreme ultra violet (EUV) images made in MgX line at 62.5 nm.

The radio observations have also shown the presence of coronal holes. At short wavelengths around 2 cm or less, which arise in the transition region, the emission in the holes is surprisingly slightly higher than the surrounding. At meter wavelengths, the holes appear as areas of reduced emission corresponding to lower brightness temperature (about 750,000 K) than the surrounding quiet corona.

Correlation of X-ray pictures with full disk magnetograms and spectroheliograms or filtergrams made in helium 10830Å line also show remarkable positional correspondence. The coronal hole region is demarked as low magnitude uni-polar magnetic area on a magnetogram, while on 10830Å He I picture, the coronal hole is delineated as 'bright patch' surrounded by conspicuous absorption region. Using the National Solar Observatory's full disk daily magnetograms and 10830Å spectroheliograms, late Dr. Karen Harvey had systematically identified locations of coronal holes from ground based observations and prepared daily coronal hole maps, which have been found very useful for synoptic study of coronal holes. Comparing the X-ray pictures with surface magnetic maps, it was also found that coronal hole regions correspond to *uni-polar* magnetic field; suggesting an open magnetic field structure on Sun. It was further revealed that field lines are not only open, but are actually diverging rapidly through which the solar wind flows out in the interplanetary space.

With the discovery of coronal holes, the long standing mystery of 27-day recurring 'M-regions' in solar–terrestrial physics was also solved. Actually, the connection between geomagnetic disturbances and the activity on the Sun had been suspected since long, but **no** particular visible feature/s on the Sun could be identified to explain the recurrent 27-day terrestrial geomagnetic events, until the X-ray pictures were available. Although, in 1939 Bartel had proposed that the geomagnetic terrestrial effects can be explained and mentioned that if, "streams or cloud of solar particles – atoms and molecules in neutral or ionized state – are emitted from certain active regions on the Sun, and when these particles interact with terrestrial magnetic field, can produce geomagnetic disturbances or storms". But in 1939 Bartel was unable to correlate the observed geomagnetic terrestrial effects with any particular visible feature on the Sun, and therefore called them as 'M-regions' (M for magnetic).

Coronal holes as the source of high speed wind was confirmed by

Krieger *et al.*, (1973) using *in situ* measurements of the solar wind at 1 AU and mapping the out-flow back to the Sun, and assuming spiral trajectory along the magnetic lines of force. It has been graphically established from Skylab and OSO-7 data that the open field coronal holes are in fact source of high speed wind streams. Coronal holes change shape and size during a solar cycle.

During the Skylab period (1972-73), a near equatorial coronal hole appeared extended to high latitudes, forming a large N-S feature as shown in Figure 4.19. To study the formation and decay of coronal holes, Timothy *et al.*, (1975) used the large hole observed between May to September 1973 from the Skylab. Before the formation of the coronal hole, two large active bipolar regions, one in the northern and other in the southern hemisphere emerged on the disk in April 1973. One

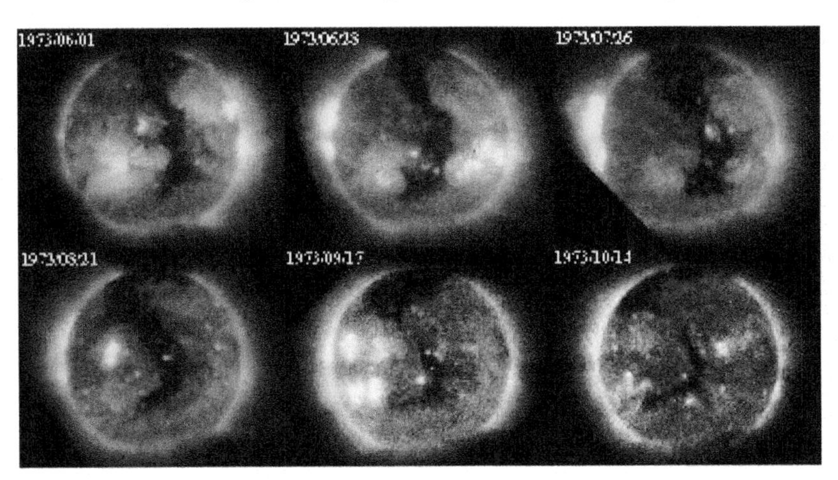

Figure 4.19 Showing North-South oriented coronal hole on X-ray pictures taken from Skylab, beginning in May 1973 at an interval of 27 days. Showing almost persistent co-rotating coronal hole for 5 rotations.

rotation later in May the magnetic fields of the following region in the northern hemisphere diffused outwards, merging several uni-polar magnetic regions. In the southern hemisphere, the leading polarity having the same sign as the following polarity as in the northern hemisphere, diffused and spread to merge and dominate the central portion of the solar disk, from the north to the south pole as shown in Figure 4.19. This coronal hole was visible for more than five solar rotations.

The life time of coronal holes vary from several rotations to few hours depending on the size and extend of the hole. Coronal holes finally vanished as bright X-ray emission appearing associated with the emergence of new active regions near the boundary of coronal holes or even in the location of the old holes.

Coronal holes show very little or zero differential rotation, although the photospheric magnetic field in which they are supposed to be anchored display significant differential rotation. From a quantitative study of Skylab data Timothy *et al.*, (1975) found the following expression for synodic rotation rate of coronal holes:-

$$\Omega_{hole} = 13.25 - (0.4 \pm 0.1)\sin^2\phi \deg/ day, \tag{4.4}$$

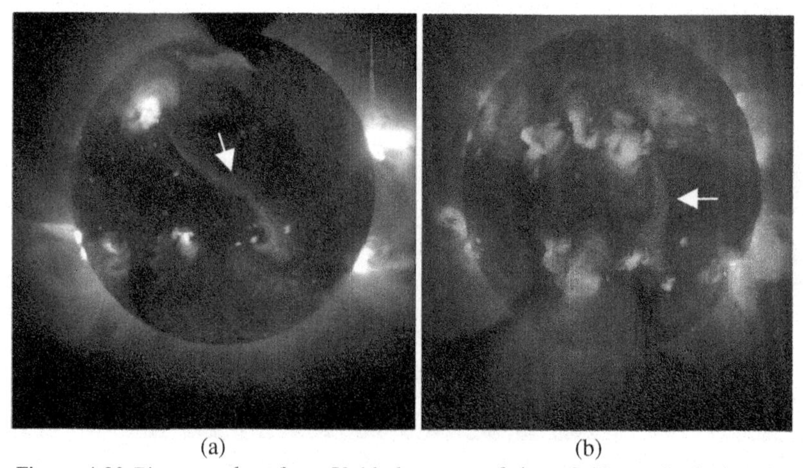

(a) (b)

Figure 4.20 Pictures taken from Yohkoh spacecraft in soft X-ray displaying, loops over active regions, coronal hole, bright points, large coronal arches joining active regions in the northern to southern hemispheres over the equator, as indicated by an arrow. (a) Picture was taken on 29 May 1998 and (b) taken on 7 July 1999. Blue &

where ϕ is the heliographic latitude. While for the sunspots, the rotation rate is given by Newton and Nunn (1951) as:-

$$\Omega_{hole} = 13.39 - 2.7\sin^2\phi \deg/ day. \tag{4.5}$$

Comparing the latitude dependent terms in Equations (4.4) and (4.5), the differential rotation for photospheric sunspots in Equation (4.5) is significantly higher as compared to coronal holes, which shows extremely small latitude dependence, suggesting almost *rigid* rotation of coronal holes. This result is conflicting as the solar surface magnetic fields display substantial differential rotation, while the coronal holes which are due to the surface magnetic fields display rigid rotation.
How this could happen?

To explain this discrepancy, Wang *et al*., (1988) have shown that the rigid rotation of the coronal fields results from the rapid decline of higher order multi-poles with height. Nash *et al*., (1988) concluded that the coronal holes rotate rigidly, results from the current-free nature of the outer corona, that is, the coronal field and its foot-points must approximately co-rotate, otherwise the field would 'wind up' and violate the current–free condition required in the corona. Wang and Sheeley (1993) have studied the specific case of the Skylab prototypical Coronal hole-1, and successfully modeled its evolution.

The other striking features observed on YOHKOH X-ray pictures are bright compact active regions, X-ray bright points, loops over active regions, and fainter and more extended arch system joining active regions and some may even extend over the solar equator, as shown by arrow in Figure 4.20. On higher resolution soft X-ray pictures large number of small bright points of few arc seconds sizes are seen, which are identified as emerging bi-polar regions. Similar features are also seen in pictures taken in EUV lines of Mg X and Si XII.

4.4.5 *Temperature and Density Profile of the Corona*

As we have seen in the preceding sections that even the quiet corona is highly non-uniform both in temperature and density, and varies in time, therefore no *one* unique temperature or density value can be assigned to the corona. For estimating coronal temperatures there are mainly four methods; and each one yields slightly different temperature.

For example; the Doppler broadening of emission lines gives about 2 million degrees K; from the density gradient resulting from intensity measures, gives coronal temperature about 1.60 million degrees K; from

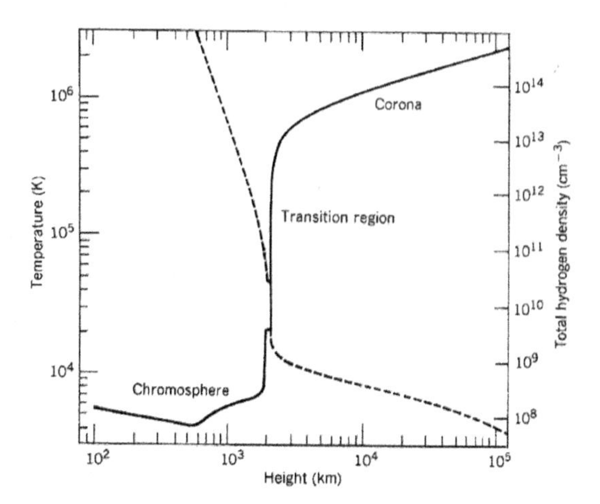

Figure 4.21 Run of temperature (solid line) and density (dotted line) in the chromosphere, transition region and corona over quiet region as given by Gabrial (1976).

radio emission at 10-cm wavelength, the brightness temperature turns out to be about 1.5 million; while from the ionization theory; the corona corresponds to about 800,000 K only. The electron densities can be calculated from the intensity and the polarization measures made during an eclipse. Coronal holes first seen in the X-ray pictures taken by Skylab satellite are dominant features of the quiet corona. Models giving conditions in coronal holes to distances of about $5R_o$ indicate that the temperature in the hole is about 500,000 K less than the quiet corona and so is the density which is about 2-3 times lower.

4.4.6 *Coronal Bright Points*

Another notable discovery made from the early X-ray images was of numerous X-ray *bright points*. These occur over the entire surface of the Sun, although the number in the equatorial region seems higher. They also appear in the coronal holes area. Their sizes range from a tiny (less than about 7000 km) to small spot sizes and have life time of only a few hours. During a day more than thousand bright points may appear on the disk. These bright points are closely related to the photospheric magnetic field and perhaps are associated with ephemeral bipolar regions. These X-ray bright points evolve in time by spreading out, increasing the separation and correspondingly weakening of the brightness.

4.4.7 *Radio, EUV and X-ray Emissions from the Corona*

Radio waves emanating from the Sun were first recorded during the World War II in February 1942 by British radars as radio 'noise' while monitoring the sky for German planes. It was then thought that Germans were jamming their radars. James S. Hey investigated this effect and found that the noise never occurred during the night time and the source moved through the day following the position of the Sun. It was concluded that the source of noise was the Sun. The signal levels at meter wavelengths were 100,000 times more intense than expected from the Sun, assuming the Sun as a black body at a temperature of 6000 K. This discovery was kept secret until 1946 due to war restriction. Earlier to this there are some references also suggesting that Sun could be a source of radio waves. Ebert (1893) in early 1890 had suggested that the solar corona may manifest visible electric discharge and concluded that, "if the Sun is really the seat of electromagnetic disturbances, then it must necessarily be a source of electromagnetic radiations". Since then number of scientists, notably; McCready, Pawsey, Payne-Scott (1947) and others have shown that the radio disturbances commonly originate in the localized regions over large sunspots. Ryle and Vonberg had shown that the radio radiations are circularly polarized. Solar radio astronomy has enormously progressed since its discovery in 1942.

The solar radio observations yield information which is quite different from that obtained from the visible light, this is because the emission and propagation of the two types of radiations substantially differ. The coronal gas which is transparent to the visible light is opaque to radio waves. In the case of corona, the ionized gases prevent radio waves to escape from the lower levels in the solar atmosphere, which become progressively higher and higher for long wavelengths. Depending on the wavelength used this restricts the radio observations only to the upper chromosphere and the corona. Another important difference and a handicap is that, only continuum radiations are generated in radio waves as they are produced by changing electric current and oscillating electrons in the solar atmosphere, unlike as in the case of optical and EVU emissions, where line emission and absorption takes place due to atomic transitions. Beside these limitations, the radio observations have another severe disadvantage as compared to optical one, due to the low angular resolution of radio telescopes as compared with the optical telescopes. As will be seen by the following argument:-

The angular resolution, R, a measure of angular separation of two features is given by:-

$$R = 1.22x \frac{wavelength}{aperture} radians, \tag{4.6}$$

$$R = 1.22x \frac{\lambda}{D} \frac{180^o}{\pi} \deg. \tag{4.7}$$

where R is the smallest angular separation of the two objects that can be resolved, D is the linear aperture of the telescope and λ the wavelength in the same units. In the case of visible light, say for λ=5000Å, a 15-cm aperture telescope would theoretically yield a resolution of 2/3 seconds of arc, while at typical radio wavelength of say 1-meter, to attain the same resolution, as in the optical region would require a radio telescope of 300 km! However, some progress has been made to achieve better spatial resolution in radio region through the use of interferrometric techniques.

There are three main mechanisms for generation of radio waves in the solar corona:-

1. Solar radio emission at very short wavelengths <1cm, is reasonably well represented by thermal emission, corresponding to a black body temperature of 60,000° K,

2. At longer wavelengths > 1cm, the radio emission rises above the black body curve, corresponding to an effective temperature of 10^6 K. Such high temperatures can not be conceived arising from a thermal source. Ginzburg (1946) had shown that such radio emissions could arise from thermal *bremsstrahlung* (breaking radiation) from the hot coronal plasma, which is permeated by fast moving electrons through coronal

matter. *Bremsstrahlung* is a process of emission of radiation, which occurs when a fast moving electron passes near a nucleus; it decelerates and jumps from a higher to a lower energy level. In astronomy this process is also called as 'free-free' transition giving rise to continuum emission,

3. Due to the high temperature and extremely low density, particles especially the electrons, which are easily accelerated due to small mass and high velocity, can travel thousands of kilometers without collision. The corona is permeated by magnetic fields, which could be quite high on the order of hundred or even thousand gauss over active regions and sunspot. In 1904 Heaviside (1904) had proposed a theory, that when an electron moves in magnetic field, it accelerates and gyrates around magnetic lines of force due to which it produces a radiation at a certain frequency, called the *gyro-frequency* or *cyclotron frequency,* ν_g. In such an environment an electron in a magnetic field would '*gyrate*' around the magnetic lines of force, at a characteristic frequency called the 'gyro-frequency', with a radius of gyration *R* and the gyro-frequency is given by the expression:-

$$\nu_g = \frac{v}{2\pi R} = \frac{qB}{2\pi mc}, \tag{4.8}$$

where m is the relativistic mass (γm_o) of the particle having charge q, v the velocity of the particle perpendicular to the magnetic field B, and c is the velocity of light. Numerically the gyro-frequency for a particle of charge can be given by:-

$$\nu_{gq} = 2.80B\frac{m_e}{m_q}.\frac{1}{\gamma}, \tag{4.9}$$

where B is measured in gauss and ν_{gq} in MHz and subscript q applies to electrons and other ion species separately. For mildly relativistic electrons (γ = 1) and assuming that the magnetic field strength over sunspots as 1500 gauss, the electrons would produce a gyro-frequency of ν_{gq} = 4.2 GHz. Thus by measuring the radio gyro-frequencies, in principle one can estimate the magnetic field in the corona.

Radio waves in the corona can propagate if their frequency exceeds a certain value know as the *plasma frequency*, which is related to the density of the coronal gas and is equal to approximately $9\sqrt{N}$ in Hz, where N is the number density of electrons (in electron per cubic meters). This could be another method to estimate the electron density in the corona, besides from the total solar eclipse observations. Any electromagnetic wave passing through the coronal plasma with frequency less than the plasma frequency is suppressed. As a consequence of this effect, the low frequency radio waves with wavelengths longer than 1-meter must come from relatively high in the corona, where the density is low about 3×10^8 per cubic cm, since the plasma frequency of 150 MHz corresponds to this density, and brightness temperature of about million degrees K. High frequency radiations come from lower regions in the solar atmosphere; a centimeter wave originates where the plasma frequency is about 3 GHz or 1 cm in wavelength, corresponding to the chromosphere and a brightness temperature of about 10,000 degrees K. In Figure 4.22 is shown the distribution of intensity of radio waves across the solar disk. These plots show a steady decrease of brightness

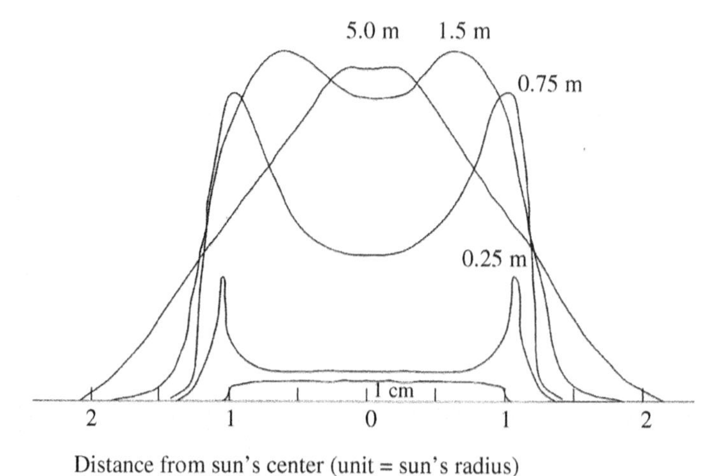

Figure 4.22 Radio intensity distribution across the quiet Sun at various wavelengths, ranging from 1-cm to 5.0-m, showing 'limb brightening' at certain wavelengths. (After M. R. Kundu).

temperature with decreasing height in the solar atmosphere and the corresponding increase in the plasma frequency (decrease in wavelength).

A limb brightening at wavelengths between few cm and 1.5 m is also observed, this arises due to increase in temperature with height. But at very long wavelengths > 5 m, limb darkening occurs, this is because of the fact that the rays which originate at the limb suffer appreciable refraction, starting their trajectories at much higher in the corona where the optical thickness is very small as compared to the optical depth of a ray from the disk center, thus giving the effect of limb darkening.

The quiet sun's meter wave emissions have brightness temperatures of the order of a million degrees. Radio emissions in the range of 50-cm and 1-m wavelengths arise from the *transition* region at temperature between 10^5 and 10^6 K. In the quiet corona, mostly the radio radiation arises due to thermal *bremsstrahlung*, the free-free transition. But over intense magnetic field regions, like sunspots and active regions, radio waves originate through the gyro-cyclotron mechanism, where thermal and relativistic electrons play a dominant role. Radio brightness

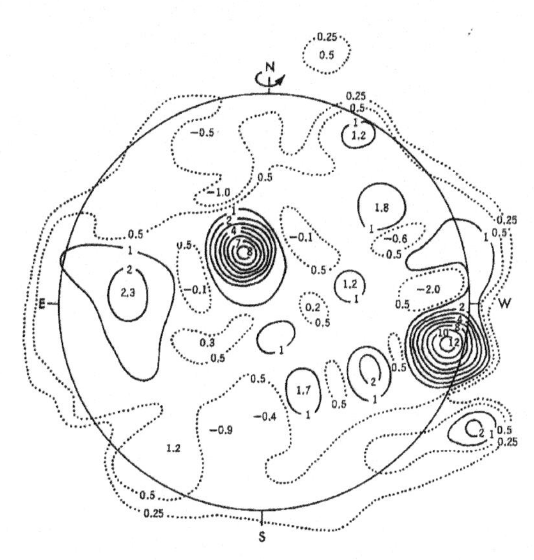

Figure 4.23 Contour maps of solar emission at 9-cm observed with Stanford radio spectro-heliograph. The peak contours correspond closely with active regions on the disk.

temperatures in the range of 10^7-10^{11} degree K have been observed in the solar corona during maximum solar activity, coronal transient events and severe radio noise storms. Such high temperatures can not arise from thermal sources, therefore the earlier mentioned non-thermal mechanisms must be at play in the corona. In Figure 4.23 is shown a radio emission map at 9-cm observed with Stanford Radio Spectroheliograph. The peak contours correspond to locations of active regions on the disk and brightness temperature of about 800,000 K.

In Figure 4.24 are shown a white light picture taken on July 3, 2004, and corresponding radio images made from the Nobeyama Radioheliograph at 17 GHz (~2-cm) wavelength and a soft X-ray image taken from SXI/NOAA satellite. All the 3 images have good positional correspondence. The radio emission closely matches with the locations of sunspots as this come from the low chromosphere, but the X-ray emission comes from upper coronal layers over active regions.

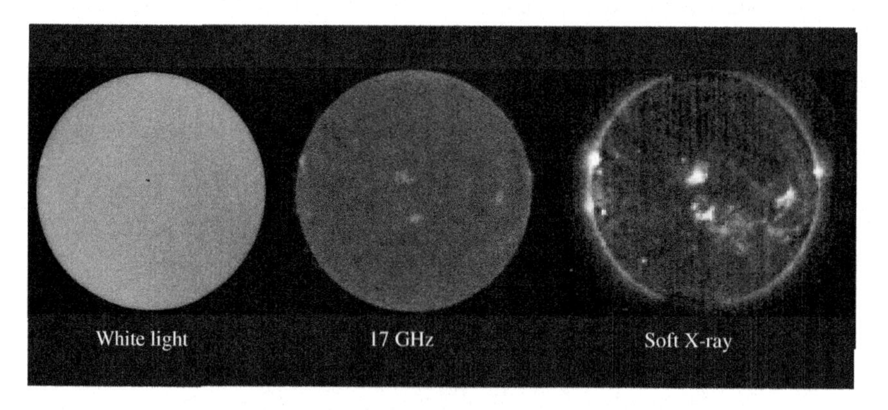

Figure 4.24 Left to right: White light picture of the Sun, image taken from Nobeyama Radioheliograph at 17 GHz wavelength, image from Solar X-ray Imager in soft X-ray on July 3, 2004. There is a close positional correspondence over active regions between all three images.

4.4.8 *Coronal X-ray and Extreme Ultra Violet Emissions*

The wavelength range between X-ray and the extreme ultraviolet is arbitrarily defined to be from 10 nm to 100 nm (100-1000Å). The coronal X-ray radiations are generated both in emission lines and in continuum. The emission lines come from highly ionized atoms through atomic transitions, similar to those emitting the coronal *forbidden* lines in the visible region. The two *permitted* resonance lines of FeXIV at 5.90 nm and 5.96 nm are emitted in the corona, since the temperature is high enough to excite FeXIV from the 3*p* ground level to *4d* state. The Fe XVII ion gives two strong lines at 1.5 and 1.7 nm. Other resonance lines also occur in the X-ray spectrum due to other abundant ions in the corona such as; ions of carbon (CV), oxygen (OVII), neon (Ne IX), emit lines at 4.03, 2.16, 1.345 nm, respectively. These lines are due to highly ionized ions and arise only from very high temperature plasma in the range of 2 to 4 million degrees K, such high temperatures are seen over sunspots and active regions in the corona.

In the extreme ultraviolet range of spectrum, several ions emit emission lines; for example, carbon (CIV) at 155 nm, oxygen (OVI) doublet at 103.2 and 103.8 nm, magnesium (MgX) at 61.0 and 62.5 nm and silicon (Si XII) at 49.9 and 52.1 nm, originates in the transition region. All these emission lines correspond to temperatures of the order of million degrees and more. Both the forbidden and permitted coronal emission lines are observed only in a relatively low region of the corona, since their intensity depends on the square of the ion density.

Besides these emission lines in the X-ray and extreme ultraviolet, the corona also emits continuous radiation in this spectral range. Similar to the chromospheric Lyman continuum spectrum, in X-ray region also a continuous spectrum is formed by recombination of electrons with nuclei of carbon and oxygen. In addition to this, the continuum is also formed through *bremsstrahlung or free-free transition*, which has been described earlier in this Section. Such emission corresponds to a temperature of about 2 million degrees K and peaks around 4 nm in the soft X-ray region.

4.4.9 *Coronal Magnetic Fields*

The general appearance of a white light eclipse photographs or images taken in soft X-rays immediately tells us that magnetic field pervades the corona. Due to high degree of ionization and low plasma density, the electric conductivity of the corona is extremely high which implies that the magnetic lines of force are *frozen* in the plasma. This would means that motion of electrons and ions are largely determined by the ambient magnetic field. It is known from the fundamental principle that a particle moving perpendicular to the magnetic field lines would describes a circular motion. A particle with an initial velocity not exactly perpendicular to the field will describe a helical motion around the field line. As we have seen earlier that the radius of gyration motion and the frequency depends on the field strength and the mass of the electron and ion. Because the electron mass is much less, it will perform spirals with smaller radius of gyration and spiraling very rapidly, while the heavier ions will perform helical motion with larger radius but slowly. In a typical case with 1mT (10 gauss) magnetic field in the corona and ion velocities corresponding to temperature of 2,000,000 K, the electron gyro-radius would be about 30 mm, emitting gyro-frequency of 30 MHz, but in the case of proton, the gyro–radius will be 1.3 meter and 15 KHz gyro frequency.

Due to the extremely low density of the corona, the protons and electrons travel thousands of kilometer without 'colliding', and confining within magnetic loops in the corona. Actually, it needs a very small magnetic field strength to *control* the coronal plasma which depends on the ratio (β) of the gas pressure to the magnetic pressure. The gas pressure even at the *base* of the corona is extremely small and a field strength of even 1mT (10 gauss) would yield β less than unity. Hence, the plasma motion is '*controlled*' by the ambient magnetic field, thus all the coronal plasma is confined along the magnetic lines of force. Now let us consider a situation; when an electron traveling along the field lines arrives at the foot points of a loop. If there were no collisions along the path, the electron would reverse on itself, as if the foot points are 'mirrors'. Near the foot points there would be greater density of other particles, therefore collisions will take place more frequently. The particles which traveled along the loops would impart much of their kinetic energy through collisions to the surrounding particles near the foot points. This would result in heating of the denser cooler chromospheric layer at the foot points, and evaporation of the gas will take place along the loops. The heat conductivity perpendicular to the

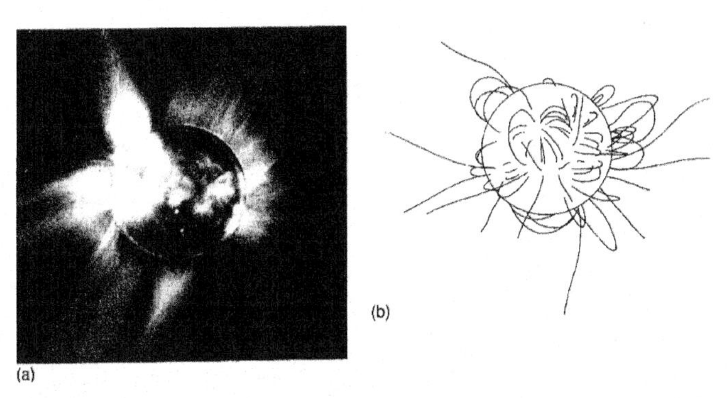

Figure 4.25 (a) Solar corona at the time of July 7, 1970 total eclipse superimposed on a X-ray picture taken by a rocket borne X-ray telescope of American Science & Engineering Corporation, shortly after the eclipse. Nearly all white light streamers have their bases marked by X-ray bright regions; (b) shows the calculated potential magnetic field lines by Newkirk (1971) for the same day.

field lines is strongly inhibited, because electrons are constrained to travel **only** in helical paths along the field lines. It will be realized that the solar corona at all levels is permeated by weak magnetic field, but is sufficiently strong to constrain the coronal plasma, which plays an important role in defining it structure. The large scale arches and active region loops seen on X-ray images actually delineate the magnetic field structures, which are filled with coronal plasma. In the case of coronal holes, as we have discussed earlier, the magnetic field lines are open to the interplanetary space thus the plasma is not confined in close loops, but is free to travel out in space.

To measure the magnetic field in the corona, we can not make use of the standard method of Zeeman Effect, as in the case of photospheric fields because the splitting of Zeeman components by very weak coronal magnetic field is too small for coronal lines. Hence, other methods have been used with limited success. The magnetic field affects the polarization properties of some spectral lines which could be measured with some fair degree of accuracy. The gyro-frequency of radio emission is a function of density and the magnetic field strength, knowing the density through other methods, estimates of the magnetic field has been made.

H.U. Schmidt had proposed a method which is widely used to

calculate the coronal magnetic field configuration. This method is based on the assumption that the magnetic field in the corona is a *potential*, and is entirely determined from the distribution of the longitudinal magnetic field strength in the photosphere, no electric current flows in the corona, since current would produce its own magnetic fields, and that would change the magnetic fields. In the case of the Sun, the surface photospheric magnetic field is assumed to have a distribution of *magnetic charges* as indicated on full disk longitudinal magnetograms, obtained from various observatories. Beyond a certain radial distance, the field lines are assumed to be stretched out radially by a steady expansion of the outer corona through solar wind. Based on this method the High Altitude observatory's group had been constructing coronal magnetic field structure. In Figure 4.25 is shown a map of coronal field, calculated from the photospheric magnetograms around the time of solar eclipse of 7 July 1970, and is compared with an X-ray image taken around the same time from a rocket borne telescope, by American Science and Engineering Corporation group. In this picture a close correspondence is clearly seen between the X-ray bright regions on the disk and white light streamers. In the south side, near a coronal hole a long streamer is also seen, which may be originating from the far side of the Sun. The computed magnetic field line pattern [Figure 4.25(b)] for this day delineates fairly well the active regions, coronal streamers and coronal hole.

4.4.10 *Coronal Heating*

One of the major unsolved problems in astrophysics is to explain how the solar corona is heated to temperatures of million degrees. Several mechanisms had been proposed, based on heating by sound waves which are generated in the convection zone, or magneto-sonic waves which would give rise to shock waves, or the Alfvèn waves traveling along coronal magnetic field. Due to one reason or other, none of these mechanisms were able to explain the heating of the corona. Therefore, scientists started looking for some other plausible explanation. As we now know that the corona is a highly dynamic region of the solar atmosphere, where numerous tiny *micro-flares* (with energy of 10^{10}

megawatts) and nano-flares activities occur all the time. Eugene Parker suggested that such small flares, which are produced by reconnection of slow twisting and braiding of magnetic fields, may supply enough energy to heat the corona. In 1990, Toji Shmizu and Hugh Hudson looked very carefully at the YOHKOH X-ray data and concluded that there is not sufficient number of micro- or nano- flares on the Sun to heat the corona to such high temperatures. From the analysis of the temperature distribution of X-ray coronal loops seen in YOHKOH data, Eric Priest *el al.*, (2000) and his colleagues came up with an idea that the source for heating the corona may be distributed uniformly along coronal loops.

Figure 4.26 Magnetic carpet. Extreme ultraviolet and magnetic field observations made from EIT and MDI instruments aboard SOHO space craft, show numerous magnetic loops of all sizes, rise up into the solar corona from opposite magnetic polarity regions (shown as black and white), scientists call it a 'carpet of magnetism'. White and dark lines are computed magnetic lines of force, assuming a potential field configuration

Recent observations made from high resolution MDI magnetograph and Extreme ultraviolet Imaging Telescope (EIT) and Coronal Diagnostic Spectrometer (CDS) instruments on board the SOHO spacecraft have revealed that thousands of magnetic bi-poles constantly appear and disappear on the solar surface, and display vigorous heating and motion at the feet of coronal loops. As seen in Figure 4.26, small magnetic loops extend into the corona from magnetic bi-poles (shown in black and white regions) in the photosphere, creating a sort of *'magnetic carpet'* in the quiet Sun. The white and black lines represent the computed magnetic lines of force, assuming a potential field configuration. The SOHO team from Stanford University and from the Lockheed-Martin Solar and Astrophysics Lab., reports that the magnetic loops are constantly reconnecting all the time all over the Sun, with release of magnetic energy. The (Extreme ultraviolet Imaging Telescope) EIT images made in extreme ultraviolet show that wherever the magnetic field in the photosphere converges there is always brightening just above in the corona suggesting that energy flows from the magnetic loops, where ever reconnection takes place. It is proposed that the strong electric currents generated in these magnetic short-circuits can heat the corona to several million degrees.

High resolution images taken from TRACE (Transition Region and Coronal Explorer) satellite in Fe IX/X line at 171 Å, show constant emergence of magnetic field from the photosphere, resulting in reconnection and rearrangement of magnetic fields in the corona. A sample example of such emerging fields is shown in Figure 4.16. From these observations it is suggested that the heating of coronal loops may occur near the base in the low corona, and that the loops are not heated uniformly along the entire length. From these recent TRACE observations, it appears that at least a large part of coronal heating is most likely due to energy release through magnetic reconnection of numerous small scale magnetic regions on the Sun. This idea seems to be gaining ground to explain the coronal heating.

Chapter 5

The Active Sun

5. 0 Introduction

In the previous Chapter we discussed the '*quiet*' Sun. Now we turn to the '*active*' part. Solar activity is not only fundamental to understanding the physics of the Sun, but it is the basic driver of Sun-Weather and our terrestrial climate. Ultimately the genesis of our life on this planet lies in the Sun. Activity occurs in all layers of the solar atmosphere. There is the photospheric activity, chromospheric activity and also coronal activity.

5.1 Photospheric Activity

At the lowest levels activity manifests itself in a variety of ways. On a white light image of the Sun one sees sunspots, and along with them, near the limb, there are bright patches, the white light 'faculae' (in Latin it means *little torches*). Near disk center faculae are not discernable, indicating that these features are located slightly higher in the photosphere. Figure 5.1 is the photosphere in white light near the limb, showing a large sunspot group and bright faculae. Both sunspots and faculae have strong magnetic fields. These are the two main photospheric activities and both vary in time, magnitude and location. Appearance and disappearance of sunspots are the main source of solar activities. The temporal variation of sunspots range between few hours to several weeks and years, for example small spots may appear just for few hours while the well known 11-year cycle lasts for decades.

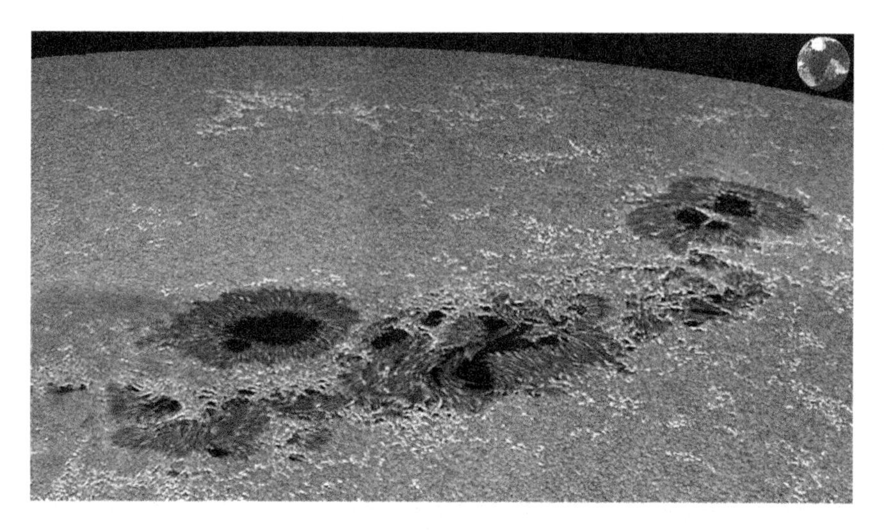

Figure 5.1 A high resolution photograph taken near the solar limb with the Dutch Open Telescope at La Palma on April 2 2001, showing white light faculae and a large sunspot group. On the right corner is shown the size of the Earth on this scale.

5.1.1 *Sunspots*

5.1.1.1 *Pre-telescopic Observations of Sunspots*

The earliest reference to naked eye sunspot observations dates back to the fourth century BC by Theophrastus of Athens (c.370–290 BC), a pupil of Aristotle. Thereafter a reference is found in Einhard's book *'Life of Charlemagne'*, that a spot on the Sun was seen around 807 AD. At the time it was interpreted as a transit of the planet Mercury. Another record is found around 840 AD, that Abū-l-Fadl Ja'far ibin al-Muktafi (c.906-977 AD) reported that the philosopher al-Kindī observed a spot on the Sun. Similarly this was attributed to a transit of Venus. Other records of sunspot sightings in Europe date back to about 1200 AD by Ibn Rushd, and around 1450 AD by Guido Carrara and his son Giovanni. In Russia, during the fourteenth century, sunspots were seen through the haze of a forest fire. The most systematic early pre-telescopic records of naked-eye observations of sunspots, however, were kept by Chinese , Japanese and Korean astronomers. These ranged from 28 BC to 1638 AD. From a study of early oriental sunspot records more than 139 sightings have been cited by Clark and Stephenson (1978). To this list Wang (1980) has added 12 more sightings. In these early Chinese reports it was generally mentioned that; "the Sun appeared yellow and there was a black vapor as large as a coin at its center". Or as "the Sun appeared orange in colour and within it there was a black vapor, like a flying magpie. After several months it gradually faded away". Or simply it was mentioned that; "within the Sun there was a black spot, as large as a hen's egg, or as large as a pear, and even the sighting of two black spots had been reported, "as large as pears". In many records it is mentioned that the "Sun had no brilliance", implying that due to haze and dusty conditions in the atmosphere, the Sun's intensity must have decreased. Under hazy conditions it becomes possible to look at the Sun and see sunspots with the naked eye. In one of the Chinese records made in 1638 on December 9th, it was stated that "within the Sun there was a black spot, and black and blue and white vapors". This may be a significant observation indicating that the observer witnessed a white light flare and the dark and bright surges which normally accompany such intense events. We may conclude that this ancient Chinese observation was perhaps the first report of a white light flare, much earlier than Carrington's famous white light flare of 1859.

It is rather surprising that no records of sunspot sightings are available during the early times from Greece, Persia, Turkey, Arab countries, nor from the Indian subcontinent. This despite the fact that during this ancient epoch, from c.200 BC–1600AD, science, astronomy, mathematics, medicine and philosophy were all well developed in these countries. What could be the reason for such non-reporting of naked eye sunspots? One of the reasons may be that in these geographical regions the sky remains very clear and the 'dimming' of the sunlight from atmospheric haze, dust storms etc., seldom takes place. Therefore the naked eye sighting of sunspots was difficult. But in China, Japan, Korea and adjoining regions, the atmospheric conditions were, and still are, such that atmospheric haze persists for quite some time during the year. It is likely that sightings were made during hazy hours in the morning and evening when the Sun's intensity is attenuated. As reported in Russia in 1365 and 1371 AD, forest fires and the resulting smoke also helps to decrease the sunlight for naked-eye observations.

Another reason for the non-reporting of sunspots may arise from the fact that great respect was paid to the teachings of Aristotle, who maintained that the Sun was a perfect body without blemish. This belief, which actually became a part of orthodox Christian theology during the Middle Ages, holds that any blemish reported on the Sun would be sacrilege. The concept of the Sun as a perfect body without any markings persisted from the time of Aristotle and continued even into the telescope era, and might be responsible for the lack of sunspot records. Even the great Johann Kepler (1571-1630 AD), intentionally or un-intentionally, assigned a sunspot seen on May 8, 1607 as a transit of the planet Mercury. In India, too, reports of rare and unusual celestial events like eclipses, comets, sunspots etc., are very fragmentary. Even though the science of astronomy and mathematics had made great strides. This non-reporting of such unusual rare celestial events in India may be due to the fact that it was generally believed that the sighting and recording of such celestial events ware 'inauspicious'.

5.1.1.2 *Early Telescopic Observations of Sunspots*

Sunspots have attracted the attention of astronomers ever since the first telescopic observations in Italy by Galileo Galilee in 1611, by Christopher Scheiner (1575-1650) in Germany, Thomas Harriot (1560-1641) in England, and Goldsmid (1587-1650) in Holland. From his systematic work, Galileo showed that sunspots belong to the Sun are not small planets revolving around the Sun. Galileo also noticed a westward drift of spots on the solar surface. This might indicated that the Sun rotates. For the next two hundred years solar studies were confined to visual telescopic observations of the Sun in white light, mostly to determine sunspot motion, the number of sunspots, their growth and latitude variation. These observations led to several important discoveries.

The extensive study of sunspots, beginning in 1611 AD by Christopher Scheiner and independently by Galileo, provided the first measurements of *solar rotation.* It was found that the Sun rotates on its axis in about 27 days. As Galileo, Scheiner also found that sunspots occurred only in two narrow belts, extending some 30° on either side of the equator. Thus was discovered the *latitude distribution of sunspots.* Scheiner also made observations of sunspots for an extended period of time. His collected observations were published in 1630 AD in a volume entitled *Rosa Ursina sive Sol.* The sunspot drawings contained in *Rosa Ursina* covers 1625 to 1627 AD, which represents Scheiner's most important contribution to solar physics as it served to trace back the sunspot cycle to the first telescopic observations. Eventually this led to the discovery of *Maunder Minimum* by John Eddy in 1970. Scheiner also noted that sunspots on either side of the equator at higher latitudes had longer periods of revolution than those near the equator. This was the first indication of *solar differential rotation.*

Heinrich Schwabe (1789-1875) was also a keen observer and recorded carefully with the help of a small telescope the occurrence of sunspots for 43 years. In 1843 he first announced a periodicity of 10 years in the numbers of sunspots, what we now know as the *11-year sunspot cycle.* At the time, his discovery attracted little attention. But he continued to observe and finally in 1851 published a treatise entitled *Kosmos,* giving a table of sunspot cycle statistics from 1826 onwards. From this work the world immediately became aware of this important solar phenomenon which had escaped the notice of telescopic observers for nearly 200 years! Schwabe was awarded the Gold Medal of the Royal Astronomical Society for this discovery.

Schwabe's discovery profoundly interested an amateur English astronomer Richard Christopher Carrington (1826-1875). He had been observing the Sun with his 4.5-inch refracting telescope from 1853-1861. In 1863 he published his solar observations in the classic monogram *Observations of the Spots on the Sun.* Carrington found that the average latitude of sunspots decreased steadily from the beginning to the end of each cycle, a phenomenon we now we know in the form of the *butterfly diagram.* He also determined very accurately the solar rotation rate and the inclination of the Sun's axis i to the ecliptic plane from the apparent path of sunspots traversing the solar disk. He assigned i a value of $7^{\circ}.25$. This quantity is still being used in astronomical ephemeredes for calculating heliographic co-ordinates. Recent investigations made by Balthasar *et al.* (1986), based on much larger data sets obtained from Greenwich between 1874 and 1947 and between 1947 and 1984 at Kanzelhöhe, Austria, suggests that: -

$$i = 7^{\circ}.137 \pm 0^{\circ}.017. \tag{5.1}$$

5.1.1.3 *Evolution of 'Pores' and Single Sunspots*

A high quality white light photograph of the Sun shows fine granular structure, the granulation, and often small dark regions of only few arc seconds in size and surrounded by granulation. These dark regions are called as *pores*. Pores may be seen for tens of minutes to hours. In Figure 5.2 is an example of pores, small and large sunspots. Pores are the simplest form of sunspots without penumbral structure. Many pores do not develop beyond this stage. Under very good seeing conditions one can see that the edge of a pore is notched by the surrounding granules. In common with spots, pores have magnetic fields. Field strength is comparable to a small spot, about 2000 gauss. Large and well developed spot groups usually contain many pores. In most cases pores have a tendency to occur in an area following the leading (western) spot rather than the preceding (eastern) spot of the group. The occurrence of pores is not restricted only to spot groups, but they sometimes form either alone or in small clusters far away from spot groups. Since the pores could be seen only on high resolution white light pictures, and usually such observations are taken of only a small limited region on the Sun, very fragmentary information is available about their distribution and variation with solar cycle.

Most pores have diameters in the range of 2 to 5 arc sec, though some are comparable to granules of 1 to 2 arc sec. Pores of 5-7 arc sec have the tendency to develop into small spots with at least some rudimentary penumbral structure. The brightness of pores is on an average less than the intergranular surrounding region, but is greater than the umbra of large spots. Pores have much longer life times as compared to photospheric granules and may remain unchanged for hours.

The first indication of a new pore forming is the appearance of a dark

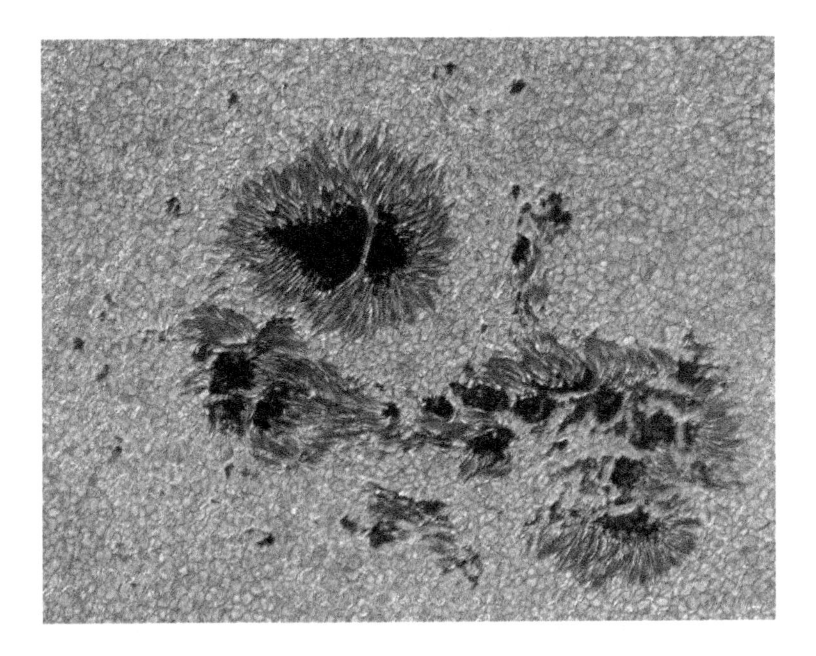

Figure 5.2 Photograph of sunspot group showing pores, sunspots with and without penumbra, light bridge. This picture was taken on 6 May 2003 in g-band from VSST.

region about the size of a granule in the granulation pattern. The growth of a new pore occurs through the gradual disappearance of surrounding granules. This may be interpreted as due to the opening up, or 'lifting' of magnetic flux tubes from beneath the surface. Time lapse observations of the initial development of pores can throw some light on the mode of penetration of magnetic flux tubes. During this period the underlying field is presumed to be pushing the flux tubes upwards and thus one may expect to see changes in the granulation pattern. Loughhead and Bray (1961) have shown that significant changes in shape and area of pores occur during their birth and that their development period requires 30 to 45 minutes. Similarly, the dissolution of pores which fail to develop into a spot takes place by individual granules pushing into the dark area of pores and then gradually covering the whole pore. The majority of pores never develop into sunspots, but sometimes isolated clusters coalesce into a sunspot. Extremely high resolution time lapse pictures of solar surface is now available from the Dutch Open Telescope (DOT) and the 1-m New Swedish Solar Telescope (NSST) and these will contribute to the understanding pores.

A typical sunspot consists of a dark central region called the umbra, derived from the Latin word meaning 'shadow', surrounded by a less dark penumbra. The penumbra consists of fine elongated bright fibril structures, less than 0.5-1 arc sec (350-700 km) wide and several arc sec (1500-2800 km) long, directed towards the umbra. A high resolution picture of a sunspot taken in white light by Rimmele from the Dunn Telescope at the Sac Peak Observatory is shown in Figure 5.3(a). Here we see umbral bright 'dots' and light bridges, while the penumbra consists of fine filaments. In Figure 5.3(b) is shown a drawing of a sunspot made by Father Secchi around 1870, using a small aperture telescope. It is remarkable that he recorded similar details including the umbral structures and the extension of penumbral filaments inside the umbra. Secchi's visual observations of 130 years ago are comparable to those of modern days!

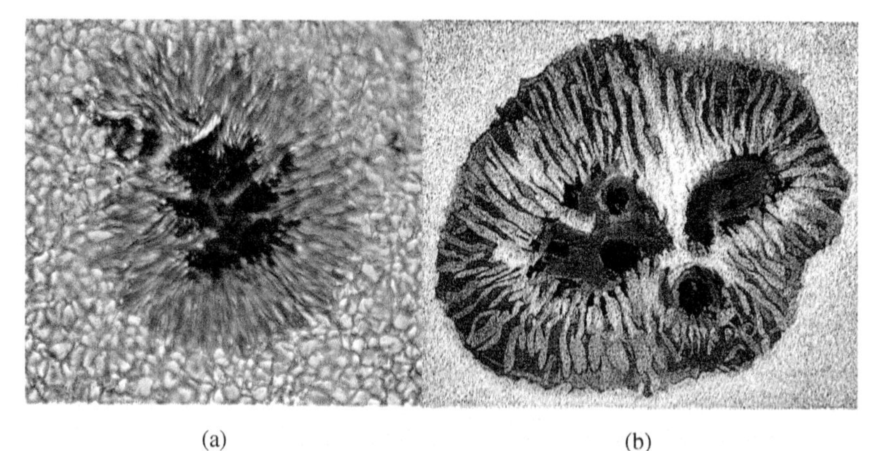

(a) (b)

Figure 5.3 (a) White light photograph of a sunspot, showing high angular resolution of 0.2 arc sec or 150 km. Taken with the Dunn Telescope at the Sacramento Peak Observatory, (b) Drawing of a sunspot by Father Secchi in early nineteenth century. Compare the fine details of both these pictures.

5.1.1.4 *Evolution of Sunspot Groups*

A satisfactory understanding of the physics of sunspots is perhaps one of the most challenging problems in astrophysics. Several theoretical models have been proposed to explain the phenomena of the development of sunspots, which range from tiny umbral details of less than 100 km in size to active regions whose enormous dimensions exceed 100,000 km. In the temporal domain, we see transient umbral flashes of less than 5-10 seconds duration to possible decade–long changes connected with the activity cycle. In spite of decades of concerted efforts by many observers and theoreticians, we still do not understand many of the basic processes occurring in sunspots.

Extremely complex sunspots display a large variety of physical activity. As previously indicated, sunspot groups grow from pores in the quiet undisturbed photospheric. During the initial stages the group consists either of a few pores in an area of about 5-10 square degrees of heliographic coordinates. A single spot at the preceding westward end may subsequently develop. Most of groups never evolve beyond this stage and disappear the following day. However, those that do survive may follow the evolutionary trend as reported by Waldmeier in 1955. Based on sunspot evolution he classified sunspot groups into 9 classes, which are now known as *Zurich Sunspot Classification*. In the following, the day-to-day development sequence of a sunspot group is given:-

2nd day after the appearance of a pore: the group shows an increase in area that becomes elongated. Individual spots concentrate near the preceding and the following ends of the group. Usually one spot at each end develops much more than the others.

3rd day: The spot continues to grow and the main preceding spots develop penumbrae.

4th day: The main spot also develops a penumbra between the two main spots. Several small spots may appear and the total number of spots may reach 20-50.

5th -12th day: The group attains its maximum area during this period, often around the 10th day.

13th–30th day: Small spots between the two main spots tend to disappear, and then the following (eastern) component disappears,

usually by breaking into smaller spots, which gradually decrease in size. The preceding (western) single spot assumes a roundish shape.

30th-60th day: the preceding spot gradually becomes smaller until it completely disappears, unlike the following spot, which generally breaks into small spots before disappearing.

The above mentioned approximate stages of development and decay of sunspot groups gives a good idea of the growth of sunspot activity. The exact details of the underlying process is still lacking. Extremely high resolution uninterrupted observations, taken over long periods of time would improve our understanding of group development.

In Figure 5.4 is shown pictorially the Zurich classification of sunspots.

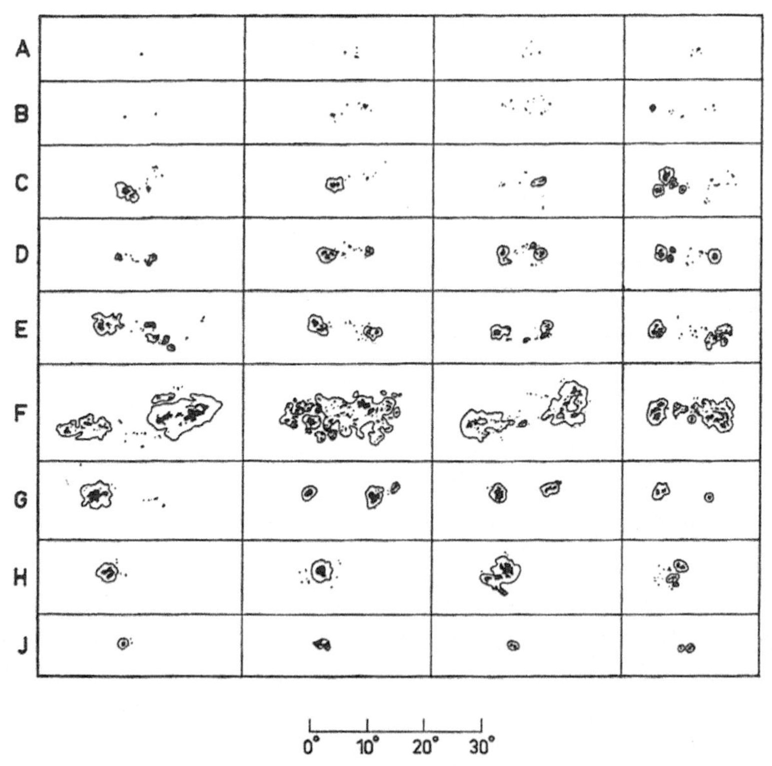

Figure 5.4 Zurich classification of sunspot groups, showing four examples of each of the 9 classes from A to J. The preceding (western) spot is left. The scale at the bottom indicates degrees of heliographic longitude

5.1.2 *Sunspot Penumbra*

In the previous section we have seen that sunspots begin their lives as pores, if they do not die out, they transform into spots with penumbrae. The transition takes place slowly from pore to a spot with well defined umbra, surrounded by the penumbra. Normally the umbra-penumbra boundary appears rather sharp, but under high resolution a system of fine penumbral filaments project onto the dark umbra, as shown in Figure 5.5.

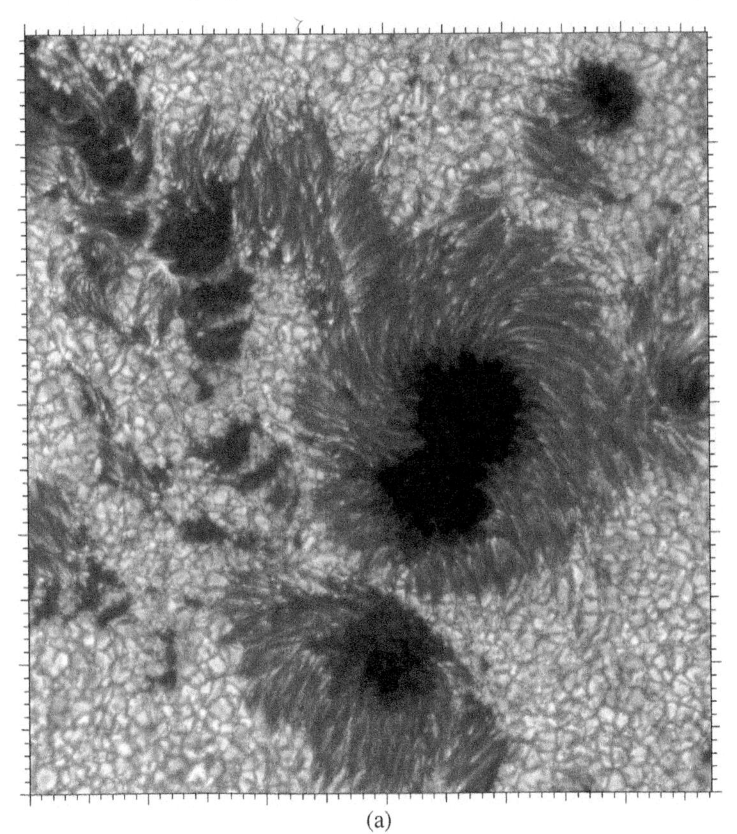

(a)

Figure 5.5 (a) Shows part of the sunspot group near disk center taken on July 15, 2002 from SST. Tick marks are 1000 km.

Thin bright penumbral filaments are seen on a darker background. These are directed towards the umbra. The outer penumbral-photospheric boundary ends abruptly with a sharp but jagged boundary. Recently,

Scharmer *et al.*, (2002) have achieved spatial resolution of better than 0.12 arc sec with the 1-m New Swedish Solar Telescope (NSST). On their best pictures they report *dark cores* inside penumbral filaments as in Figure 5.5b. The lifetimes of these dark cores are about the same as that of the penumbral filaments themselves, on the order of at least one hour. The foot points of these cores seem to be centered or adjacent to umbral brightenings.

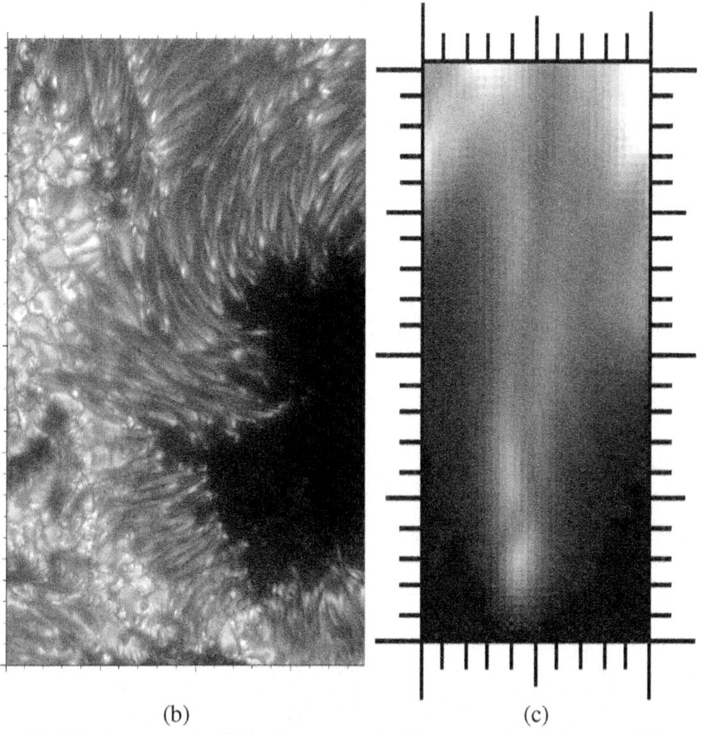

(b) (c)

Figure 5.5 (b) shows part of the larger spot, (c) a sample dark-cored filament is shown where tick marks are 100 km.

5.1.2.1 *Bright Ring around Sunspots*

Waldmeier (1939) and Das and Ramanathan (1953) have reported the appearance of enhanced bright zones, or rings, around sunspots on photographs taken in violet and blue violet light. From a detailed study of 41 spots, Waldmeier measured the ratio of the diameters of the penumbra (p) and the diameter (d) of the bright ring and obtained a mean value of 0.72 which increases with p. Observations made at 4000Å, the brightest part of the rings was about 3% higher than the surrounding photosphere. Das and Ramanathan (1953) also obtained similar results from spectroscopic determinations in the continuum near 3842Å, but in absorption line centers of λ 3820.5 (FeI), λ3937 (K-line of Ca II) and λ3971 (H line of Ca II) the ring in all cases was about 10% brighter than the background. It has been proposed that the bright rings represent the radiation deficit from the dark spots. Spruit (1982) has modeled the case, however, and finds that it takes about a year for the radiation balance from spots to be restored to the solar disk. It seems likely to us that the bright rings result from unresolved changes in granulation near spots–the so-called "abnormal" granules affected by magnetic fields. Recently interest in the investigation of bright regions around spots has been revived. Rast *et al.*, (1999) have made high precision photometric observations of bright rings around eight sunspots. The rings are about 10°K warmer than the surrounding photosphere. One plausible explanation that has been proposed for the observed bright ring is that, due to partial suppression of the convective energy transported from the underlying layers by the sunspot magnetic fields, this blocked energy shows up as a bright ring around the spot penumbrae. About 10% of this radiative energy missing from the sunspots is perhaps emitted through the bright ring.

It has been found that the flux deficit in sunspots is much larger than what can be accounted for by any enhanced bright zone. Also, not all sunspots display bright zones, even on pictures taken under exceptional good seeing condition. How can we then explain the sunspot's missing energy? If Spruit is correct, the process is so slow as to be undetectable.

5.1.2.2 *Motions in Sunspot Penumbrae*

At the Kodaikanal observatory in India, Evershed (1909) used Doppler spectroscopic observations of sunspots to demonstrate that solar gases show a predominantly radial outflow in the penumbrae at the photospheric level. Such a flow appears to be parallel to the surface and is directed outwards from the umbra-penumbra boundary towards the surrounding photosphere. This mass motion is now known as the Evershed Effect. In this early work, velocity amplitudes from 1 to 2.5 km per second were measured, though on rare occasions, velocities as high as 6 km/s were also reported.

Since this initial work, many authors (Abetti 1929, Kinman 1952, Maltby 1960, Holmes 1961, Servajean 1961, Makita 1963, Bhatnagar 1966) have worked on the Evershed effect. Some observers report tangential components to change this horizontal outward radial flow to a spiral pattern, while others measured only vertical components to this material motion. In one early major study at the Mount Wilson observatory, St. John (1913) observed the Evershed effect in several hundred photospheric and strong chromospheric lines. He concluded that weak lines show outward motion while the strong lines of CaII H_2 & K_2, H_3 & K_3, Hα, and Hγ show an inward motion towards the umbra. The amplitude and direction of the mass motion seemed to be related to the line strength. The schematic shown in Figure 5.6 represents St. John's measurements using photospheric, temperature minimum and chromospheric lines.

Since these early and classical studies, a number of further investigations have been carried out on this enigmatic outflow. Much of the earlier work is summarized in Bray and Loughhead's (1964) book on *Sunspots,* while Thomas (1994) provides a review of later work in his 1994 article in *Solar Surface Magnetism.* Since this time, the advent of adaptive optics and high spectral resolution imaging has significantly improved observations of the Evershed flow. It is now reasonably well established that this flow occurs only in the dark fibrils of the penumbra, and that material in the brighter filaments is stationary. This correlation is particularly striking when source depth of the continuum brightness is close to that of the line showing the Evershed flow. Furthermore, the magnetic fields in these dark areas appear to be horizontal, whereas fields outside these regions appear to be tilted to the horizontal. Recent infrared observations by Penn *et al.*, (2003), of molecular CH lines originating specifically in the cooler gas from deep in the photosphere, has demonstrated that Evershed velocities of up to 8 km/s are possible in dark fibrils deep within the penumbra. Evershed imaging by Clark and colleagues at the McMath-Pierce solar telescope at Kitt Peak with CO infrared lines of varying strengths has shown a change from radial outflow in the dark fibrils deep in the solar atmosphere to a spiral motion, as the source depth approaches the temperature minimum layers.

In spite of the large amount of observational and theoretical work on the Evershed effect over the past last hundred years, the dynamics of the flow are not yet fully understood.

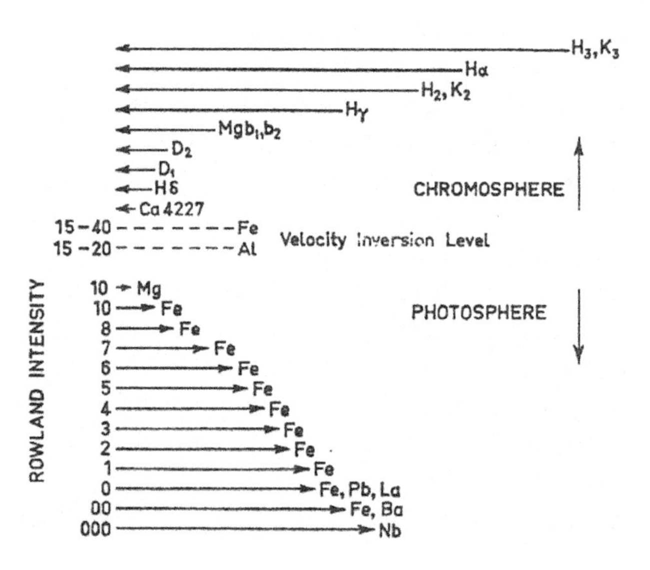

Figure 5.6 Schematic showing the Evershed velocity as a function of line strength. The lengths of the arrows are proportional to the measured velocities. Arrows to the right indicate an outflow and arrows to the left inflow.

The most reasonable theoretical model by Thomas and his associates (e.g. Montesinos and Thomas (1997)) proposes that the flow is like a siphon with the material being driven along a magnetic flux tube by the pressure drop between the end points of the tube. In this model, the pressure gradient is postulated as being maintained by a magnetic field gradient between the weak fields within the penumbra and those existing in regions of higher concentration outside the penumbra. The inverse Evershed flow higher above the penumbra is explained in this model by a field gradient between the strong umbral fields and the weak fields of the photosphere over which this inverse flow extends. The requirement of small field concentrations within the photosphere near to the penumbral boundary to produce penumbral Evershed flow is perhaps a weakness of this siphon model. A second model has recently been suggested by Schlichenmaier and his co-workers (Schlichenmaier *et al.*, 1998) that likens magnetic flux tubes to a tethered balloon. The motion of the region of the tube at the source height of the line being used for the observation produces the apparent outward flow as the tube moves like a tethered balloon under convective heating.

In contrast to the general material flow within dark fibrils, motions of specific features within brighter filaments have been seen occasionally. High resolution time-lapse white light movies of sunspots made from the balloon-borne telescope Stratoscope by Danielson and Schwarzschild in 1957-58, and recent observations made in 2000-2002 through DOT and NSST telescopes, have revealed that the bright penumbral filaments or *dots* appear to move *inwards* towards the umbra, with speeds on the order of 0.5-1km/s. The dark penumbral material or the background represents the outward Evershed flow. These motions are very clearly seen on time lapse movies produced by DOT and NSST.

5.1.2.3 *Asymmetric Evershed Flow and 'Flags' in Sunspot Spectra*

In addition to general line shifts that represent Evershed flows, more detailed examination of spectral lines show several other features that probably originate in other kinds of material motion in sunspot penumbrae.

Kinman (1952), Abetti (1957), Bumba (1960), Bhatnagar (1966) have observed that the magnitude of the radial component is not always the same on opposite sides of a sunspot. One of the hypotheses to explain this ambiguity is that the Evershed flow follows the lines of magnetic field, which fan out from the umbra and therefore present different angles on the two sides of a spot. Observations at high spectral resolution show several interesting phenomena of spectral line asymmetry, line broadening and the appearance of a so-called 'flag' on spectral lines. Figure 5.7 shows a line drawing of the 'flag' phenomenon as observed by Bumba (1960). It will be noticed that, at the penumbra-photosphere boundary in the penumbral region, a 'satellite' line appears displaced in opposite directions on the two sides of the umbra with corresponding velocities of greater than 5-6 km/s. If the satellite line is not resolved then the line appears asymmetric. No satisfactory theoretical explanation has been yet given for this interesting phenomenon. Perhaps localized rapid flows or turbulence are the cause for the observed spectral line profiles.

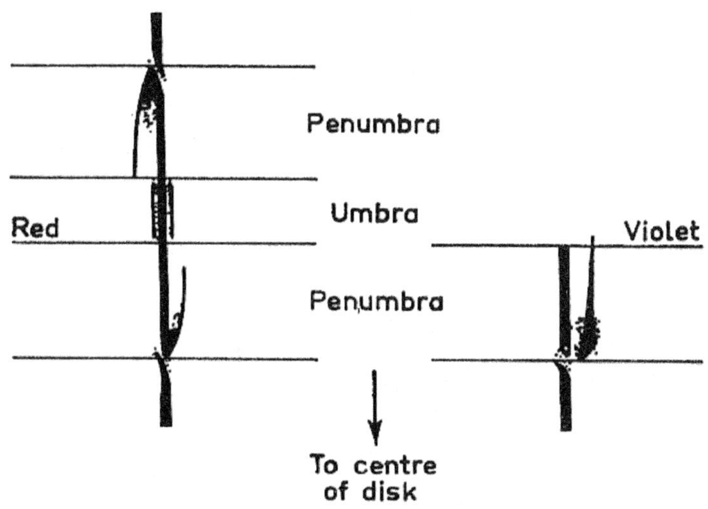

Figure 5.7 Showing 'Flag' phenomenon in sunspot spectral line as reported by Bumba (1960).

In 1974 Larry Webster of the Mount Wilson Observatory came across George Hale's observing diary of 1892 in the Hale Lab at Pasadena. Two pages of Hale's dairy dated May 26, 1892 and July 14 and 15, 1892, nos. 138 and 139, are reproduced in Figure 5.8. These drawings were made by Hale himself at his Kenwood private observatory in Chicago. The spectral line 'C' line is Hα, with the slit crossing a sunspot's umbra and penumbra. It is clearly seen on May 26th drawing made at 9:53, that the C-line shows Doppler shifts and an asymmetry on the two opposite sides of the penumbra, towards the long and short-ward sides of the spectrum. The drawing made at 9:57 shows even the *'flag'* phenomena as

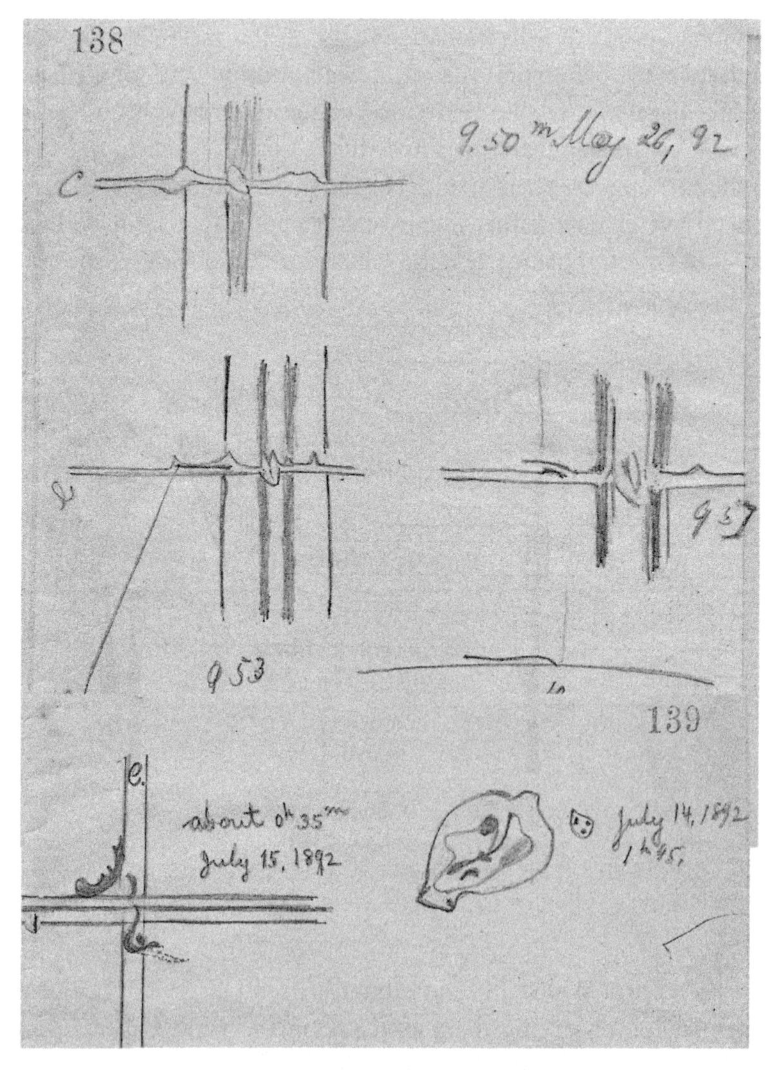

Figure 5.8 Reproduction of the original George Hale's drawings of sunspot spectra made on May 26th 1892 and July 14th and15th July 1892, at his Kenwood observatory. These drawing clearly show the Doppler shifts of the C-line in penumbral region and also the 'flag' phenomenon.

reported by Bumba in 1960. This drawing, made on July 15th 1892 at O^h 35^m, shows that the line shift was seen even beyond the penumbral boundary. These observations demonstrate that Hale had indeed first observed the 'Evershed Effect' in 1892, some 17 years before Evershed. But he did not interpret this phenomenon as material motions.

5.1.2.4 *Wave Motions in Sunspots*

From time lapse Hα filtergram movies taken at the Big Bear Solar observatory, Zirin and his undergraduate student Stein (Zirin and Stein 1972) discovered outward running penumbral waves emanating near or from the penumbral-umbral boundary in the chromosphere. In the chromospheric layers, the 3-minute standing oscillations were found dominant in the umbra, while 5-minute running waves were strong in the penumbra and super-penumbra. These waves were seen both in the line center and wings of Hα. The average speed of these penumbral running waves was estimated as about 15-25 km/s. Giovanelli (1972) also observed independently similar umbral oscillations and wave phenomenon. Recently, in 2000-02 this phenomenon has been studied in great detail by Rutten and colleagues using the DOT and NSST telescopes. They have observed outward moving waves in penumbra with speeds ranging from about 10 to 25 km/s. These penumbral waves seem to be correlated with bright umbral flashes. The best way to appreciate this phenomenon is to see the DOT and NSST movies of sunspots taken in white light and K-line available on web sites of DOT and NSST. These waves have been also observed in TRACE extreme ultra violet data, extending out from sunspot *moats* horizontally and vertically into the corona.

The interest in the running penumbral waves has continued. In recent papers by Brisken and Zirin (1997), Georgakilas *et al.* (2000), and Schlichenmaier (2002), these authors have suggested a few probable mechanisms for the penumbral running waves, but there is still doubt as to their true nature. Running penumbral waves seem to be associated with umbral oscillations and propagate outward as spherical waves which are triggered by umbral flashes. These in turn excite the overlying and surrounding penumbral photospheric, chromospheric and even coronal regions as evident from extreme ultra violet TRACE data analyzed by Muglach (2003), by Georgakilas *et al.* (2002), and also by Brynildsen *et al.* (2002). These data indicate that over the inner penumbra the motion is *inwards* towards the umbra, while in outer regions of the penumbra the wave motion is *outwards*. The speed of the waves in the corona is also much less than in the chromosphere and photosphere, on the order of 0.5–1.0 km/s, and the distance between the crests of the waves was found to be about 2500 km.

5.1.2.5 *Proper Motions of Sunspots*

In a bipolar spot group the preceding spot lies at a lower latitude than the following spot. Differential solar rotation then causes the higher latitude spots to rotate slower than the low latitude spots, and this results in a longitudinal drift between the spots. If this differential rotation is accounted for, still in the case of some spot groups, both the preceding and the following spots show inherent *proper motion* in longitude and latitude. Such proper motions have been reported first by Waldmeier (1955). From a physical point of view this proper motion reflects the dynamical behavior of the magnetic flux tubes underneath the surface. Such motions could be a source of energy build-up; a mechanism which is released in the form of solar flares and other activities.

5.1.3 ***Structure of Sunspot Umbrae***

5.1.3.1 *Umbral Granules or Dots*

Under high resolution, bright sub-arc second umbral structures have been observed. Bright umbral dots are seen in Figure 5.3a picture taken through the Dunn Telescope. Father Secchi's sunspot drawing (Figure 5.3b) also show filamentary structures. Rösch in 1956 made the first successful attempt to photograph umbral granulations and found that these are smaller in size as compared to the photospheric granulation. Beckers and Schröter (1968 a, b) measured the intensity, size and life times of umbral dots and gave a value of the average intensity as ~0.13 times the photospheric intensity, their diameter as ~ 0.16 arc sec, with a life time of about 1500 seconds.

In Figure 5.9 is shown an over exposed photograph taken by Beckers,

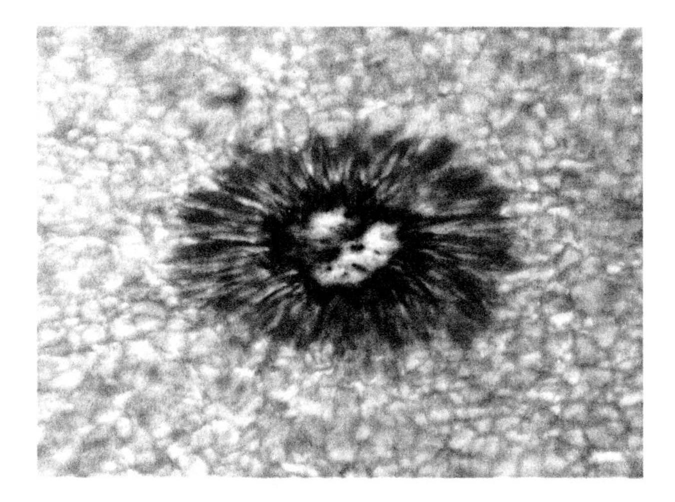

Figure 5.9 Showing inhomogeneous structure of the umbra, this picture is over exposed to bring out the umbral details.

showing the inhomogeneous structure of sunspot umbra. We note that no corresponding fine structure has yet been recorded in the magnetic field. Beckers and Tallant (1969) observed, on filtergrams taken in the K-line of CaII, short lived bright umbral flashes in the chromosphere lasting from few seconds to minutes. It is not known whether there is any connection between these chromospheric flashes and the photospheric umbral details. The life times of the two phenomena are quite different.

5.1.3.2 *Umbral Light Bridges*

On white light photographs of sunspots, umbrae often display complex structure termed *light-bridge/s*. They display diversity in shape, size and intensity. In Figures 5.1 and 5.3(a, b) light-bridges are evident. Most of the light bridges span an umbra and may cover an appreciable part of it. Some light-bridges are only 1 arc sec in width, extending into the umbral region. They may continue for hours and even days, implying that there exists remarkable thermal and dynamic stability in the umbra. Lites *et al.*, have measured the height of bridges to be 200-450 km above the umbra.

Light-bridges play a role in the final stages of sunspot evolution. The appearance of a light-bridge can be a sign of impending division or final dissolution of a spot. Sometimes an irregular spot is transformed into several smaller spots, as a result of divisions made by light-bridges.

5.1.4 *Bipolar Characteristics of Sunspots*

Hale in early 1900 noticed on chromospheric Hα spectroheliograms that sunspots generally appear as pairs, and that the Hα fibril structure around the spot group, as seen in Figure 5.10, resembles the distribution of iron fillings around a bar magnet. This inspired Hale to look for magnetic fields in sunspots. Using the Zeeman Effect, which had been discovered only a few years before, he indeed found (Hale 1908 a,b) strong magnetic fields of several thousand gauss.

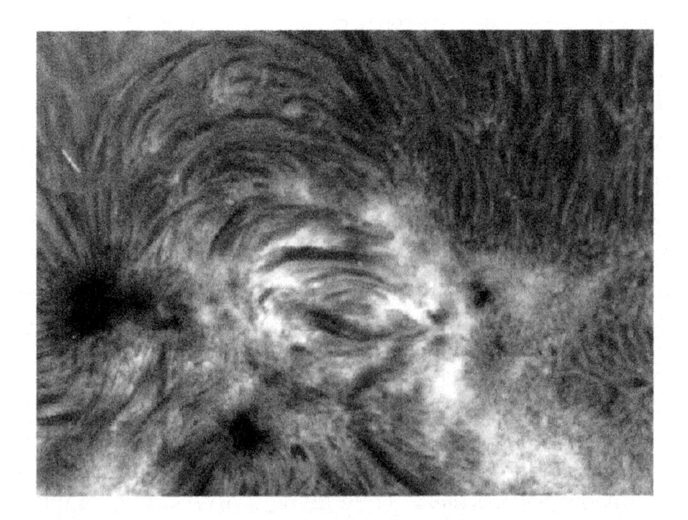

Figure 5.10 A bi-polar sunspot group taken in Hα showing filamentary structure joining the polarities of the two spots.

Generally a bi-polar group develops initially with the appearance of a single spot on the preceding (westward) side and then, a little later on the following side (eastward), small spots or a number of siblings appear. The result is a bi-polar sunspot group as shown in Figure 5.10. Hale observed that the leading or western most spots of any group in the northern hemisphere, have the same polarity, say positive, while the following or eastern spots of the group have opposite or negative polarity. In the southern hemisphere the situation is reversed, the leading spots will have negative polarity and following spots will have positive polarity. Every eleven years the polarities change sign and return to the original polarity in 22-years; this is known as *'Hale's Law'*.

Several years later Hale *et at.*, (1919) discovered that the axis, or line, joining the preceding and the following spots of a bipolar group in the beginning of solar cycle is almost parallel to the solar equator. But with time and as the solar cycle advances, the tilt of this line to the equator increases. This is now known as the *Joy's Law* and may be telling us something about the generation of solar magnetic fields inside the Sun.

5.1.4.1 *Magnetic Fields in Sunspots*

The discovery of magnetic fields in sunspots opened up a new chapter in solar physics. According to the simple Zeeman Effect, an atom which is radiating a spectral line in a magnetic field is split into a number of components. These are either linearly polarized parallel to the field, or circularly polarized perpendicular to the field. The un-displaced line is referred to a π–component and the displaced are σ-components. These

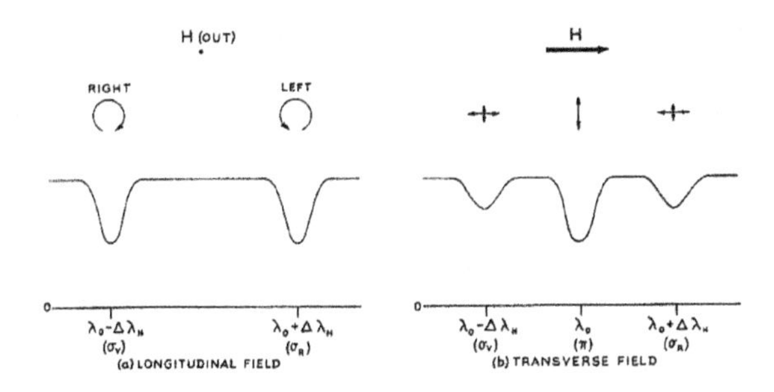

Figure 5.11 Polarization and splitting of an absorption line in the case of normal Zeeman triplet. (a) Longitudinal field directed towards the observer: Only two σ-components are present and both are circularly polarized in opposite directions. (b) Transverse field: Both π and σ-components are present, the former is plane polarized perpendicular to the field and latter plane polarized parallel to the field.

Zeeman components are always symmetrically displaced about the normal position of the spectral line. In Figure 5.11 is shown a 'normal Zeeman triplet' splitting diagram with the π and σ components.

The splitting of spectral lines is giving by the expression:-

$$\Delta\lambda_B = \frac{eB}{4\pi m_e c^2}.\lambda_o^2, \quad \text{or} \tag{5.2}$$

$$\Delta\lambda_B = 4.67 \times 10^{-5} g.B.\lambda_0^2, \tag{5.3}$$

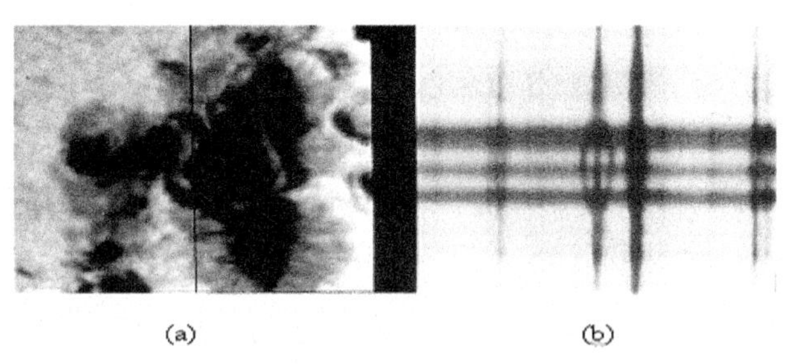

Figure 5.12 (a) Slit position marked on a sunspot, and (b) Zeeman spitting of λ 5250 over this sunspot.

where e the electron charge in e.s.u., m_e the mass of electron in gm, c the velocity of light in cm/s, g is the Landè g factor, B the magnetic field in gauss and λ_0 the wavelength of the line in the same units as $\Delta\lambda_B$.

In Figure 5.12(b) is shown an example of Zeeman splitting in the iron spectral line λ5250 where the slit of the spectrograph is positioned over a large sunspot.

5.1.4.2 *Measurements of Sunspot Magnetic Fields*

The simplest method to determine the strength of a sunspot magnetic field is to measure the separation $2\Delta\lambda_B$ of the σ-components of a suitable absorption line with the help of a high resolution spectrograph and using Equation (5.3). The Zeeman splitting can be measured either photographically or visually. The dispersion of the spectrograph should be large enough to yield sufficient separation for measurement of smallest field strength; for example a field of 1000 gauss would correspond to 7.7 mÅ separation for the λ5250.2 line of FeI.

Hale devised a very simple method for visual measurement of fields using the Mount Wilson tower telescope and its large dispersion Littrow spectrograph. This technique is still in use for daily measurements of sunspot fields at the Mount Wilson Observatory. The device consists of a polarizing-analyzer placed in front of the slit of the spectrograph, which is essentially a λ/4 plate followed by a linear polaroid for measuring the longitudinal fields. For transverse fields a λ/2 plate is used. The λ/4 analyzer consists of a grid of mica strips of suitable width and cut so that the principal axes of the adjoining strips are at right angles, followed by a polaroid sheet with its axis at 45° to the axes of the mica strips. When the analyzer–polarizer combination is placed in front and close to the slit of the spectrograph, the sunspot lines appear as the 'zigzag' pattern shown in Figure 5.13. One of the two σ-components in the alternate λ/4 mica strips is suppressed, and thus the accuracy in measuring the separation of the two components increases.

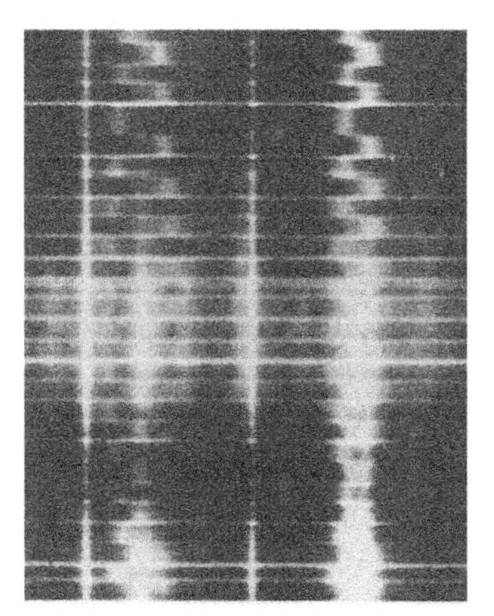

Figure 5.13 Zeeman splitting of a triplet line FeI 6302.5 Å in a large sunspot near the center of the disk. This spectrum was taken with a compound analyzer-polarizer device as proposed by Hale. The zigzag pattern of the line is clearly seen over the sunspot umbra and penumbral spectrum.

In the case of transverse field measurements, the $\lambda/2$ analyzer plate is used with the principal axes of the alternate $\lambda/2$ mica strips making 0 and 45 degrees with the axes of the polaroid sheet. In this case the two σ components are suppressed by one strip and the π component by the next, and so on. The direction of the field is determined either from the measurement of relative intensities of the individual components or from a laboratory calibration.

5.1.4.3 *Distribution of Magnetic Fields in Sunspots*

As a result of the pioneering measurements of spot magnetic fields began by Hale and his colleagues at the Mount Wilson Observatory during the early twentieth century and by later workers, the following picture of the magnetic field in sunspots emerged:-

1. That the magnetic field is generally symmetrical around the axis of the spot in the umbra.
2. The maximum value of the field is at the center of the umbra, and the lines of force are perpendicular to the solar surface.
3. Away from the umbral center the field decreases and the lines of force near the penumbral boundary are inclined to the vertical. In recent times much more precise field measurements have been made using vector magnetographs.
4. All sunspots have detectable magnetic fields and the strength increases with spot size. The magnetic field ranges from about 1800 gauss for the smallest pores to 4000 gauss for the largest spots.

The darkest position in a spot corresponds to the highest field strength and, statistically, there is a monotonic relation between field strength and umbral intensity.

In Figure 5.14 is shown a recent measurement using the infrared line at 15648 Å, made by Livingston at the National Solar Observatory, of the total magnetic field strength at the darkest place in the umbrae in the two complex sunspot groups on 24 October, 2003. The polarities of sunspots were obtained from visual measurements at Mount Wilson Observatory.

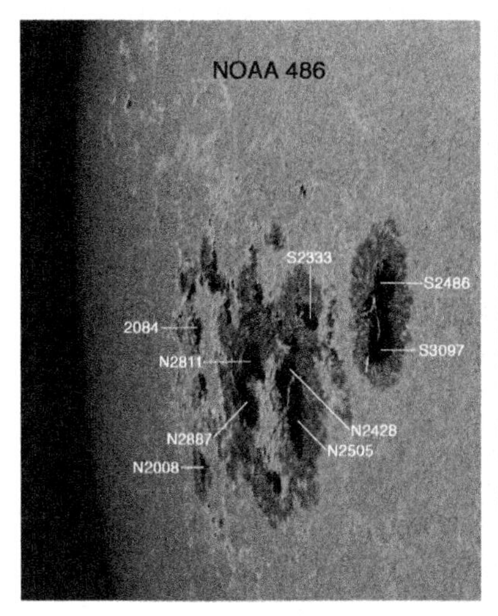

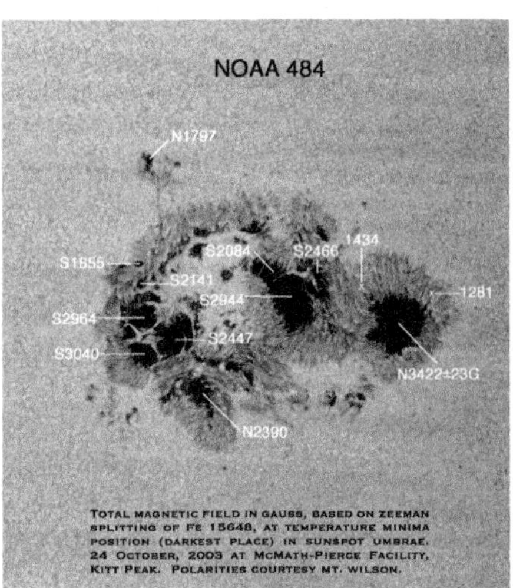

Figure 5.14 Showing total magnetic field strength & spot polarities, measured in sunspot groups NOAA nos. 486 and 484 in gauss on 24 October 2003. These are based on the Zeeman splitting of Infrared line of Fe I, 15648 Å at the darkest place in the umbrae using the McMath-Pierce facility.

5.1.4.4 *Center-limb Variation of Magnetic Fields in Sunspots*

There is an apparent systematic decrease of field strength as a spot approaches the limb. This is a consequence of the vertical divergence in the field over the spot.

5.1.4.5 *Variation of Magnetic Field across a Sunspot*

The magnetic field decreases across a spot from a maximum value near the center of the umbra to a small value beyond the outer boundary of the penumbra. The following empirical relation gives the magnetic field strength B with radial distance in a spot:-

$$B = B_m\left(1-\frac{r^2}{b^2}\right), \tag{5.4}$$

where B is the strength of magnetic field at a radial distance r from the spot center, B_m is the maximum field strength and b is the radius of the spot. This is only an approximate empirical relation.

5.1.4.6 *Direction of Lines of Force in Sunspots*

The direction of lines of force varies across a sunspot was first proposed by Nicholson and Joy at the Mount Wilson observatory in 1938. Their measurements have now been improved by modern vector instruments. But their general result is still as indicated in Figure 5.15.

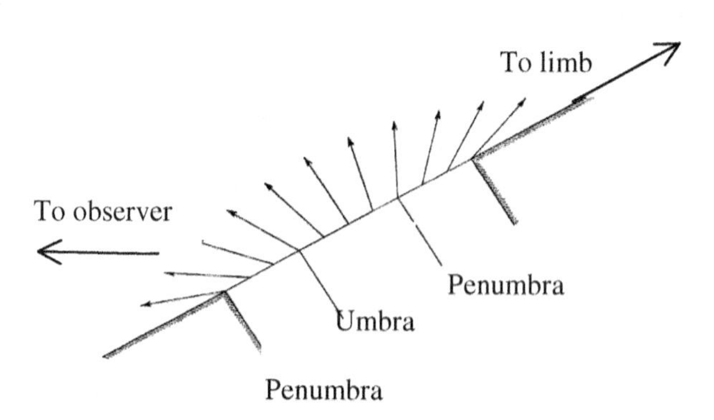

Figure 5.15 Showing change in the direction of lines of force across a sunspot. When a spot is near the limb, the lines of force are directed towards the observer on the side away from the limb, while in opposite direction on the side towards the limb. This results in the apparent observed difference in polarity on the two sides of the spot.

5.1.5 *Sunspot Models*

Physical conditions as a function of depth in sunspots, or the *sunspot model*, are obtained from the measurements made in the continuum and in spectral lines. Photometry in the continuum gives us the effective temperature, while photometry at selected wavelengths provides us with measures of the radiation temperature at *unit optical depth.* The temperature at unit optical depth in a large spot could be 2000 K less than the photospheric temperature at the same optical depth.

5.1.5.1 *Umbral Model*

Measurement of the ratio of umbral and photospheric intensities as a function of cos θ, where θ is the heliographic angle of the spot, can yield a sunspot umbral model. The physical conditions in umbrae are given by the variation of temperature, gas and electron pressure, and density with optical or geometric depth. We have seen earlier that neither sunspot umbrae nor penumbrae are of uniform in intensity. However, as a first order approximation several authors have attempted to give observational and theoretical sunspot models, assuming sunspots as homogeneous. Umbral models have been constructed by Michard (1953), Mattig (1958), Makita and Morimoto (1960), Deinzer (1965), and Yun (1968).

The simplest method to obtain the *effective* temperature T_{eff} of a umbra or penumbra is to compare the total radiation, i.e., integrated over all wavelengths, with the corresponding figure for the photosphere. If I* and I are the measured radiation fluxes, then Stefan's law gives

$$T^*_{eff} = T_{eff}\sqrt[4]{\left(\frac{I^*}{I}\right)}\,, \tag{5.5}$$

where T^*_{eff} and T_{eff} are the effective temperatures of the umbra and photosphere respectively. It may be noted that for an atmosphere in radiative equilibrium, the effective temperature is equal to the actual temperature at an optical depth of 0.67. Scattered light and seeing effects introduce errors and several authors have attempted corrections to the spot intensity. Corrected values of I*/I for spots range from 0.27 to 0.77. Although some authors applied small corrections, which reduced the measured value of 0.31 to only 0.30, but occasionally large corrections have been necessary, ranging from 0.49 (uncorrected) to 0.28 (corrected). From a large number of intensity measurements for spots of different sizes, several authors have derived a statistical relation between the umbral diameter and the temperature. The spot temperature decreases with the size of the spot as shown in Figure 5.14. Considering T_{eff} as 5785 K for the Sun, and the corrected value of I* = 0.27 for umbra,

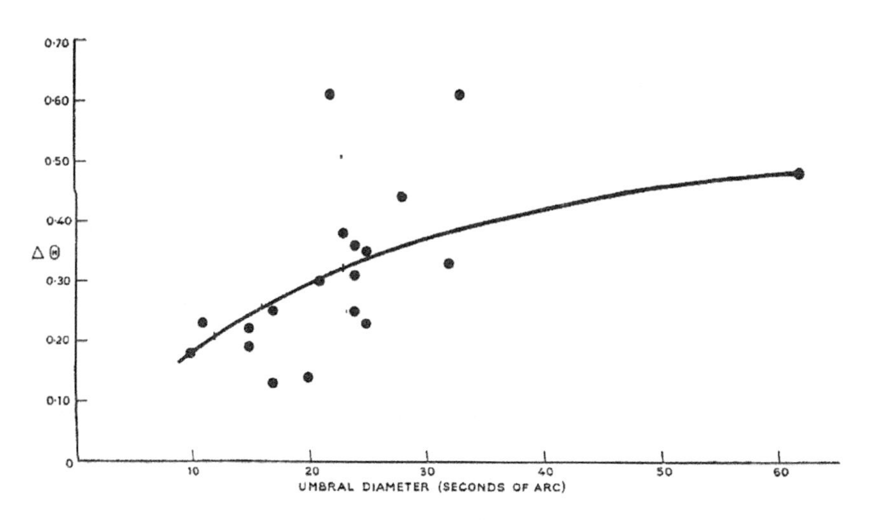

Figure 5.16 Showing temperature, ΔΘ of umbra as a function of diameter, where T*, is the umbral temperature obtained from the expression: $\Delta\Theta = 5040(1/T^* - 1/T)$, and T is the photospheric temperature and umbral diameter in seconds of arc.

yields T^*_{eff} as 4160 K, indicating a temperature difference between the photosphere and the umbra of 1625 K. It will be noted from the Figure 5.16 that larger spots are darker and cooler than the smaller ones. It is difficult to make accurate estimates of the scattered light and seeing contributions and these could give considerable errors in determination of the true umbral intensities and temperatures. In deriving the sunspot model, the next step is to determine p*, the gas pressure. This is accomplished by measurements of certain iron and titanium lines, as the equivalent width of weak Fraunhofer line formed by pure absorption, depends on the electron pressure. Michard derived the value of p_e^* at optical depths of 0.08 and 0.15 and p_e^* at $\tau^* = 4.0$ was derived from the intensity measurement of the Balmer discontinuity.

The electron pressure p_e^* depends on the gas pressure p*, the temperature T*, and the ratio of the ionized metals to hydrogen 1/A. Assuming p_e^*, T* to be known and taking value of A from the Strömgren's ionization tables as 10^4, and the helium abundance to be negligible, Michard obtained the values of p* as function of optical depth as given in Table 5.1. This is an empirical sunspot umbral model proposed by Michard (1953).

The geometrical depth in kilometers h^*, corresponding to an optical depth τ^*, is calculated by the following expression:-

$$h^* = 4.16x10^6 \int_{\tau_o^*}^{\tau^*} \frac{d\tau^*}{k^*\Theta^* p^*}, \tag{5.6}$$

where in the relation $d\tau^* = k^* \rho^* dh^*$, k^* is the continuous absorption coefficient due to negative hydrogen ion (H^-) and ρ^* is gas density, and $\Theta^*=5040/T^*$. In Table 5.1 the origin of the h-scale for both spot and photosphere was taken arbitrarily to be $\tau_o^* = \tau_o = 0.02$.

Table 5.1 Comparison between Empirical Models of Photosphere and Umbra (Michard, 1953)

T_{5000}, T^*_{5000}	T	T*	$\Delta\Theta$	P dyn/cm^2	P* dyn/cm^2	P_e dyn/cm^2	p_e^* dyn/cm^2	h km	h* km
0	4270	3550	0.24						
0.02	4520	3720	0.24	1.29 X 10^4	6.17 X 10^3	0.89	0.13	0	0
0.05	4760	3880	0.24	2.34 X 10^4	9.12 X10^3	1.70	0.25	77	413
0.10	5010	4060	0.23	3.47 X 10^4	1.32 X 10^4	2.88	0.45	132	733
0.15	5200	4180	0.23	4.37 X 10^4	1.59 X 10^4	4.17	0.63	167	927
0.25	5490	4340	0.24	5.89 X 10^4	1.91 X 10^4	7.76	0.93	206	1190
0.50	5930	4590	0.25	7.76 X 10^4	(2.51 X 10^4)	19.10	(1.59)	256	(1620)
1.0	6450	4870	0.25	9.77 X 10^4	(3.55 X 10^4)	58.20	(2.69)	300	(2120)
2.0	7100	5250	0.25	1.15 X 10^5	(5.25 X 10^4)	214	(5.43)	332	(2660)
3.3	(7620)	(5490)	(0.26)	(1.23 X 10^5)	(6.17 X 10^4)	(562)	(8.32)	(345)	(2975)
4.0	(8080)	(5720)	(0.26)	(1.26 X 10^5)	(6.76 X 10^4)	(1230)	(12.9)	(353)	(3190)

Note: Less reliable values are given parentheses.

Comparing the photospheric and umbral models given by Michard, two interesting results emerge:-

(1) That the umbra is much more *transparent* than the photosphere, at unit optical depth. Optical depth unity corresponds to 2120 km, that is

one could see up to this depth in the umbra, but in the photosphere only to 300 km.

(2) That in the umbra the gas pressure is less than in the photosphere by a factor of two to three.

Mattig (1958) also derived an empirical umbral model based on the same temperature distribution, but calculated the gas pressure assuming hydrostatic equilibrium and a different chemical abundance. Because of these assumptions, gross differences in the gas and electron pressures, and the corresponding geometrical heights, are noticed in the two models. Mattig found the umbra to be somewhat less transparent than the photosphere, whereas Michard found umbra to be more transparent. In Table 5.2 is shown a comparison of the two models.

As is it is well known that the sunspot umbrae are *not* homogeneous uniform structures, hence these sunspot models refer to only an average umbra.

Table 5.2 Comparison between Michard's (1953) and Mattig's (1958) Umbral Models

T^*_{5000}	T*	P* dyn/cm^2 Michard	Matting	Pe* dyn/cm^2 Michard	Matting	h* km Michard	Matting
0	3100						
0.001	(3330)		8.13 X 10^3		0.037		0.00
0.002	(3400)		1.18 X 10^4		0.056		25.00
0.005	(3520)		1.91 X 10^4		0.1		57.00
0.01	(3620)		2.75 X 10^4		0.16		83.00
0.02	3720	6.17 X 10^3	3.89 X 10^4	0.13	0.26	0.00	110.00
0.05	3880	9.12 X 10^3	6.17 X 10^4	0.25	0.49	413.00	144.00
0.1	4060	1.32 X 10^4	8.17 X 10^4	0.45	0.89	733.00	171.00
0.2	4270	1.76 X 10^4	1.23 X 10^5	0.79	1.70	1070.00	206.00
0.3	4400	2.03 X 10^4	1.48 X 10^5	1.08	2.40	1290.00	224.00
0.5	4590	2.51 X 10^4	1.91 X 10^5	1.59	3.80	1620.00	246.00
0.7	4720	2.94 X 10^4	2.24 X 10^5	2.04	5.01	1840.00	261.00
1	4870	3.55 X 10^4	2.63 X 10^5	2.69	6.76	2120.00	278.00
1.5	5080	4.45 X 10^4	3.31 X 10^5	3.88	9.77	2440.00	299.00
2	5250	5.25 X 10^4	3.80 X 10^5	5.43	12.90	2660.00	314.00

Corresponding to the effective temperature of 4160°K, the spectral type of a sunspot umbra turns out to K0 while that of the Sun is G2.

5.1.5.2 *Penumbra Model*

The temperature, pressure and density structure in penumbrae can be derived from intensity measurements by identical methods as used for umbra. Just as umbrae are *not uniform*, the penumbrae are also manifestations of superimposition of bright and dark features of varying dimensions and life times; however there is a gradual decline of intensity from the photosphere to the umbral boundary. So an attempt was made by Makita and Morimoto (1960) to derive a consistent penumbral model, referring to an average penumbra. It was found that over quite a large part of the penumbra, the intensity lies in a fairly narrow range of 0.72-0.80, corresponding to an effective penumbral temperature of 5500 K compared to 5785 K for the photosphere. Looking at the high resolution pictures of sunspots, it will be obvious that these numbers are in great error and at present no satisfactory sunspot penumbral model is available.

5.1.6 ***Wilson Effect***

As the Sun rotates on its axis, sunspots appear to traverses across the solar disk from the east to west limbs of the Sun. During its disk passage the appearance of the sunspot changes from an oblong - elongated shape near the east limb, to become round near the disk center, and again oblong near the western limb. These changes are foreshortening effects as the Sun is a spherical body and the spots are '*embedded*' on its surface. Alexander Wilson (Wilson and Maskelyne 1774), a professor in the Glasgow University, proposed that sunspots are indeed depressions in the solar atmosphere. Wilson concluded this, from his observations made of a sunspot seen near the western limb of the Sun on 22nd and 23rd November 1774. This apparent depression is now known as the Wilson Effect. In Figure 5.17 is shown a schematic of the apparent variation in the shape of a sunspot as it traverses across the disk. A controversy arose as to whether the sunspots are depressions on the solar surface or they appear as saucer-shaped 'depressions' because of lower density and higher transparency over the spots. This is still not resolved. During the early days of solar physics, the great astronomer Sir William Herschel

(1795) supported Wilson's idea and believed that luminous clouds form above upper layers of the bright surface of the Sun. And that the sunspots are regions where the clouds are temporarily displaced by the solar atmospheric currents, so that the inner dark cool body of the Sun is revealed in sunspots. Herschel speculated that the "*true surface*" of the Sun is protected from the heating effect of the luminous envelope by the low-lying clouds, visible as penumbra; hence, inside of the Sun could be inhabited! This picture of the Sun as a relatively cool solid body surrounded by layers of luminous clouds gradually disappeared with the advent of spectroscopy applied to solar astronomy. However, this idea of cool solar interior lingered on for more than 80-90 years as mentioned by Sir John Herschel's (son of Sir William Herschel) standard textbook – *'Outlines of Astronomy'* – in which Sir William Herschel's scheme is reproduced even at the end of 1860.

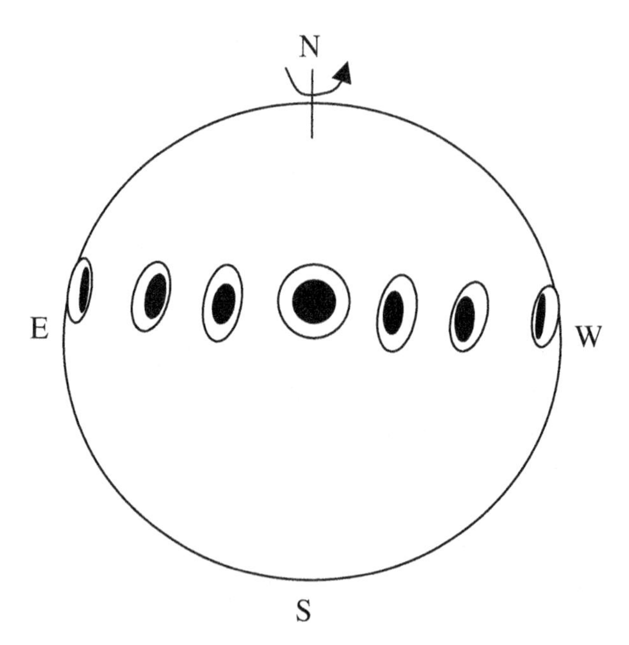

Figure 5.17 Variation in the appearance of a sunspot as it traverse the solar disk, showing the Wilson effect.

5.1.7 *Life-times, Number and Latitude Variations of Sunspots*

The lifetimes of sunspots vary from a few hours to several days, even months. Their size may range from a few thousand square kilometers to several millionths of the solar disk. The number of sunspots also varies with time. As mentioned earlier, Heinrich Schwabe (1789-1875) announced the existence of a sunspot cycle with a period of about 10 years. Rudolf Wolf (1822-1896) further improved this figure to 11.2 years.

To have some kind of quantitative estimate of sunspot activity on the Sun, Wolf, a Swiss astronomer, in 1848 introduced the now well known empirical relation, known as – *Wolf relative sunspot number,* R which is defined as: -

$$R = k\,(10g + f), \tag{5.7}$$

where f is the total number of spots on the visible disk at a particular time, irrespective of size, and g is the number of spot groups. The factor k is a personnel reduction coefficient, which depends on the observer's method of counting the spots and subdividing into groups, and on the aperture of the telescope, magnification employed and the 'seeing' conditions. Wolf determined the sunspot number, using the 8-cm aperture Fraunhofer refractor of the Zurich observatory of 110-cm focal length and a magnification of 64, and took k = 1, thus fixing the scale of relative numbers. For example, if there is only one spot on the Sun, R = 11, in a group of 5 spots, then r = 15, and if there are 5 independent individual spots, R would be = 55. In 1882 Wolf's successors at the Zurich observatory changed the counting method and derived a new value for k as 0.60, which reduced the new observations to the old scale. This value of k has been in use at the Zurich observatory since then to the present day. In order to fill in gaps in the records, Wolf initiated an international collaboration. Since 1981 many solar observatories and large number of amateur astronomers around the globe have been cooperating by sending their daily sunspot statistics to the Sunspot Index Data centre in Belgium. Earlier the Zurich Observatory compiled the sunspot numbers which were published in the Mitteilungender

Eidgenossischen Sterwarte, Zurich, in the Quarterly Bulletin on Solar Activity, in the Zeitschrift für Meteorlogie and in the Journal of Geophysical Research.

Using a more or less arbitrary method, but quite consistent, Wolf was able to estimate the relative sunspot number from the earlier observations dating back to 1749 until 1893. Between the period 1894 and 1926, the data were collected by A. Wolfer, and from 1927 to 1944 by W. Brunner. M. Waldmeier thereafter continued observations until 1979 at the Zurich Observatory. The most extensive compilation of sunspot numbers is due to Waldmeier (1961) who has given the following sunspot data in his important monogram:-

(1) The years of minimum and maximum sunspot activity during the period 1610-1960.
(2) The annual mean sunspot numbers for the period 1700-1960.
(3) Monthly mean sunspot numbers, for the period 1749-1960.
(4) Daily sunspot numbers for the period 1818-1960.

In Figure 5.18 is shown the 11-year sunspot cycle dating back from 1610 to 2000 AD.

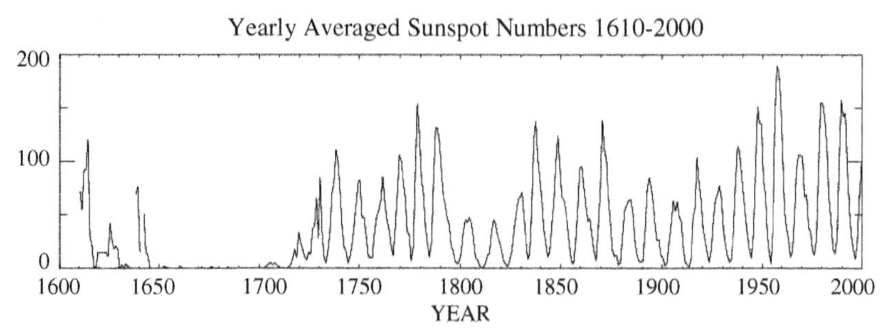

Figure 5.18 Sunspot number with time beginning from 1610 to 2000 AD, This plot clearly shows the 11-sunspot cycle and the 'Maunder minimum' during 1640 to 1715 AD.

It was Richard Christopher Carrington (1826-1875) who made the discovery that the average latitude of sunspots decreases steadily from the beginning to the end of each solar cycle. This latitude drift was further investigated by Gustav Spörer (1822-1896) and is now referred to as *Spörer law*. Since the early observations made by Carrington and by

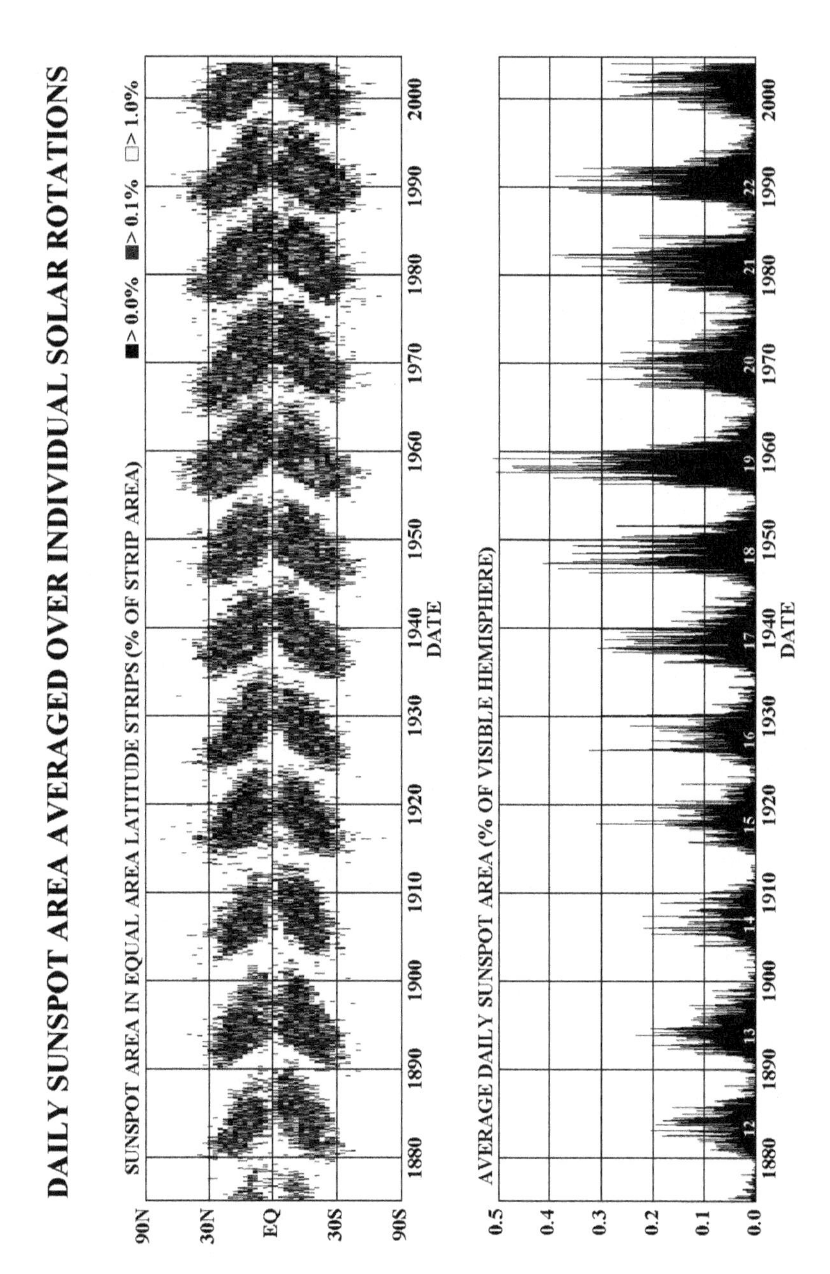

Figure 5.19 Left panel shows the 'Butterfly diagram' indicating the latitude drift of sunspots during 11-year solar cycle from 1875 to 2003 for nearly 12 solar cycles. Right panel shows the average daily sunspot numbers during this period.

Spörer, it is known that sunspots are confined to a relatively narrow latitude belt in each solar hemisphere, and that the vast majority of spots occur between the equator and latitude zone of $\pm$ 35°. Spots are rarely seen at latitudes higher than 40° and these are small and short-lived. However, occasionally spots have been observed at 50° and even at 60°. Waldmeier observed pores up to 75°. The average latitude distribution of spots depends on the phase of the solar cycle. At the beginning of the cycle the spots appear at high latitudes and steadily decrease until the end of the cycle. It will be noticed that at any one epoch there is considerable spread of sunspots in latitude; however the drift in latitude is well marked for each solar cycle. According to Waldmeier the overlapping of two cycles lasts for about 2 years. During the overlap period there are four sunspot zones. This diagram shows the latitude drift of sunspots which resembles as a butterfly and thus is called the '*butterfly diagram*'.

5.2 Faculae

5.2.1 *Photospheric Faculae*

Faculae are visible in white light only near the solar limb as patches of bright areas, see Figure 5.1. They are inseparable companions of sunspots; they become visible even before spots are seen and outlive them by several rotations. This association with sunspots distinctly shows a close correspondence with the photospheric magnetic field. In fact faculae map the photospheric field in the upper photosphere where it has expanded somewhat.

A very detailed study of photospheric faculae had been made at the Greenwich Royal Observatory (1923) from its vast collection of daily photoheliograms. The main findings from this study are summarized in the following:-

1. The centers of the main zone of the faculae have well defined progression with solar cycle,
2. Compared to the sunspot zones, the faculae zones are on the average about 15° (in heliographic coordinates) broader and

extend mainly towards the pole-ward sides,
3. There is a zone of *polar faculae* in both hemispheres, showing a slight concentration above 70° latitude. Polar faculae differ from the low latitude ones, as they appear as small, short-lived, detached flecks,
4. Faculae frequently act as a connecting link between successive spot groups they tend to cluster around the preceding and the following member of a bipolar spot group,
5. Faculae frequently appear as streaks roughly at right angle to the rotational axis of the Sun and spread several degrees in latitude, invariably towards the poles. This behavior is quite different as compared to sunspot groups, which generally spread in longitude and with hardly any latitude drift. This feature may have some relation to the magnetic field diffusion on the solar surface.
6. Excluding the polar faculae and about 10% of the total facular area in sunspot zones, all other faculae are always related to spots. There are no spots *without* faculae, but there could be extensive areas of faculae without spots.

The center–limb variation of the intensity ratio of faculae to photosphere shows that the faculae are brighter near the limb than the surroundings and disappears near the disk center. This indicates that the facular emission must be enhanced in the upper layers of the photosphere. The temperature of faculae is higher by about 100°K than the photosphere. Considering the contribution due to limb darkening the temperature of the faculae increases with height.

5.2.2 *Chromospheric Faculae or Flocculi or Plages*

The transition from photospheric to the chromospheric faculae is continuous; the photospheric faculae correspond one-to-one in position with chromospheric features. As seen on CaII K-line or Hα–spectroheliograms or filtergrams, the chromospheric faculae, which are generally known as *flocculi* or *plage*, are much brighter and have higher contrast than the photospheric faculae. Chromospheric flocculi are mainly confined to sunspot zones. Occasionally small round flocculi may appear where there was no sign of previous activity. Once flocculi are formed, the characteristic tendency of its west-end, or the preceding end, is to stretch and expand with time, towards the equator in both the hemispheres. After attaining a maximum area, predominantly elongated shaped flocculi tend to disintegrate into irregular shapes. The orientation of their east-west axis may range from 0° to 40° and occasionally could reach as high as 90°. As with photospheric faculae, in no case sunspots have been found without flocculi, but flocculi could exist without visible spots. There is a close correspondence between the photospheric magnetic field and the chromospheric flocculi.

The flocculi do not disintegrate suddenly, but gradually appear to disperse until finally the area become indistinguishable from the solar rèseau. The photospheric faculae and the flocculi resemble each other, both show granular structure of comparable scales. Most of the facular and flocculi granules have diameters in the range of 1-2 arc sec and life time times of about 2 hours. In view of the fact that both seems to be closely correlated in shape and position, some scientists believe that both these phenomena represent the same structures, seen at different heights (de Jager 1959, Bray and Loughhead 1964, Zirin 1966).

5.3 Chromospheric Activity

The chromosphere is the seat of the most spectacular displays of dynamic solar activity, ranging from emerging new active regions to network structures, plages, filaments, prominences, flares, mass ejections surges, ephemeral regions, and so on. A large number of astronomers have enormously contributed towards observational and theoretical studies of chromospheric activity. In this Section we shall discuss some of the salient results obtained during the last few decades. In the optical wavelengths, generally two main spectral lines are used to observe the chromosphere; one is the Hα line due to hydrogen and other the K and H-lines at 3834Å and 3868 Å due to singly ionized Calcium atom. Full disk images in these wavelengths have been taken regularly at number of solar observatories around the globe since the early twentieth century.

5.3.1 *Ellerman Bombs - Moustaches*

When W.M. Mitchell at the Harvard College Observatory in early 1900 centered the slit of a spectroscope at a point in an active region near a sunspot, he observed in Hα line transient appearance of two bright streaks extending outwards in the continuum. These streaks span more than 10Å along the spectrum and cross many Fraunhofer absorption lines.

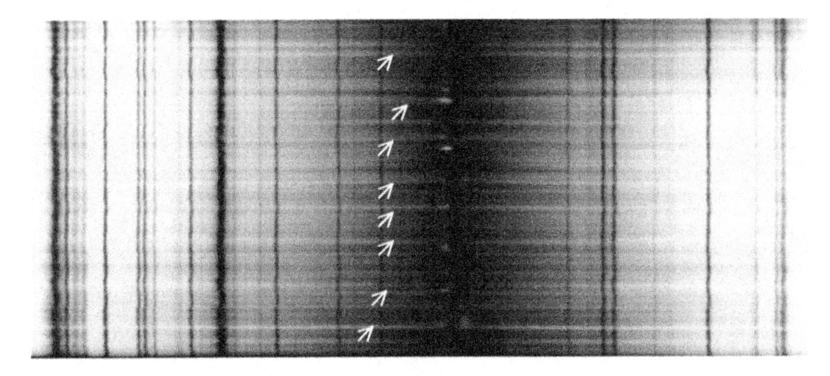

Figure 5.20 Spectrum taken in the region of CaII K-line near an active region showing bright emission streaks in the continuum. Moustaches are marked by arrows in several locations.

Following this discovery, Ferdinand Ellerman (1917) undertook a detailed study of this interesting phenomenon at the Mount Wilson's 150-foot tower. He found in the spectra taken near active regions, between a spot group and the outer boundaries of penumbrae, that there was a narrow band of bright emission extending over several Ångstroms on either side of Hα and in other Balmer lines, but in each case the emission did not seems to cross the line center. He concluded that the average duration of such streaks was only for a few minutes. Severny (1956) at the Crimea observatory studied this phenomenon in great detail and named it as '*Moustache*' from its similarity with the human moustache. He also found that moustaches are seen in all members of Balmer series up to H_{10} and also in the H and K-lines of CaII and several strong and weak lines of iron and other metal lines.

Tamara Payne (1993) from a study in 1993 at Sacramento Peak was able to say that Ellerman bombs are confined to a 500 km layer, 600-1100 km above the photosphere. There is nothing above the low chromosphere. She determined that the energy release was about 10^{27} ergs.

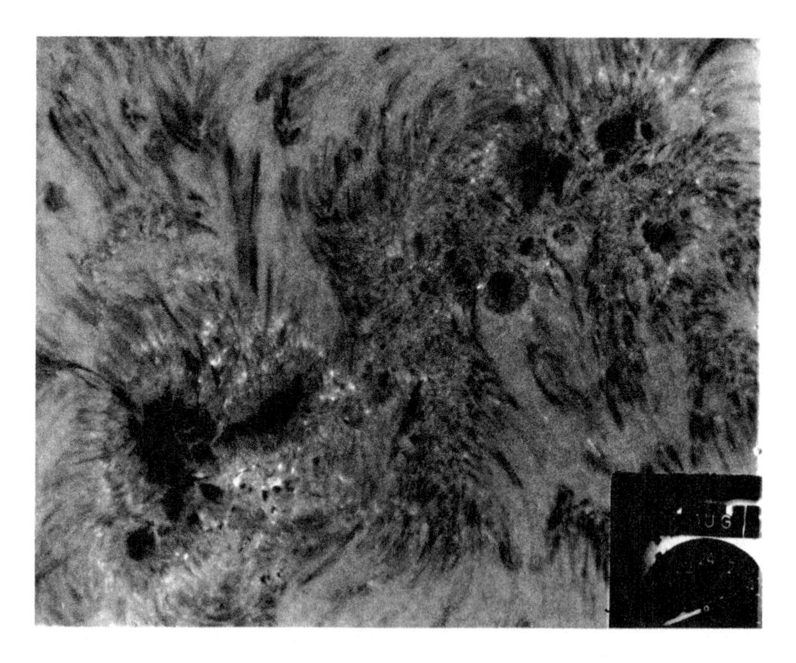

Figure 5.21 Large number of bright Ellerman bombs on a Hα filtergram taken from the Big Bear Observatory on August 25 1971, at 0.7Å away from the line center in the red wing. Ellerman bombs are invariably seen at the outer boundary of the penumbrae.

In Figure 5.20 is shown a high spectral and spatial resolution spectrum taken around the CaII K-line displaying several moustaches. Bright steaks are visible extending to several Ångstrom in both wings of the line in the continuum, but avoiding the line K-line center. Some moustaches show asymmetry of the intensity in the blue and red wings. In some rare cases a moustache can display an *inclination* to the direction of dispersion, a kind of tilt as seen in the case of spicule spectrum. Perhaps this indicates a spiraling motion in the Ellerman Bomb structures. Ellerman bombs are also seen as bright points in high resolution *off band* filtergrams taken in Hα line at about 0.5-0.7Å away from the line center, as shown in Figure 5.21. All around the sunspot active region, especially near the outer penumbral boundary, bright points are visible in this picture; these are in fact Ellerman bombs or moustache seen in two dimensions.

5.4 Evolution of Chromospheric Active Regions

The first indication of activity at the chromospheric level is marked by the disturbances in the Hα –fibril structure. Flocculi or plage then forms a bright patch, and a small spot may also develop. Subsequently a spot of opposite polarity to the existing one develops on the eastern or following end of the plage region. This bi-polar configuration generally attains its maximum area between 6th and 13th days from the first appearance of the disturbance in the quiet fibril structure.

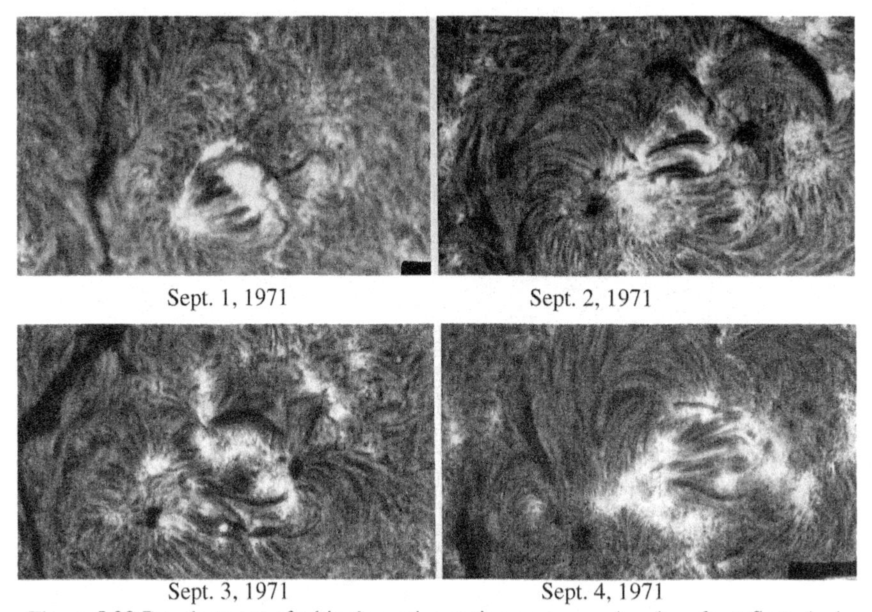

Figure 5.22 Development of a bipolar active region on successive days from Sept. 1- 4, 1971, pictures taken from Big Bear Solar Observatory.

Throughout this development phase the size, brightness and complexity in surrounding fibrils of the plage region continues to increase. Between the 14th and 30th day, the chromospheric plages begin to shrink in area and at the same time the following spot or portion of the plage may disappear, leaving only the western or leading spot of the plage. Filaments (prominences) may appear in the active region. From the 30th to 60th day the last spot slowly dissolves and the brightness of Hα and CaII plages also decreases. Any dark filaments which sometimes appear during the declining phase of the region elongate and tend to align almost parallel to the solar equator.

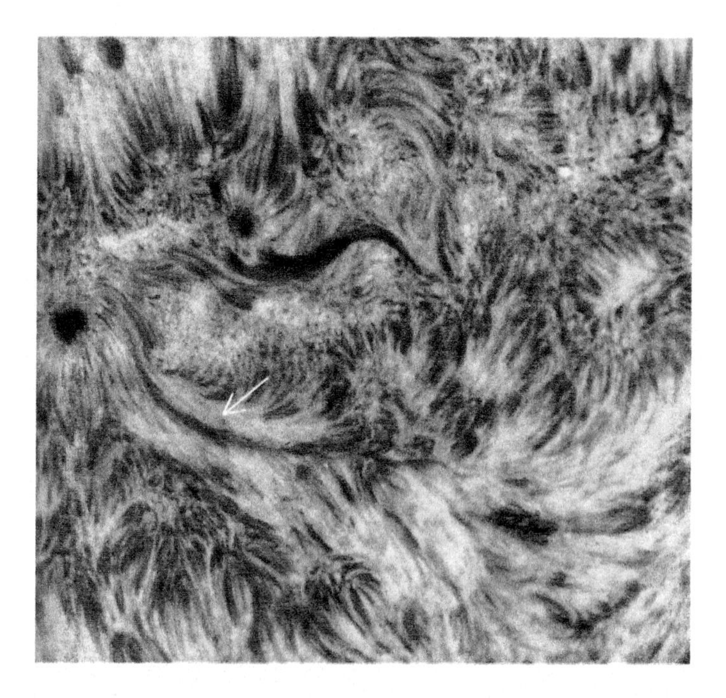

Figure 5.23 Showing fine scale fibril structure near a dark filament, very near to the filament the fibrils are almost parallel shown by black arrow, while slightly away they tend to bend towards the filament shown by white arrow.

Studies have been made of the chromospheric fine features in active regions by Bruzek (1967), Weart and Zirin (1969), Martres and Soru-Escaut (1971), Zirin (1971), and Frazier (1972). Figure 5.22 illustrates the development of a chromospheric active region. It will be noticed how dark fine arch systems develop between the bipolar configuration.

Bruzek refers to these fine structures as an *Arch Filament System* (AFS) while Zirin likes to call them *Emerging Flux Regions* (EFR). The dark AFS between the preceding–*p* and the following–*f* spots follow the *transverse* magnetic lines of force, connecting regions of opposite polarity. Zirin (1974) and several other authors have reported that material in the AFS rises to the top of the arches and drains down at the edges. As the bipolar spot group separates and expands more and more arches appear and stretch, and more horizontal fields emerge. There is also a tendency for new flux to emerge near the active region.

As an active region grows in size and complexity the associated fibril structure also increases. Generally fibrils display remarkable stability and may show little change over period of an hour or more. In a bipolar spot group a complicated fibril pattern is often visible, resembling the characteristic 'iron filing' pattern formed by a bar magnet as shown in Figure 5.10. Filaments near an active region also seem to influence the fibril arrangement in their vicinity. Fibrils adjacent to a filament tend to run parallel to length of the filament while those a little distant may appear to bend away from it. Sometimes fibrils at the end of filaments are seen lying perpendicular to its axis. Martin and co-workers have studied fibril structures and their orientation. In Figure 5.23 is shown a high resolution picture of an Hα filament and surrounding fibrils.

5.4.1 *Magnetic Fields and Chromospheric Flocculi*

It has been well established now that a one-to-one correlation exists between the Ca II plages and magnetic fields. All plages and chromospheric network display this close correspondence with magnetic field strength and are almost linearly related to the CaII plage intensity. H-α emission is also correlated with magnetic fields as shown in Figure 5.24. However, over sunspot umbrae with field strength of 2000 gauss or more the CaII and H-α emissions are missing while in regions with medium to weak magnetic flux there is one-to-one correlation.

Hα bright emission is not seen over sunspot umbrae, while in regions of medium to weak flux there is one-to-one correlation. If magnetic energy is responsible for heating the chromospheric plages, why does it fail to heat regions of stronger field strength? At present no plausible

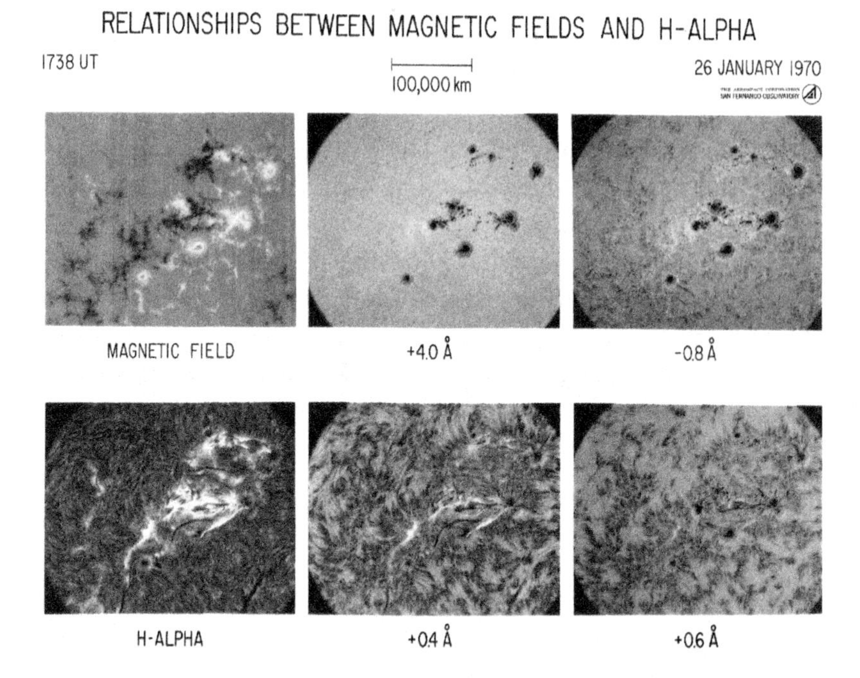

Figure 5.24 Showing spectroheliograms taken in off band and line center of Hα and corresponding to a longitudinal field magnetogram from the San Fernando Observatory on 26 January 1970. Black & white regions show the two polarities. Notice the close correspondence between Hα emission and the magnetic field, dark filaments indicated by arrows are seen lying between opposite polarity regions and AFS *joining* regions of opposite polarities.

explanation has been given for this lack of emission over strong magnetic field regions. In Figure 5.24 the neutral magnetic lines are also clearly demarked by filaments lying between the two opposite polarity regions, and arch filament systems are seen joining regions of opposite polarities. Away from the active region the field displays the network structure of both polarities.

5.5 *Large Scale Magnetic Fields*

Since the mid eighteenth century the Sun was suspected to have general global magnetic field, as suggested from the appearance of coronal polar

polar plumes seen during a total solar eclipse. It was only after almost 200 years that the father and son team of Babcock & Babcock (1952), Babcock (1953), and independently Kiepenheuer (1953) devised photoelectric techniques to detect weak magnetic fields beyond sunspots. Since then enormous progress has been made both in spatial and spectral resolution allowing the detection and measurements of field strengths on the order of few gauss. From full disk high resolution magnetograms it is now well established that the magnetic field is not confined only to active region zones, but it is distributed all over the Sun and in all latitude and longitude zones. Weak fields appear everywhere, while the boundaries of network cells show enhanced field strengths.

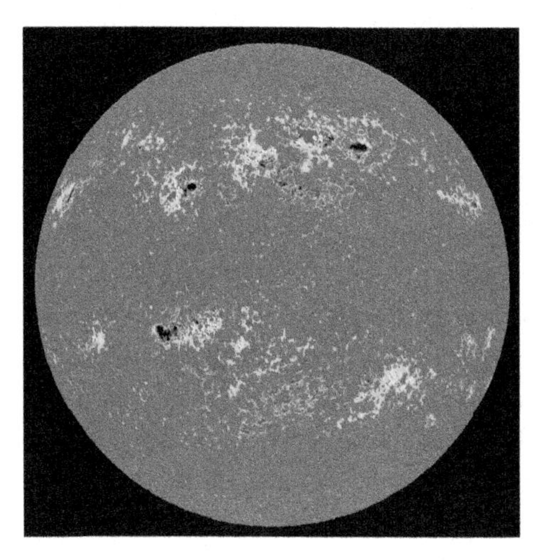

Figure 5.25 Showing full disk magnetogram taken from National solar observatory on 12 February 1989, Heliographic north is at top, east at left. Positive (north) polarity is shown in green to yellow to red: negative (south) polarity is in blue to dark blue to black. Hale's law of sunspots polarity is revealed in the sunspot zones of the two hemispheres.

In Figure 5.25 is shown a full disk magnetogram taken during high solar activity period on 12 November 1989, at the National Solar Observatory. It provides a nice over all picture of large scale magnetic flux pattern. On such magnetograms one can identify a number of characteristic features. For example, Hale's polarity law is clearly seen in the two sunspot zones. In the northern hemisphere the preceding (p-westward) spot's polarity is negative, while in the southern hemisphere the preceding spot's polarity is positive. Young and old active regions, the magnetic network, and back-ground fields can be identified.

The dominant bipolar regions are found in the two intermediate latitudes bands between 10°-35° in the two hemispheres. These bipolar regions are essentially pairs of opposite magnetic polarity regions; with one directed out of the Sun and other pointing into it. The magnetic field lines of force emerge from a sunspot of one polarity and loop through the low solar atmosphere above it and enter a neighboring sunspot of opposite polarity. These loops constrain and are filled with hot plasma above a sunspot group and active region. They are among the brightest features seen in X-ray and EUV pictures, as shown in Figure 4.14. These bright loops also delineate the magnetic lines of force in the corona.

A new bipolar magnetic field emerges in latitude zone of about 30°-40° as a compact region. Its polarities are oriented approximately in the east-west direction. As time progresses, the axis of the bipolar region tilts towards the equator (Joy's Law), the region grows towards the following portion, and new siblings of spots appear near the p-spot. Over a time scale from a few weeks to months, the region decays as indicated by dispersal of flux over a large area. This flux becomes diffuse, consisting of tiny fragments which mingle with background fields. Leighton in 1969 proposed that the diffusion of magnetic field is carried out by 'random walk' process through supergranulation motions. In this process the flux tubes (fragments of magnetic flux) are swept towards the boundaries of the supergranulation cells and thereby form a network pattern with typical cell size of 30,000 km. It is assumed that the individual magnetic flux elements that make up the network have a typical size of 100 km. The dissipation or annihilation of magnetic flux is proposed by Martin *et al.*, (1985) to occur by the process of cancellation of magnetic fields of opposite polarities.

Newly born bipolar sunspot groups always follow the Hale's polarity law, and the polarities of the preceding and the following groups are reversed with the commencement of a new 11-year cycle. In the case of older, partially decayed magnetic regions, the Hale's law is not always evident as fluxes merge with the remnants of other magnetic regions and the background fields. During the course of the 11-year cycle bipolar regions migrate towards the equator as seen in the 'butterfly' diagram shown in Figure 5.19. A new solar cycle starts when a sunspot group with reversed polarity (as compared to the preceding cycle) first appears at high latitude. To get back to the original polarity configuration it takes 22-years, and this is known as the *22-year solar magnetic cycle.*

5.5.1 *Fine Scale and Ephemeral Magnetic Regions*

Full disk magnetograms reveal that magnetic features have a wide range in pattern size, extending from about one sixth of the solar diameter to few arc sec or even fraction of an arc sec. As the spatial resolution of the telescope-magnetograph combination is improving, more and more fine structure of magnetic features is emerging. Extremely high spatial resolution magnetograms have been obtained from the Swedish Solar Vacuum telescope at La Palma displaying magnetic features of nearly 300 km size and field strengths of more than 1000 gauss, Figure 5.26.

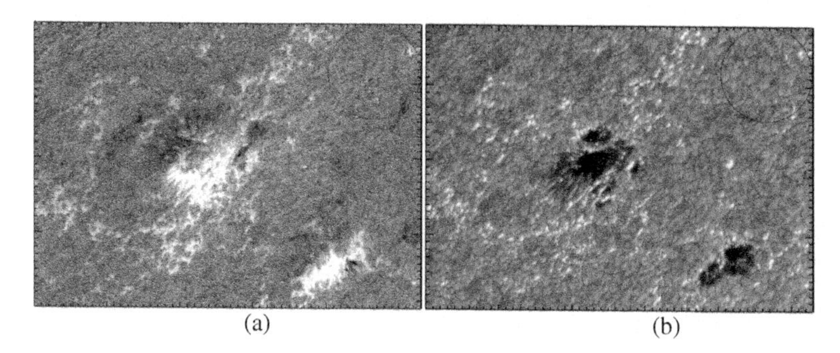

Figure 5.26 (a) High resolution longitudinal magnetogram taken from the New Swedish Solar Vacuum telescope, showing 0.5 arc sec magnetic elements, the white and dark regions indicate the two polarities, (b) Same region taken in G-band on 6 June 2003. Circle on left image indicates size of the Earth.

Figure 5.27 is a high resolution magnetogram of a quiet region at the disk center. This picture reveals that the Sun's disk is covered with small scale intermittent flux fragments of mixed polarities, giving a 'salt-pepper' appearance. The bright and dark areas represent magnetic flux of opposite polarities. Here mixed polarity network fields are preferentially located at the network boundaries which we judge to be supergranulation cells. Such high sensitivity magnetograms disclose the weak mixed polarity fluxes inside the cells. This is the *intra-network* field. The enhancement of fields at the network boundaries is a result of horizontal motion in supergranules which 'pushes' the field lines to the boundary to manifests this flux increase. A schematic diagram explaining this idea is shown in Figure 4.7. Stenflo (1973) proposed that the basic building blocks of the network fields are *magnetic flux tubes* of 100 km or smaller dimension, which is beyond the present day observational capability to resolve.

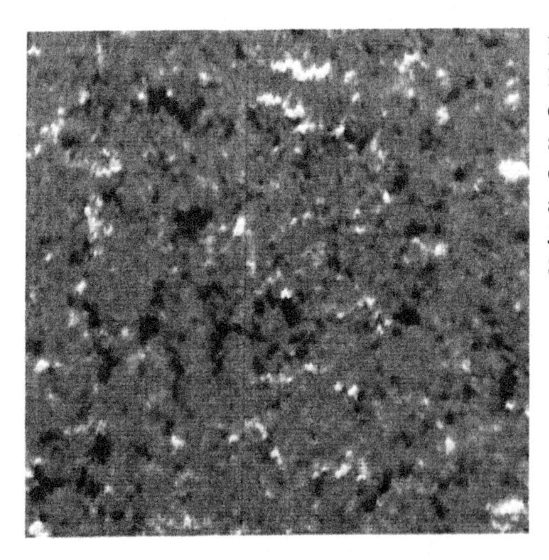

Figure 5.27 High resolution longitudinal magnetogram of a quiet region near the disk center showing fine features of 1arc sec or less of both polarities, indicated as white and black areas. Taken by Jack Harvey from the National Solar Observatory.

Using the 'line-ratio' (FeI 5247.06/FeI 5250.22 Å) technique, Stenflo (1973) was able to estimate the magnetic field strength in the quiet photospheric network and arrived at a value between 1000-2000 gauss. Similar field strengths were found also in active plages regions, where the total flux is much larger by an order of a magnitude. Stenflo believes that although the basic building blocks of solar magnetic field are magnetic flux tubes, the main difference between the network, quiet and active regions (outside sunspots) is primarily due to the number density and size distribution of flux tubes. The whole Sun seems to be permeated by such thin flux tubes. Theoretically, Stenflo (1976) has estimated the sizes of these flux tubes to be nearly 100 km, as this is approximately the horizontal mean free path of the visible photons in the photosphere. Recently Keller (1992) has used the technique of speckle polarimetry and resolved flux tubes of 200 km sizes.

Today we can directly measure the field strength of flux tubes using the favorable line of FeI 15648 Å. As we discussed in the Chapter 7, the Zeeman splitting of this line and its temperature insensitivity, means we no longer must resort to indirect methods like line ratios. Indeed, Stenflo was correct and the fields range 1000-2000 gauss.

From first appearance of a full disk magnetogram, it may appear that there are only large scale magnetic active regions (AR) which dominate the Sun's magnetic field. But on closer examination, smaller and intermediate scale magnetic features are also seen. These are called '*ephemeral regions*' (ER) and have been extensively studied by Harvey and Martin (1973), and Martin and Harvey (1979). ERs are compact small bipolar regions of typical 5-10 arc sec sizes with life times ranging from few minutes to hours.

5.5.2 *Dispersion and Annihilation of Magnetic Fields*

In the previous sections we have discussed how the new magnetic flux emerges. After attaining a maximum area, the active region (say a bipolar region) appears to first start fragmenting from the following portion of the active region. The fields seem to considerably weaken and disperse away from 'axis' of the bipolar region. Finally it merges with the background fields. High resolution time lapse movies of magnetic fields made by Vrabec (1973), Michalitsanos and Bhatnagar (1975) and others at the Big Bear Solar Observatory showed that fields are 'nibbled' away or driven out as small fragments from the active regions towards the surrounding photosphere. Dispersal speeds are about 1-2 km/s. These observations give some support to the idea of diffusion and cancellation of magnetic field over the solar surface.

5.5.3 *Polar Magnetic Fields*

The background field develops a global pattern, which is clearly seen on magnetograms taken during the minimum period of the solar cycle, Figure 5.28. In this figure, all over the solar disk are tiny bipolar magnetic regions. But near the polar regions specks of a single polarity is predominant. At Mount Wilson, H. W and H.D. Babcock (1955), using their new magnetograph, detected polar magnetic fields. They found that one polarity was dominate and opposite, around the two poles.

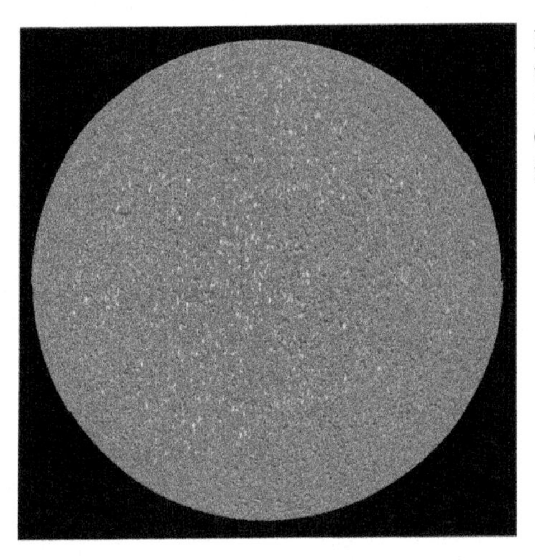

Figure 5.28 Full disk magnetogram taken at the time of solar minimum on 12 October 1996 from National Solar Observatory. Yellow and blue colors indicate the two polarities.

H. W and H.D. Babcock (1955) also observed a reversal of polarity near the time of solar maximum in 1957-58. Further work by H. D. Babcock and Livingston (1958) confirmed the reversal of the polar field. Howard (1972, 1974) measured the polar fields for a full solar cycle from 1960-73, and observed reversal of polar fields. These polar fields gradually change polarity, but at times both poles may show the same polarity or mixed polarities of very small magnitude. Recently, it has been reported by Pete Riley of Science Application International Corporation in San Diego, that beginning in March 2000 for nearly a month, the Sun's south magnetic pole faded and the north polarity emerged in its place. Thus during this period both poles of the Sun had the same polarity. According to Riley, the south pole never really vanished; it simply migrated north and for a while became a band of south magnetic flux, smeared around the Sun's equator. By May 2000 the south magnetic polarity returned to its usual location near the Sun's southern spin axis, but not for long. In 2001 the solar polar magnetic field completely flipped polarity, the south and the north poles then swapped positions. In 2003-2004, the polarity of north pole is positive; the field lines are directed outwards, while in the

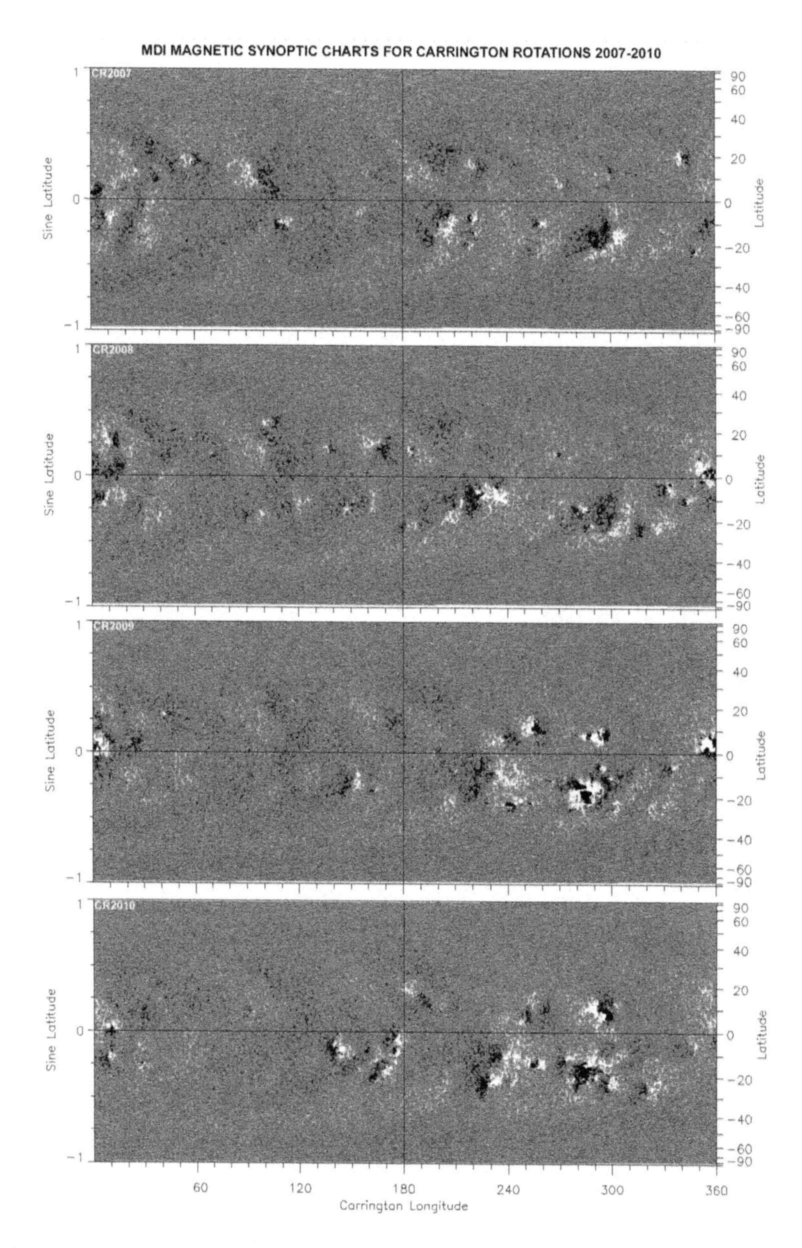

Figure 5.29 Synoptic maps of longitudinal magnetic field over four Carrington rotations, during August to December 2003. These plots clearly show emerging & decaying magnetic regions from one rotation to the next. White & black indicate the two polarities.

south polar region, the field lines are directed inwards. After the change-over of polarities, the polarity of the polar caps becomes the same as the polarity of the eastern (following) portion of the bipolar magnetic regions of the same hemisphere.

The above picture does not mean that the Sun has a bi-pole like the Earth. Rather, polar zone polarities are a consequence of flux surface migration (Sec. 5.5.4) with dominate polarities following Hale's Law.

The emergence and decay of magnetic regions continuously takes place during the 11-year cycle. This is clearly demonstrated from synoptic observations of longitudinal magnetic fields. Figure 5.29 shows synoptic maps made during four Carrington solar rotation periods of 27.27 days from August to December 2003. On these magnetic field synoptic charts one notices the emergence of new flux and the decay or diffusion of old magnetic regions from one rotation to the next.

5.5.4 *Migration of Fields*

As the 11-year solar cycle progresses from the minimum to maximum, two types of field migration occurs. One is towards the equator; the preceding p-region of the bipolar group moves towards the equator in both hemispheres. The second one is towards the poles; the following f-portion of an active region spreads in area and also displays a lateral extension towards the poles. With time the active region declines in area, brightness and magnetic flux. Migration of large scale fields have been also observed on the solar disk. One of the techniques used to delineate the field migration is to make use of Hα dark markings, the filaments (prominences seen on the disk), which are known to be *controlled* by magnetic field. Hα filaments seen on the disk are known to lie between regions of opposite polarity. The area and location of filaments and prominences show a variation with the 11-year cycle. From Hα and K-line spectroheliograms available at the Kodaikanal observatory, an analysis of prominence areas and position, during 5 solar cycles beginning from 1905 until 1952, was carried out by Ananthakrishnan (1954). He showed from the latitude and area distribution of prominences during five solar cycles that during the minimum period prominences are mainly confined to low latitude zones.

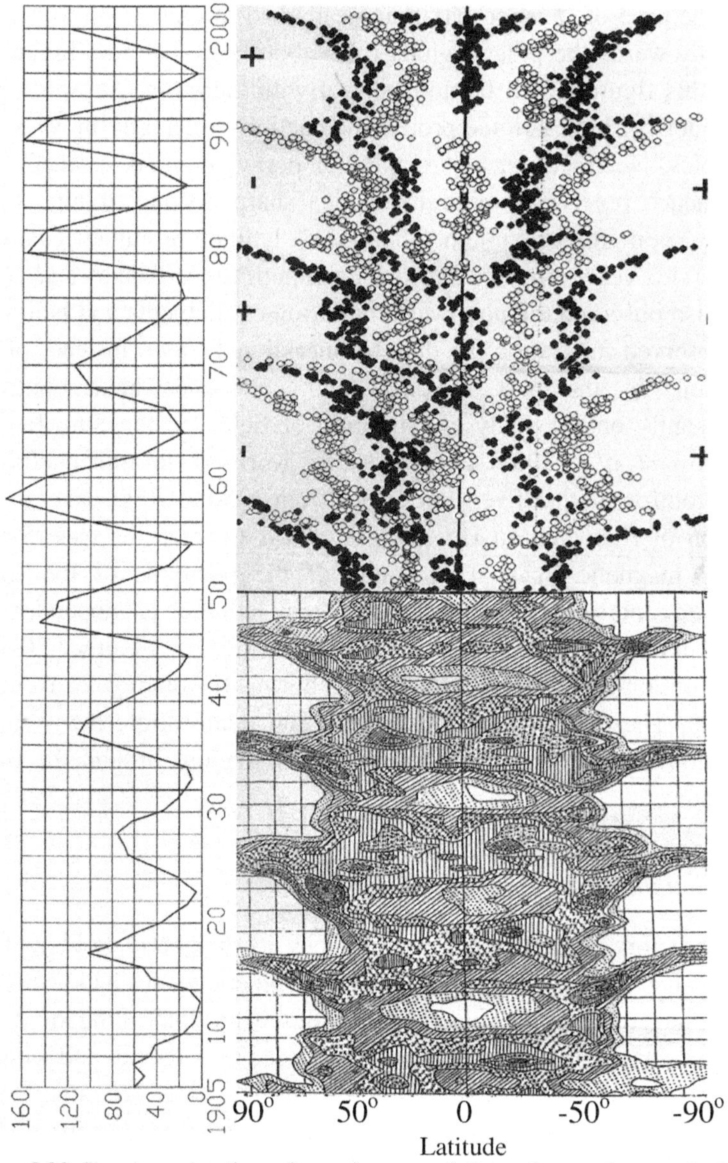

Figure 5.30 Showing migration of prominence activity and neutral magnetic field lines during 10 solar cycles from 1905 upto 2000. The data from 1905 to 1950 are from prominence positions and area measurements by Ananthakrishnan (1954) and for the period 1950 to 2000 from Makarov *et al.*, (2001) of positions of the trajectories of neutral lines from Hα synoptic charts. Left panel shows the 11-year solar cycle showing mean sunspot number with year.

Near the time of sunspot maximum a marked migration of prominences is seen towards the polar regions from about ±60° latitude, Figure 5.30. From this figure it will be noticed that with increasing sunspot activity the centers of high latitude prominence activity gradually move towards the poles. Near the sunspot maximum period there is a steep rise of prominence pole-ward migration and a sharp decline of high latitude activity soon after maximum. The highest latitude prominences are seen about 1-1.5 years after the sunspot maximum time. Topka *et al.*, (1982) have also observed the pole-ward migration of filaments and believe that the observed migration is due to meridional flow instead of field diffusion. In the past Makarov and colleagues have contributed significantly on the study of migration of fields on the Sun. Recently, Makarov *et al.,* (2001) extended this work of prominence/filament migration for another five solar cycles from 1950 to 2000. They used the position of filaments on Hα synoptic charts to trace the trajectories of neutral magnetic lines. Their plots of the positions of the 'neutral magnetic line' verses the heliographic latitude is appended with Ananthakrishnan's plot of prominence distribution. Figure 5.30 shows the migration of prominences/filaments during 10 solar cycles from 1905 to 2000. These data beautifully match and demonstrate the systematic migration of magnetic fields towards polar regions around the time of maximum.

5.5.5 *Generation of Magnetic Fields*

The question of how the Sun produces its magnetic field and goes through its 11-year sunspot and 22-year magnetic cycles had been one of the prime problems in solar physics. Based on some kind of dynamo motion, several investigators have proposed theoretical and empirical models to explain the solar magnetic field and its generation. The first conceptual, empirical and kinetic model was based on observed phenomena and put forward by Babcock (1961). Babcock's model proposes 5 stages of development through which the Sun undergoes the 22-year magnetic cycle. A schematic diagram, given in Figure 5.31, illustrates these stages of magnetic field development.

In stage 1, about 3 years before the beginning of a new cycle, at the time of minimum solar activity, there exists a north-south dipole poloidal magnetic fields of about 8×10^{21} Maxwells flux and a very weak field strength (~1-2 gauss). The magnetic field lines lie below the solar surface in the meridional N-S plane. Field lines emerge from the surface only above ± 60° latitude. These field lines emerging from the polar regions may extend well into the corona.

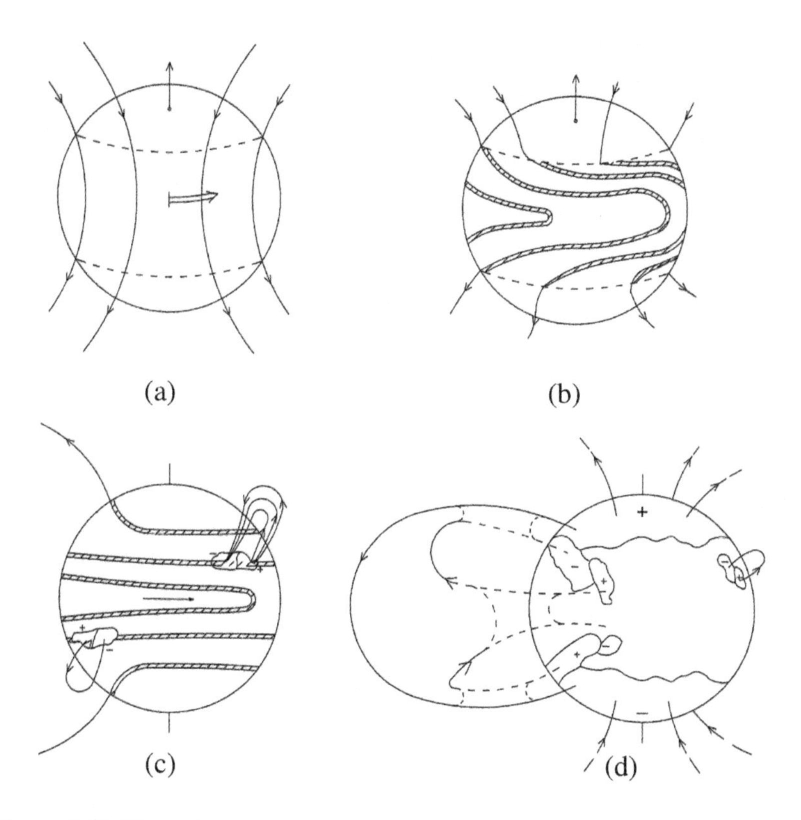

Figure 5.31 Illustrating Babcock's model (a) showing the schematic of dipole poloidal magnetic fields, (b) due to differential rotation the field lines are drawn inside the sun, (c) at the time of solar maximum, bipolar spots appear in the two hemisphere regions at higher latitude, (d) showing expanding lines of force above older bipolar magnetic regions, severing & reconnecting lines of force.

During stage 2, the submerged field gets intensified as the poloidal field lines are stretched inside the Sun's surface by rotating plasma which carries the magnetic field lines and winds them up. This occurs because the field lines are "*frozen-in*" the solar plasma. Stretching of the frozen-in lines transfers kinetic energy of the Sun's differential rotation energy and intensifies the field. As the rotation rate at the equator is higher, the poloidal field lines are progressively drawn out into an increasingly east-west direction towards the equator as shown in Figure 5.31 (b). Babcock showed that in 3 years differential rotation would wrap the low-latitude field lines about 5 times around the Sun, and the field would intensify to several hundred gauss. Further the field intensification occurs due to the gradient of the Sun's differential rotation with depth and convection, which results in twisting of flux tubes in which the field lines are embedded. This twisting results in discrete *braided flux ropes* of much higher magnitude than the few hundred gauss that may be produced by toroidal field from differential rotation alone. The flux tube ropes achieve field strengths of several thousand gauss intensity and drift towards lower latitude active region zones, where they emerge as bipolar regions. Emergence of *flux ropes* takes place by combination of *magnetic buoyancy* and the kinking of flux tube if they are twisted beyond a certain point like a twisted rubber band.

In the 3rd stage, the Babcock's model describes the formation of active regions from the emergence of flux ropes, in Figure 5.31 (c). Each of the 'Ω' shaped flux tube erupting through the solar surface will produce a bipolar active region with preceding p- and following f-polarity regions in one hemisphere and opposite polarities in the other hemisphere. This scenario explains Hale's polarity law. Babcock showed that the observed time scales of about 8 years between the first appearance of activity at about $\pm$ 30^{0} latitude and the last one at low latitude, can be reproduced using the observed differential rotation rate. The initial poloidal field of 8x10^{21}Maxwell is amplified about 3 times by the differential rotation in this time period.

In the 4th stage, Babcock's model describes the neutralization and reversal of the Sun's poloidal field. As the solar cycle proceeds, the following f-portion migrates towards the nearest pole and the preceding p-portion towards the equator. This results in neutralization of the existing polar fields of the previous cycle by the f-polarity regions of the existing (present) cycle, while the p-polarities in the two hemispheres cancel near the equatorial latitudes. Further pole-ward migration of fields leads to replacement of older fields by the new ones of opposite polarity. In Figure 5.31 (d) illustrates how the earlier cycle's polar fields are first cancelled and are replaced by the new fields. As the p- and f- active regions separate over a given hemisphere, the field lines also rise into the corona. The reconnection of field lines leads first to a cancellation of fields and then the replacement by new poloidal field lines connecting the two hemispheres.

In the 5th stage of Babcock's model, the poloidal field reverses in about 11 years after the beginning of stage 1. The pole-ward ends of the submerged fields may remain, after the emergence of the active regions, which are expected to have the polarity of the previous cycle. This would be opposite to the previous cycle's polarity orientation of the active regions. Latitudinal differential rotation will then act first to straighten out the surrounding field lines and eventually wind them up again during the second half of the 22-year cycle, but with opposite polarity.

Babcock's solar kinematics model based on the solar differential rotation explains most of the observed phenomena, such as the Hale's polarity law, Joy's law of tilt of the axis of bipolar regions, reversal of polar fields and the 22-year magnetic cycle. Since Babcock's pioneering model on generation of solar magnetic field, several models based on the solar dynamo have been proposed. From the recent helioseismology results, some scientists have proposed that around 0.7 $R_{\odot}$ from the Sun's center the solar magnetic fields are some how generated through a dynamo mechanism near the base of the convection zone. Here the sound speed and density profiles show a distinct sudden 'bump' called the *tachocline,* (see Figure 9.6). It is not yet understood how magnetic fields are generated deep inside the convection zone and emerge to the surface. Further no one has tried to explain the absence of sunspots during 70 years of Maunder minimum from 1645 to 1715 AD. How can the solar dynamo be switched off for 70 years?

5.6 Solar Prominences and Filaments

Prominences seen above the solar limb during a solar eclipse and outside the eclipse presents one of the most fascinating sights of the Sun. They have been seen for hundreds of years and systematic observations go back to atleast 125 years. Perhaps the earliest prominence observation was made by Muratori in 1239, during a total eclipse. He reported burning "hole in the moon", as mentioned by Father Secchi in his book, *Le Soleil*. This observation was interpreted as a prominence. Medieval Russian chronicles also mention descriptions of 3 or 4 prominence sightings by Vassenius during May 2, 1773 eclipse. He called them 'red flames' and believed them to be clouds in the lunar atmosphere. With the introduction of photography and spectroscopy since 1860, observations and study of prominences have come a long way.

Bright emission structures seen in strong chromospheric lines on the solar limb are called prominences and have been know for many years in the past as *'protuberance'* in French. But the same features, when seen against the disk, appear as dark filaments. To distinguish between the limb and disk features, prominences seen on the disk are called *filaments*. Prominences are sheets of relatively cool and dense material at a temperature of about 6000-10,000 K. Much higher temperature prominences are also seen in EUV radiations. They are somehow embedded in the surrounding hotter corona and have magnetic fields of the order of 5-10 gauss.

5.6.1 *Classification of Prominences*

Prominences and filaments appear in all variety of shape and sizes, ranging from a few thousand kilometers to almost the diameter of the Sun. In Figure 5.32 an extended dark filament is seen stretched from one edge of the Sun to the other and measuring nearly 100,000 kilometers. Some prominences may be visible only for a few hours, while others may persist for several months, and occasionally they suddenly erupt and magically even re-form almost in the same shape and position. Actually, the term prominence is used to describe a variety of objects, ranging from quiescent, active, eruptive, surges, sprays loop prominences etc. Prominences have been classified in several ways by number of authors, but the most salient characteristics can be briefly described by Pettit's classification, given Table 5.1 The prominence classification mentioned in this Section is based on the nature and degree of motion. Depending on the speed of the prominence material, it is described as an active or quiescent prominence.

Figure 5.32 Full disk filtergram taken in Hα at the Big Bear Solar Observatory on July 16, 2002, displaying a huge extended dark filament measuring nearly 100,000 kilometers long.

Table 5.3 Pettit's prominence classification

Class	Name	Description
I	Active	Material from prominence streaming into nearby active centers like sunspots.
II	Eruptive	Whole or part of prominence ascends with uniform velocity of even several hundred km/s. The velocity may at times suddenly increase.
III	Sunspot	These are found near sunspots often in the shapes of loops or fountain.
IV	Tornado	These are rather rare events, appears in the from of vertical spiral of a closely wound rope like structure.
V	Quiescent	These are large prominences, displaying only minor charges over period of days, and appear as thin vertical sheets consisting of fine filamentary structure.

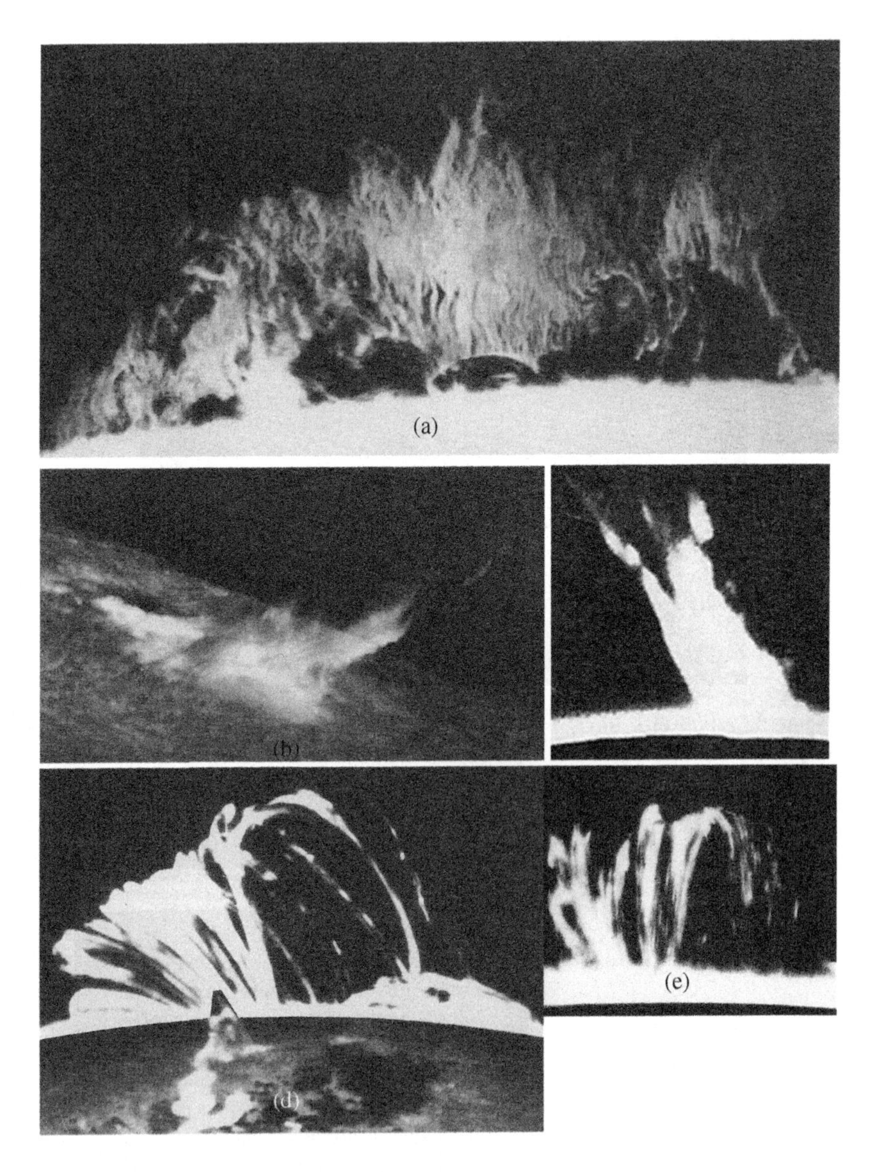

Figure 5.33 (a) A large quiescent prominence showing fine filamentary structure appears as vertical sheet of gases supported at number of places on the solar surface, (b) Surge prominence erupting out of an active region, (c) Spray prominence (d & e) Post flare-loop prominence with feet of loops anchored in the sunspot region.

5.6.2 *Filaments*

Filaments are essentially prominences seen as absorbing features on the disk of the Sun. Thus from the observations of filaments, distribution of prominences over the whole disk can be studied. Filaments are found preferentially in two latitude belts in the two hemispheres - one in higher latitudes near the polar regions and other in active mid-latitudes.

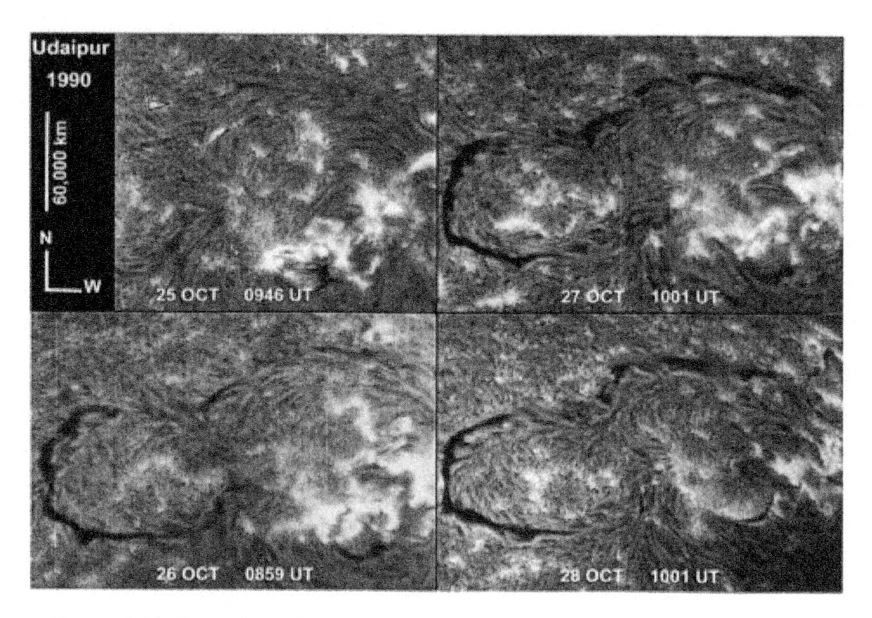

Figure 5.34 Formation of a filament in a magnetic neutral region between two opposite polarity regions.

Generally filaments are found along or within a longitudinal magnetic neutral line, between two opposite polarity regions. However not all neutral magnetic field regions are marked by filaments, but the possibility of condensing or formation of a filament in such regions is generally high. In Figure 5.34 is shown a sequence of filament formation near an active region. Filaments are formed in what is now known as *filament channel*, which is essentially a neutral magnetic field region between the two opposite polarity regions.

5.6.3 *Quiescent Prominences*

Although 'quiescent' would mean that there is no activity, in fact high resolution Hα movies show that there is considerable mass motion in such prominences. Both quiescent prominences and filaments on the disk show fine filamentary structure and Doppler shifts both to the blue and red, suggests material motion in loop like flux tubes. In prominences and filaments the solar material is concentrated in thin rope-like structures of diameters less than 300 km. Horizontal motions on the order of 10-15 km/s have been observed and also there is considerable internal motion. These motions play important roles in their stability and can be precursors of major disturbances. Valnicek (1968) has observed

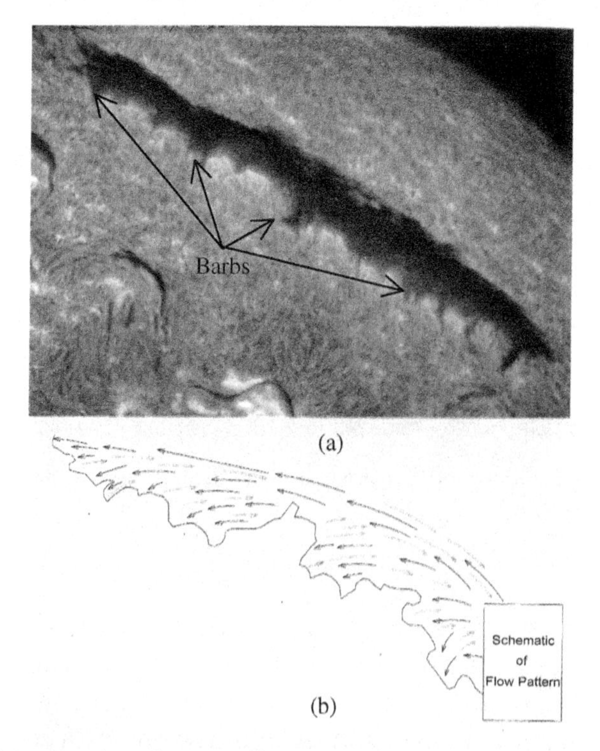

Figure 5.35 (a) Showing a large Hα filament observed on 10 April 1990 at Udaipur Solar Observatory(USO) with number of 'barbs', they seem to be 'hooked' in the chromosphere–photosphere layers. Notice conspicuous brightening underneath the filament, (b) Schematic drawing showing counter streaming motion along the spine and barbs of a filament.

internal plasma motions up to 300 km/s velocity. The observed untwisting motion of interwoven helical structures no doubt speaks of the magnetic field configuration in the prominences. Filaments also display up-down and side-ways oscillatory motion, with a periods of 3 to 15 minutes and an amplitude of about 1-6 km/s. Occasionally much longer periods and speeds of 50-80 km/s have been observed. Such filaments are called as '*winking*' filaments. This effect may arise from motion in the line of sight as the filament is Doppler shifted in and out of the pass band of the Hα filter.

Bright rims just underneath (see Figure 5.35a) the prominences were first noticed by Royds (1920) and later discussed by d'Azambuja in 1948. One of the possible explanations for such bright rims may be, as a result of local heating of the chromosphere due to magnetic reconnection under the quiescent prominences as suggested by Engvold (1988).

Filaments also display downward or rising slow streaming motion along the '*spine'*, axis of the filament and along the 'barbs' features, which are supposed to support the filaments. Engvold (1976) had observed motion of the order of 5-10 km/s. Recently, Zirker *et al.*, (2001) have reported observing such mass motions, as shown by a schematic in Figure 5.35(b).

5.6.4 *Disparition Brusque*

Occasionally a quiescent prominence, either on the disk or limb suddenly undergoes an ascending motion, with very high velocity on the order of few hundred km/s. This phenomenon was perhaps first observed by Deslandres in 1889 in France and is termed as *disparition brusque,* to distinguish it from eruptive prominence. Prior to occurrence of disparition brusque, the prominence material generally displays increased random velocities of the order 30-50 km/s, then suddenly whole or part of the prominence starts to ascend with increasing velocity, attaining velocity in excess of the escape velocity of 618 km/s at the photosphere, and to about 400 km/s at a height of 100,000 km in the corona. Recently disparition brusques are known to be associated with Coronal Mass Ejections.

The velocity of escape from the Sun is given by the expression:-

$$V_e = \sqrt{\frac{2GM_o}{h+R_o}}, \qquad (5.8)$$

where $M_\odot$ is the solar mass, $R_\odot$ is the radius, and G is the gravitation constant and h the height above the photosphere.

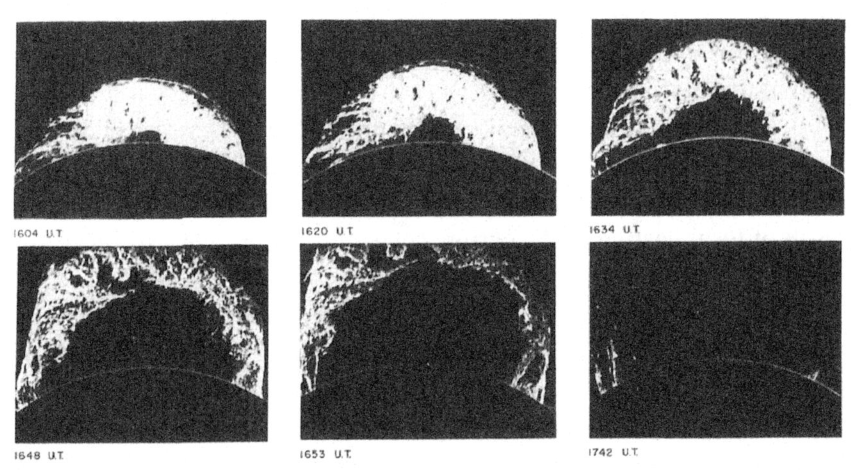

Figure 5.36 Disparition brusque observed on 4 July 1946 at the Climax station of the High Altitude Observatory. This event has been called 'Grandpa' and may be the largest prominence ever recorded.

One of the grandest examples of disparition brusque was recorded on 4 July 1946 at the Climax station of the High Altitude Observatory, the sequence of this event is shown in Figure 5.36. Initially this prominence was a small quiescent one; within a few minutes it suddenly developed into a huge eruptive prominence covering almost the diameter of the Sun, and rising with a speed of several hundred kilometers per second, and vanished within an hour and half. In addition to the violent ascending motion, the prominence material also displayed spiraling motion and helical filamentary structures. The manifestation of spiraling and helical structures reveals the interaction of magnetic field with plasma. From such observations it is well established that the photospheric magnetic field plays an important role for support and stability of prominences. How a prominence stays quiescent for a long time and then suddenly erupts, is a subject of continuing interest. In some cases a filament may disappear for a few hours or days, and then re-form in almost at the same location and shape. Disparition brusque have been also observed in high temperature EUV lines such as ionized CII, CIII, OVI, MgX etc. Mouradian and Soru-Escaut (1994) call them *thermal* disparition brusque, which may or may not be related to solar flares or triggered by some kind of instability in the solar atmosphere.

5.6.5 *Active Prominences*

All active prominences, such as surges, sprays, and loops, display fast structural changes and violent motion. They have life times of minutes to hours. Active filaments (prominence on the disk) near or in active regions are related to spots and flares, and show significant mass motion on the order of 30 km/s. These motions are often the precursor to flares.

5.6.6 *Loop Prominences*

These are among the most beautiful manifestations of magnetic flux tubes extending in the corona, as shown in Figure 5.33 (d, e) in Hα. Loop prominences are intimately related to flares and are often referred as post-flare loops. Loops are formed during the declining phase of a two ribbon flare, and appear to be connecting the two strands of the flare. The 'legs' of the loops can be traced back to sunspots of opposite polarities. From the top of the loops, material streams down along the two legs but without the spiraling motion as seen in other types of prominences. A loop system may last for several hours during which they expand and may reach typical height of 50,000 km or more. They are seen progressively forming higher and higher over post-flare active regions. Arch-loop systems have been also seen in the EUV and in soft X-ray emissions as shown in Figure 5.37. This picture was taken by TRACE at 195Å on November 4, 2003 at 22:35 UT. The foot points of the loops demark the flare ribbons near the solar limb and the tops of the loops show enhanced brightness. Another similar event (see Figure 5.43) was observed on the disk by TRACE on 28 October 2003, wherein a long chain of post-flare loop system arcades were seen extending to several hundred thousand kilometers on the Sun. This event was associated with

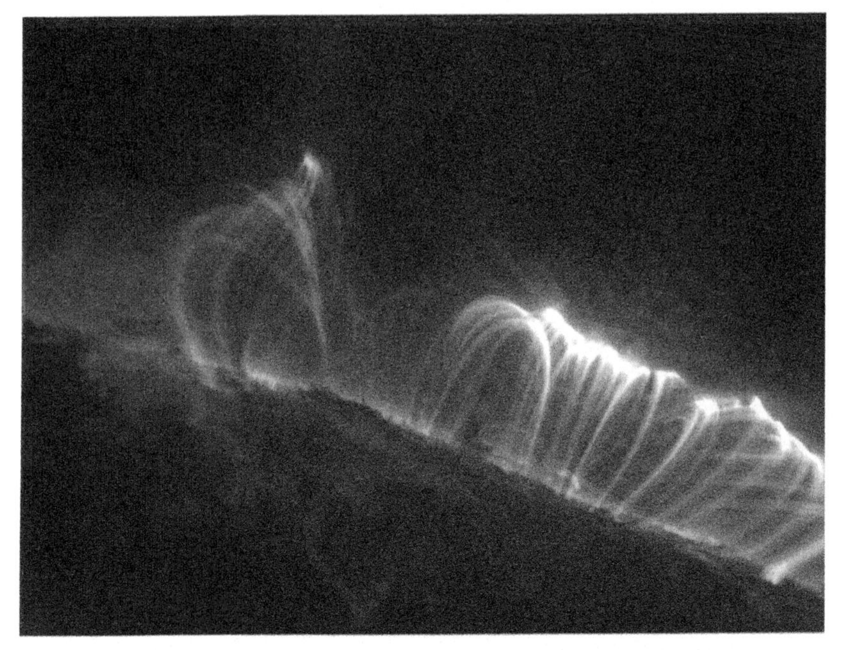

Figure 5.37 A post flare arcade-loop system on the limb observed by TRACE at 195Å on November 4, 2003 after an intense X-ray flare. The foot-points of the loop system seem to demark the flare ribbons and the material seen draining down from the loop top along the legs.

an intense X4.2 two ribbon and filament eruption. The two sets of loops, one seen in the optical region in Hα, and other in EUV, refer to two very different temperature regimes. In Hα it is about 10^4 K, while in the EUV it is nearly 1.4×10^6 K. How do we reconcile such large temperature differences in the same, or nearly same, physical features?

5.6.7 *Eruptive Prominences*

These types of prominences also come under the 'active' prominence class, as they are associated with some kind of activity on the solar disk which triggers them. Cinematographic techniques first introduced for taking prominence observations by McMath at Michigan, and Lyot in France, revealed interesting structure and dynamics of eruptive prominences. At Mount Wilson Observatory, Pettit studied eruptive prominences in detail. He and later workers have found that prominences do not ascend uniformly but in 'jerks'. Initially a prominence shows some acceleration which may change as it attaints height in time. Prominences also show twisting and untwisting motions in their structure, which is revealed by spectroscopic and 2-dimensional Doppler images taken in the wings of Hα line.

A very spectacular eruptive prominence event was observed on 14 January 1993 on the eastern limb of the Sun, in Hα filtergrams taken at the Udaipur Solar Observatory, and by the YOHKOH Soft X-ray telescope. The sequence of development of this event in Hα and in soft X-ray is shown in Figure 5.38 (a, b). The eruptive prominence started from a quiescent bright mass of material, seen a few days before the event on 14 January. Filtergrams taken around 05:04 UT show that the main mass of the prominence seen as a 'hook' was pointed towards the south as shown in Figure 5.38 (a), while a picture taken at 06:58 UT shows that the 'hook' had completely turned 180 degrees around and was pointing towards the north. This distinctly indicates a rotational or *un-twisting* motion around its vertical axis. Initially the material rose slowly with a characteristic velocity of about 100-400 km/s, but after reaching nearly 350,000 km in height, it suddenly accelerated and shot out with a velocity of more than 1200 km/s to attain a height of about 650,000 km. As the material rose it slowly disintegrated into fragments. All along during the rising phase of the eruptive prominence it displayed rotational motion. The whole phenomenon lasted for nearly 2 hours. Simultaneous observations are available from YOHKOH in soft X-rays, they reveal untwisting motion and large-scale kinks and loop structures as in Figure 5.38 (b). This event started at about 6:00 UT, and 10 images are available from YOHKOH between 06:15-07:03 UT. The rectangular shaped soft X-ray loops display a kink formation at the loop tops. As the X-ray loops grow in height they seem to 'untwist' and a bubble shaped structure appeared near the top, as seen on the picture taken at 06:55:01UT. The soft X-ray loop features delineate the magnetic lines of force in the corona. And what we see in this sequence of images is the stretching of magnetic lines of force by the eruptive prominence. The ascending prominence material pulls along with it the embedded magnetic field in the corona, as displayed by strands of X-ray loops. The build–up of magnetic energy occurs by twisting of the quiescent filament along with

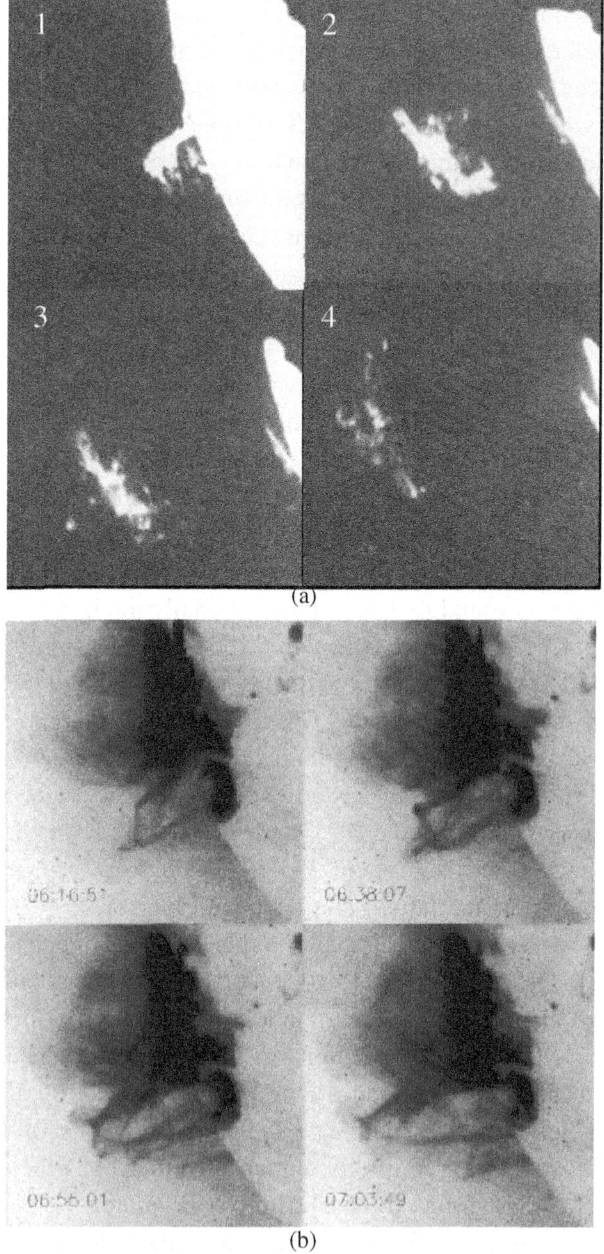

Figure 5.38 (a) Eruptive prominence in Hα on 14 January 1993 taken at USO, 1 at 05:04, 2 at 07:13, 3 at 07:25, 4 at 07:40 UT, (b) sequence in soft X-ray, observed by SXT on Yohkoh.

its embedded magnetic field. Some kind of instability develops and suddenly the prominence 'untwisted' and erupted. Somewhat like a twisted rubber band, it violently untwists when released. Comparing and over-laying the soft X-ray images and Hα filtergrams, it is clearly seen that the prominence material was *sitting* on the X-ray loop tops and rose with them, which ultimately was blown off to higher heights. Up to certain heights, the prominence material was perhaps constrained by the magnetic field. Thereafter it got loose and shot out at a high speed as the *close* magnetic loops opened up. This interesting event has been studied by Wahab Uddin and Bondal (1996) and a correlation study with YOHKOH data has been made by Bhatnagar *et al. (1996).*

5.6.8 *Surges and Sprays*

Surges are straight or slightly curved spikes which are shot out from a small bright flare mound at a velocity of more than 100-200 km/s. They can reach heights more than 200,000 km in the corona before fading out. Most disk surges initially start as an emission feature, but then turn to absorption. Many small surges originate near penumbral borders and are directed radially away from the spot. This means that surges are directed along magnetic field lines. Surges show a tendency to recur at the same place at a rate of about 1 per hour or so. This indicates that whatever the source and triggering mechanism, they reform in quick succession. Various mechanisms have been proposed for explaining the initial acceleration of surges. In all likelihood it is the underlying magnetic field which provides the initial energy. The mass and energy contained in large flare associated surges is 10^{12}-10^{13} kg (particle density 10^{11} -10^{12} cm^{-3}) and 10^{30} ergs respectively. A picture of a typical surge in Hα is shown in Figure 5.33(b).

Sprays are much more violet flare associated event, wherein the flare material is ejected out, which frequently disrupts into fragments. After an initial high acceleration of a few km sec^{-2}, sprays reach high velocities on the order of 500-1200 km/s within a few minutes, and may reach heights greater than 600,000 km. After the initial phase, the spray material decelerates and may spread over a large volume in the sky. Sprays are of two types; one, the *flare-spray* type is associated with flares which emanate from a rapid explosion-like expansion of a flare. On the solar limb, such flare-sprays appear as expanding bright mound which suddenly disrupts and material flies into space as shown in Figure 5.33c. The development of such sprays suggests that flare plasma is originally constrained in a closed magnetic field configuration, which bursts due to increased kinetic energy density, and the material escapes into the corona. Helical structure also has been seen in sprays, which indicates that magnetic field lines permeate the prominence plasma.

The other type is the *prominence–spray*. Sometimes active region filaments erupt and are rapidly driven away as a spray during the flash or maximum phase of a nearby flare. This type of spray is less energetic as compared to flare-associated spray.

5.7 Support and Stability of Prominences

Prominences and filaments can last from weeks to months on the solar disk and then suddenly erupt. How can such large volume and mass of material at the relatively low temperature of 8,000 - 10,000 K while

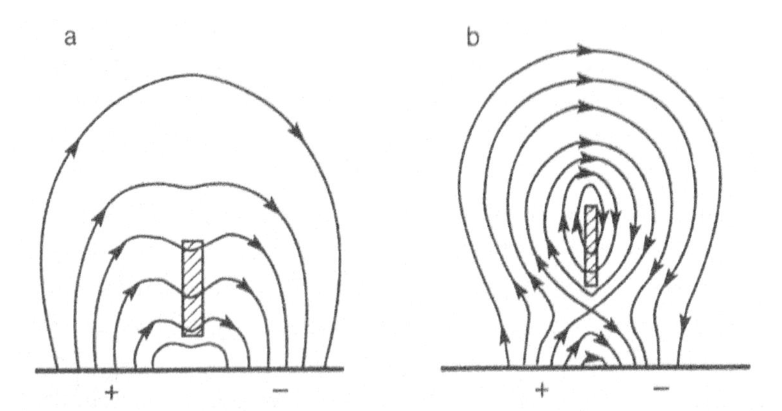

Figure 5.39 (a) As per Kippenhan-Schluter's model with embedded prominence (hatched area) in *normal magnetic field topology*, (b) as per Kuperus-Raadu's model with *inverse polarity* configuration.

embedded in a million degree coronal plasma, survive for months and balance against the solar gravity? These questions have been puzzling many theoreticians. Several plausible mechanisms have to explain the support, stability and eruption of prominences. In all theories it is the magnetic filed that plays the primary role to maintain the equilibrium of prominences. A static balance between the magnetic pressure - $B^2/8\pi$ and the weight of the overlying prominence material, maintains the equilibrium. When this equilibrium is disturbed, the filament erupts.

Two main magnetic field topologies for supporting prominences have been suggested. One is by Kippenhahn and Schlüter (1957), the other by Kuperus and Tandberg-Hanssen (1967), which was further developed by Kuperus and Raadu (1974). In Figure 5.39 (a) is shown a schematic of the magnetic field topology as per Kippenhan-Schluter's model (K-S model). In this model the prominence material sits at the top of field lines and is supported by a *normal polarity field*. In the Kuperus-Raadu (K-R) model, as shown in Figure 5.39 (b), the prominence is embedded in an *inverse polarity field*.

The other question about prominence physics concerns how the magnetic field is oriented in and around the prominence body. In the simplest case, a prominence is considered as a sheet of plasma, standing vertically in the corona, above a magnetic neutral line. The field lines emanating from one polarity region in the photosphere enters the prominence and exit from the other end to join region of opposite polarity. However, observations indicate that the magnetic field vector is not perpendicular to prominence sheet, but makes a small angle between the long axis of the prominence and the magnetic vector, as shown by Tandberg-Hanssen and Anzer (1970) and by Kim *et al.* (1988). Prominences at the limb and filaments on the disk show that they are *supported* by a series of *legs*, which appear as supporting pillars for the prominences. These legs are anchored in the low chromosphere and photosphere. Mass motions have been also observed at these 'foot-points' and along the legs. It is pointed out by Forbes (1986) that the 'legs' or '*barbs*' of the prominence are co-spatial with supergranulation cell boundaries. As we have seen in Chapter 4, that at the intersection where 2 or 3 supergranules meet, the magnetic flux is considerably enhanced, and perhaps it is this extra magnetic field that anchors the legs and supports the prominence. Bhatnagar *et al.*, (1992) have shown from observations of a quiescent filament made during its disk passage, that it remains stable until at least two legs are available to support the filament. It is seen that slowly the legs somehow disintegrate, leaving behind just one or two legs. When only one leg is seen attached to the prominence the chances of its eruption becomes very high and invariably it erupts shortly thereafter. The triggering agency for filament eruption can be due to some sort of disturbance in the magnetic field topology or by its own imbalance between magnetic pressure and the kinetic energy in the filament.

5.8 Solar Flares

Solar flares are among the most interesting and widely studied phenomena on the Sun. Flares occur suddenly, releasing enormous energy on the order of 10^{26} to 10^{32} ergs in a very short time of 100 to 1000 seconds. They emit over a wide range of radiation extending from radio, visible, extreme ultraviolet, X-rays, γ–rays and particle emission. Flare events are closely related to our Earth's environment and their effects echo throughout the solar system. A single solar flare can create an explosion equivalent to several billions of hydrogen bombs each of 100 megaton TNT destructive power, exploded simultaneously. All this energy is initially generated in a relatively small region on the Sun, occupying less than even 0.01% of the solar disk. Actually, for a short time flares can be the seat of the hottest place on the Sun, reaching tens of million of degrees. Although in the visible region, the percentage increase in the over-all integrated visible flux is hardly perceptible, but at short wavelengths and in radio waves it is indeed significantly enhanced. Flares are rarely visible in white light at photospheric level, but in chromospheric, EUV and X-ray wavelengths, they display enormous release of energy and structural changes. In the radio region, their effect is marked by various types of emissions. In white light only very energetic flares are visible. The earliest observation of a white light flare was made on 1 September 1859, independently by two Englishmen, Richard C. Carrington and Richard Hodgson, who were observing the same active region at the same time. Ancient Chinese records of sunspot sightings in the seventeenth century suggest that white light flares were perhaps observed along with sunspots, as mentioned in Chapter 5.1.1.1

5.8.1 *Flare Classification*

Flares appear in various shapes, sizes and temporal scales and release a wide range of electromagnetic and particle emissions. A flare in visible region may be defined as a sudden transient increase in brightness of a pre-existing plage region to at least two times the normal chromospheric intensity. It is accompanied by an increase in the X-ray and radio flux. In Hα flares generally appear either as a compact extremely bright region or as a two or multi-ribbon structure. The flare intensity suddenly increases within a few seconds to minutes and this is known as 'impulsive' phase. After reaching peak intensity, it declines rather gradually. However, some flares may show a gradual increase to a maximum intensity and a much slower decline. Flares are ranked in importance depending on the intensity of their emissions in optical, radio or X-ray radiation. For example, in the optical region where flares are generally observed in Hα, the *importance class* is decided by corrected (due to foreshortening) peak flare area and estimated peak Hα brightness; in the case of radio emission by the flux intensity at 5000 MHz frequency range, measured in *solar flux units (1 sfu=* 10^{-22}w/m^2/Hz = 10^4 jansky), and in the X-rays by the peak intensity recorded in 1-8Å spectral band in *watts/m*^{2}. The Hα 'importance' class of flares is designated by letter S (sub-flare), and numerals 1, 2, 3, 4, in increasing order of their importance and depending on the corrected peak area of the flare, a suffix F, N, B is added if the flare brightness is faint, normal or bright. The Hα flare area is used, as it can be easily measured and there is a fairly good correlation between the area and other flare related effects. However, this classification does not indicate the impulsive or gradual nature of a flare. Flares could be quantitatively classified better in radio and X-ray radiations as flux can is more accurately measured than area. Systematic and continuous high precision flux measurements are available in 0.5–4Å and 1-8Å ranges from Geostationary Operational Environmental Satellites (GOES) of NOAA. In Table 5.4 is shown the 3 types of flare classification from optical Hα images, radio data at 5 GHz, and soft X-ray flux at 1-8 Å.

Table 5.4 Flare classification as per Hα, radio emission & soft X-ray flux.

Hα classification			Radio flux at 5000 MH_Z** in s.f.u.	Soft X-ray class ***	
Importance Class*	Area (Sq. deg.)	Area 10^{-6} solar disk		Importance class	Peak-Flux in 1-8 Å w/m^2
S	2.0	200	5	A	10^{-8} to10^{-7}
1	2.0-5.1	200-500	30	B	10^{-7} to10^{-6}
2	5.2-12.4	500-1200	300	C	10^{-6} to 10^{-5}
3	12.5-24.7	1200-2400	3000	M	10^{-5} to 10^{-4}
4	>24.7	>2400	3000	X	$>10^{-4}$

Note: *The 5- Hα importance classes are sub-classified into 3 sub-classes as F (faint), N (normal) and B (bright) depending on the peak intensity of the flare. For example a 3B designated flare would mean that it covered an area between 12.5 - 24.75 sq. deg. & was bright (B).

**Radio flux measured at 5000 MH_Z (6-cm) are given in solar flux units 1 sfu = 10^{-2} W/m^2/Hz= 10^4 jansky.

***S X R importance class is indicated by a letter (A, B, C, M or X) followed by a number, the value of the measured flux, e.g. if the peak S X R flux is $5.2*10^{-4}$ W/m^2 it will be designated as X5.2.

5.8.2 *Temporal Characteristics of Flares*

In Figure 5.40 is shown the temporal characteristics of a typical flare giving rise to all kinds of radiation. From these curves it will be noticed that soon after the pre-cursor phase, which may last for 2-5 minutes, a very conspicuous impulsive phase occurs in all ranges except in the thermal visible and soft X-rays. This impulsive phase may last for a few seconds to a minute, consisting of energy release in burst of microwave (~3 GHz), EUV ($\geq$ 30 Kev), hard X-ray ($\leq$ 30 Kev) and γ-rays (1Mev), followed by a gradual phase lasting for several minutes to hours depending on the intensity of the flare. The visible and soft X-ray emission profiles show extended and gradual time variations as compared to emission at other wavelengths. Various types of radio emissions, e.g., Type I, Type II, Type III and moving Type IV are generated following the flash phase. Even high energetic particles are also released during intense flares.

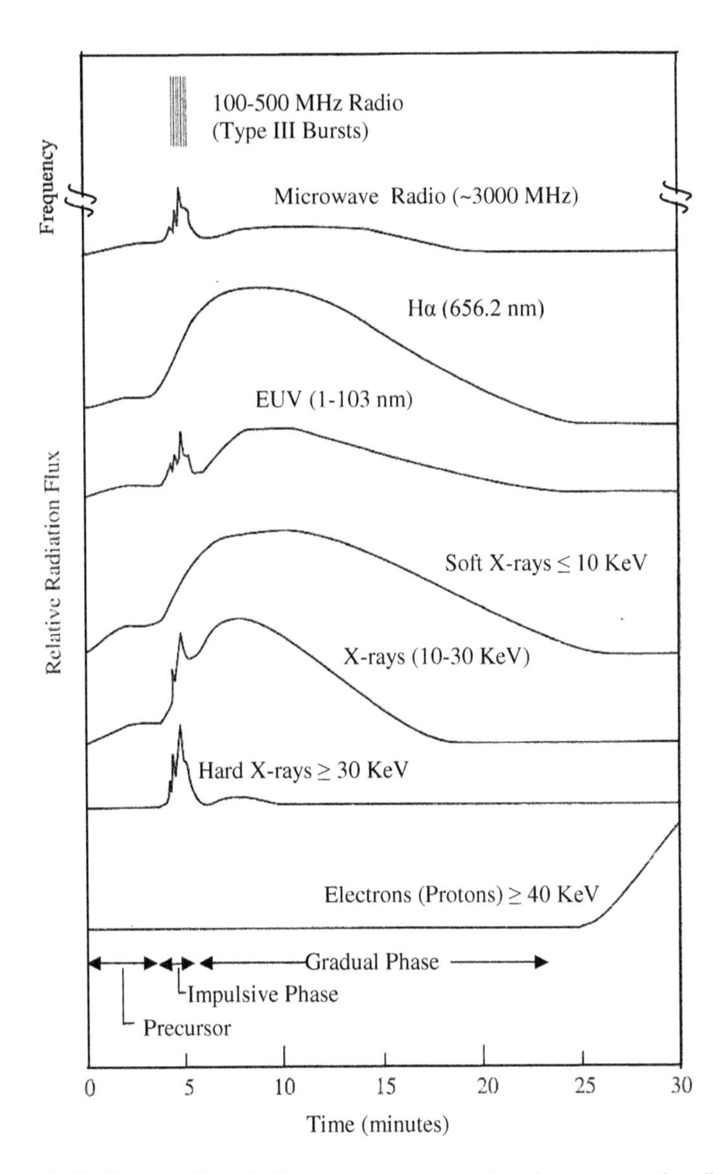

Figure 5.40 Time profile of flare energy release at various types of radiation, extending from radio, microwave, visible, EUV, soft and hard X-rays. A distinct impulsive phase lasting for a few seconds is seen in all wavelengths except in optical and soft X-rays. The optical and soft X-ray emissions may be considered as an 'afterglow' of an impulsive flare.

5.8.3 *Optical Flares*

In the optical region, flares are best seen in strong chromospheric lines such as Hα and the H & K lines of CaII. They are called chromospheric flares. They appear as a sudden brightening of a plage region. As all flares occur in strong magnetic field regions, it follows that it is the magnetic energy that fuels them. How the magnetic energy converts to flare energy is a subject of continuing study. There are several competing theories to explain the generation of solar flares and energy release. The basic idea of all theories is that the magnetic field configuration in an active region becomes 'stressed' or 'sheared' which causes the built–up of magnetic energy that is 'stored' in the corona. Through some kind of instability or perturbation in the region, the 'build-up' or 'pent-up' energy is suddenly and explosively released at a higher level in the solar atmosphere. This release of energy manifests itself as sudden brightening of a small active region on the Sun. In the optical domain this is known as the *flash* phase, which lasts for only a few minutes. Generally, this phase is preceded by a 'pre-cursor' or pre-heating phase or thermal phase, wherein the plage region may display slightly higher brightness than the surrounding region, but not bright enough to be marked as a flare. The flash phase frequently includes an *explosive* or *impulsive* phase which is a very sudden increase in brightness (within ~1 minute) of a small part of the flare and sometime is accompanied by rapid expansion of the flaring region. In the case of very energetic flares this impulsive phase coincides with hard X-ray and microwave bursts. The duration of optical flares ranges from several minutes to few hours.

Flares develop in 3 phases; first a rapid rise in intensity, then a flat plateau, followed by a gradual decline. In almost all flares the rise to maximum intensity in Hα is much more rapid as compared to their decline. Some flares may show a pre-flare or pre-heating phase also, just a few minutes before the start of the main flare. The Balmer and other strong chromospheric lines (H & K lines of CaII) are very broad in flares, up to 10 Å as seen in Figure 5. 41.

The optical flares appear either as compact bright regions or as two or multi-ribbon flares, located between regions of opposite magnetic polarities, and separated by a magnetic neutral line. Compact flares give

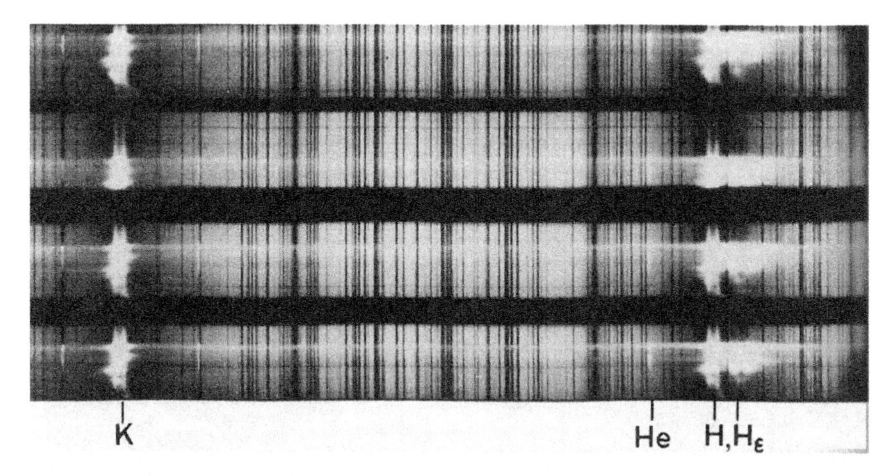

Figure 5.41 Flare spectrum in the region of H & K CaII line, showing wide emission wings of H_ε and the H & K lines with emission in many other lines (with Permission from D. Reidel Publishing Company).

rise to strong surges and sprays, and shoot out material at high speeds of the order of 100-600 km/s to large distances in space.

5.8.4 *Two Ribbon Flares*

Many intense flares develop as pair of bright ribbons; they start on either side of a main inversion line of magnetic field of opposite polarities in an active region. The flare strands spread parallel to the neutral line. As the flare progresses, the two flare ribbons move apart and during the declining phase, the space between them is filled with higher and higher bright post-flare loops. An intense two ribbon flare of 4B/X17.2 class was observed in the Hα line at Udaipur Solar Observatory (USO) on 28 October 2003, in active region NOAA 10486. In Figure 5.42 is shown a sequence of development of this energetic flare. As the flare progressed, the flare ribbons moved apart with more than 20 km/s speed and lasted for more than 4 hours. The dark filaments, demarking the neutral magnetic lines in the active region showed considerable motion and finally erupted as shown in frame taken at 11:10:09 UT. As the flare declined, several post-flare loops started appearing joining the two opposite polarity regions on the two sides of the neutral line. In EUV

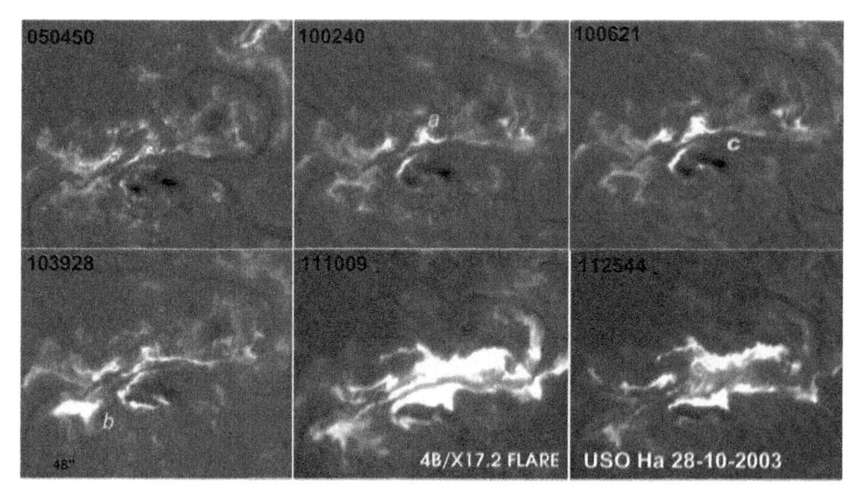

Figure 5.42 Showing development of a class 4B/X17.2 very energetic two ribbon flare observed at the Udaipur Solar Observatory on 28 October 2003, in NOAA No. 10486. Note that the flare ribbons separate as the flare progresses and the disappearance of dark filaments and appearance of loops joining the flare strand.

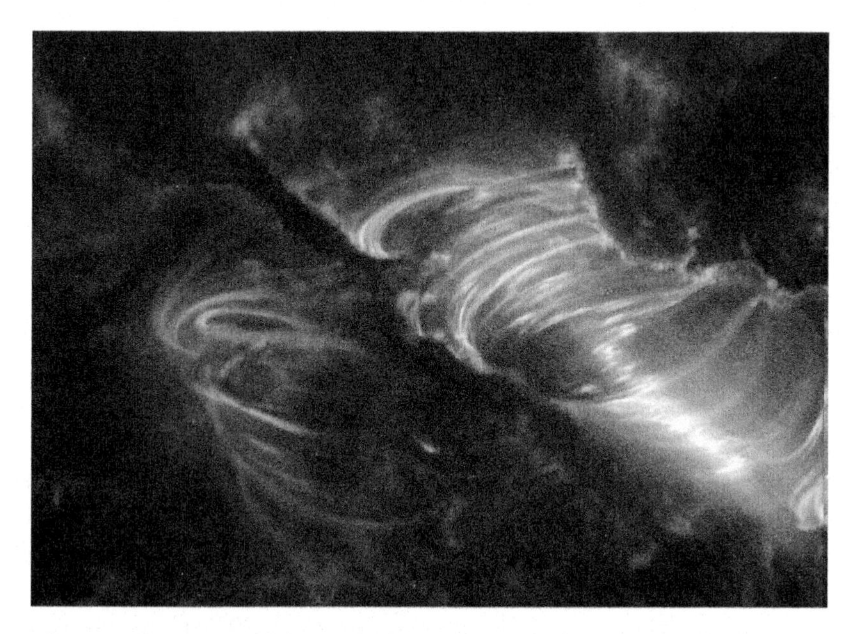

Figure 5.43 Corresponding image of the same flare (28 October 2003) observed by TRACE in 195Å wavelength showing beautiful post flare loops in EUV.

emission at 195 Å, beautiful post eruptive arcade loops associated with this flare were also observed by TRACE spacecraft, and is shown in Figure 5.43. Such intense flares release energy of the order of 10^{32} ergs. A large fraction of this energy output is distributed between the thermal soft X-rays and optical radiations, and a small part in EUV, non-thermal hard X-rays, radio emission and in accelerating charge particles up to GeV energies and kinetic energy associated with shock waves.

5.8.5 *Homologous Flares*

Occasionally certain flares occur repeatedly at short intervals at the same place and show strikingly similar pattern of structure and development. Such events are called homologous flares. As it is well established that magnetic fields and its configuration plays a major role in generation of flares, in the case of homologous flares it seems that the build-up, storage and release of magnetic energy is repeated rapidly, and the same magnetic field strength and configuration is restored by some unknown mechanism.

5.8.6 *Filament – Associated Flares*

Sometimes the disappearance of a filament (disparition brusque) in or close to a spot group results in the brightening of a two ribbon flare. Such filament associated flares include some of the largest and most energetic flares on record. Often these flares rise to a maximum intensity within 30-60 minutes and can last for several hours.

5.8.7 *Limb Flares*

Flares occurring on or near the solar limb offer a measure of their height in the corona. Generally they extend up to about 10,000 km in the corona and appear as bright cones or mounds. The higher parts of the flares have slightly different physical properties as compared to lower part visible on the solar disk. At higher heights, the electron densities are lower by an order of magnitude and electron temperature is higher by a factor of two [$n_e \approx 10^{11}$ -10^{12} cm^{-3}, $T_e \approx 15{,}000$ K (from Balmer lines)]. Sometimes such limb flares can be confused with bright prominences; however flares are rather static phenomenon as compared to prominences which display large scale motion.

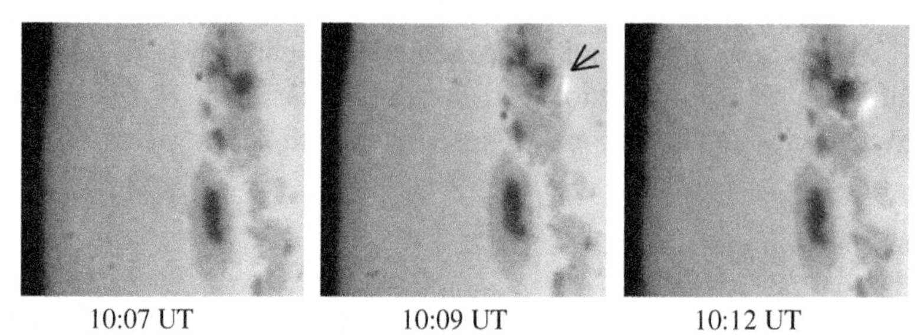

Figure 5.44 Development of a small white light flare observed on April 2, 2001 at the Udaipur Solar Observatory near a sunspot penumbra in active region NOAA No. 9393. Solar west is on the left and south on the top.

5.8.8 *White Light Flares*

In rare cases, a small portion of a flare becomes visible in white light for a short duration. Usually this is not for more than 5-10 minutes during the flash phase of an intense flare. White light flares consist mainly of one or two bright points or small areas equidistance from magnetic neutral line, close to or in the penumbrae of spots of opposite polarities. Maximum brightness in case of white light flares is about 50% higher then the photospheric intensity. In Figure 5.44 is shown a small white light flare observed on April 2, 2001 as two tiny bright points near a sunspot penumbra. Generally the energy emitted in the photospheric continuum from such flares is about 10^{30} ergs. These flares are known to be closely associated with the emission of hard X-ray and microwave bursts. It is believed that beams of energetic electrons and protons bombarding the lower solar atmosphere produce the observed continuum.

5.8.9 *Flare Associated Phenomena*

There are several optically observed chromospheric and coronal phenomena which are in different ways related to flares. Some of these are:-

1. *Filament activations.* There are 3 types of flare associated filament activations;

(i) It is observed that even before a flare begins, filaments in the vicinity or in the active region, show considerable motion and with the onset the flare may even disappear. This pre-flare activation is believed to arise from pre-flare changes in the magnetic field configuration in the surrounding active region.

(ii) Filaments in the region are frequently activated and disrupted during flash or maximum phase of the flare.

(iii) Some times distant filaments start oscillating. Known as 'winking filaments', this may be caused by the passage of wave disturbance originating in the flare.

2. *Flare ejections,* appear in the form of surge and spray prominences and fast ejecting material. In such cases the actual flare material is ejected during the flash or maximum phase of the flare. Often surges are seen as bright fast moving material which sometimes turns into absorbing features. Sprays have been observed moving out with the velocity of 100-500 km/s. Both surges and sprays are known to be associated with Type IV and Type II radio bursts.

3. *Post flare loops,* develop during the declining phase of an energetic two ribbon flare and appear joining regions of opposite polarities. They are most conspicuous in Hα, EUV, and soft X-rays. In Figures 5.37 and 5.43 is shown beautiful pictures of arcade of post-flare loops observed by the TRACE spacecraft in the EUV radiations at 195 Å. These post-flare loops may last for several hours, even after the decline of the main flare. Post–flare loops delineate the magnetic field lines and are filled with hot coronal plasma of more than a million degrees. Coronal material seems to condense at the top of loops and drain down along the legs of the loops to the foot points. The loops seem to condense from higher and

higher heights in the corona with time.

4. *Coronal Transients,* refer to sudden brightening or fading of coronal structures due to expanding, ascending or disrupting coronal arches etc. Such features have been observed in monochromatic and soft X-ray images as well as in inner and outer K-corona.

5. *Wave disturbance or Moreton waves.* Many flares in the optical region display an 'explosive phase', wherein a small area on the Sun suddenly brightens up and rapidly expands. Associated with such flares, Moreton (1960) and Moreton and Ramsey (1960) noticed in 'on-band' and 'off-band' Hα filtergram movies, made at the Lockheed Solar Observatory, that disk filaments were seen activated at even 50,000 km distance away from the flare site. Considering that the observed activation of the filament was due to some kind of traveling disturbance, Moreton and Ramsey estimated the velocity of the blast wave about 1000 km/s. Since then several workers have made observations in Hα line center, blue and red wings of Hα line, and observed conspicuous wave motions associated with flares. In Figure 5.45 is shown signature of a 'blast wave' propagating out from a flare active region with a velocity of approximately 700 km/s. This was triggered by a great flare of September 20, 1963 and accompanied by a surge and Type III and Type II radio bursts.

Recently, similar type of blast or shock waves have been also observed in images taken in EUV at 195Å observed from EIT aboard SOHO spacecraft and is shown in Figure 5.46. From these pictures it is seen that the blast wave moves and spreads out in a sector of about 90° or more with velocities ranging from 700-1000 Km/s and no apparent sign of deceleration. What could this phenomenon be? Is it a shock wave? But shocks dissipate their energy rapidly and would considerably decrease before reaching distances on the order of 50,000 km or more. These waves could be transverse magneto-hydro-dynamic (MHD) waves, traveling across the vertical magnetic fields in the corona. As the MHD wave traverses the chromosphere and low corona, it excites the solar plasma, which manifests as a moving wave of emission, perhaps with no mass motion of material. When this blast wave hits filaments and other

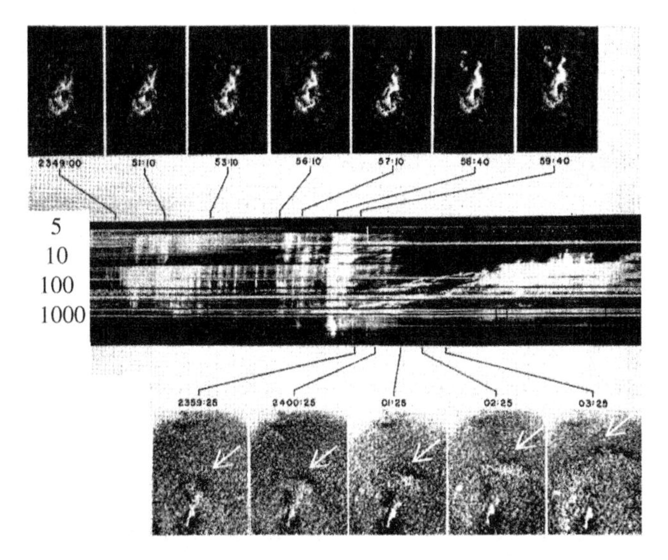

Figure 5.45 The top panel shows development of Hα on September 20, 1963 observed at the Lockheed Solar Observatory. Timings are given in UT. Middle panel shows radio emission recorded from Ft. Davis, where the y-axis shows the frequency in MHz and the bottom panel shows propagation of a blast or Moreton wave, as a dark sector boundary followed by a white area, moving away from the flare region as indicated by arrows. These five pictures are made by photographically canceling the filtergrams taken in the two wings of the Hα line (with permission from Blaisdell Publication Company).

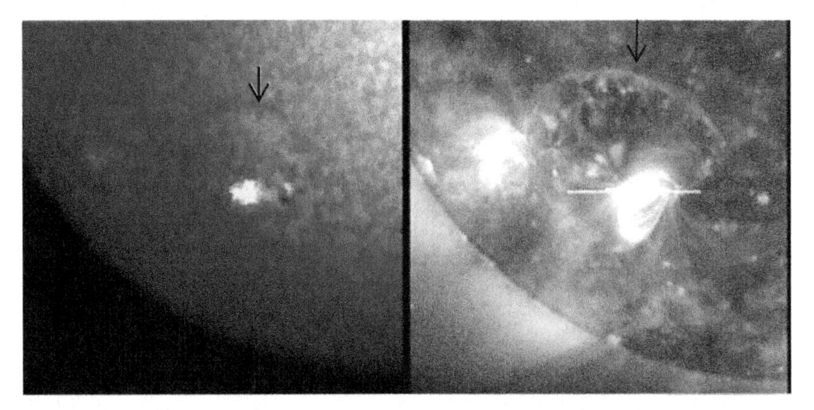

Figure 5.46 Blast or Moreton wave seen associated with a flare on images taken in the EUV at 195 Å from the EIT instrument onboard SOHO spacecraft. The corresponding picture taken in Hα line also shows the blast wave as marked by arrows.

features on the disk, they display activity and are even blown off or show up and down motions. The wave travels between the chromosphere and the transition region as a spherical traveling wave; it excites the solar plasma which is displayed as a traveling bright feature across the disk.

6. *'Halo' around flares.* Zirin and Tanaka (1973) have reported observing a 'halo' of diffuse bright emission in Hα line centre in the case of intense flare of August 2, 1972. It is interpreted that such halos may arise due to Hα radiation being scattered in the surrounding chromosphere.

7. *Solar flare effects (Sfe) in geomagnetic field.* The solar flare associated short-wave UV, EUV and soft X-ray electromagnetic radiations cause significant and detectable effect on the geomagnetic field. Initially sudden increase in the geomagnetic field is recorded with a gradual decline to pre-flare condition, such sudden temporary variation due to flares are called *crochets*. The flare radiations enhance the ionospheric conductivity in the lower Ionosphere and temporarily alter the ionospheric *Sq*-current in the sunlit side of the Earth. The resultant effect on the geomagnetic field is called Sfe (Solar flare effect) and appears in the form of kink or impulsive change in the geomagnetic H-component (horizontal) of the field of few tens of nano tesla or less than about 0.1% of the total field strength, lasting for nearly 30 minutes.

Much larger than the crochets are the *magnetic storms.* They are associated with solar wind disturbances, such as coronal mass ejections and high speeds solar wind streams, which have much higher plasma densities and speeds. When a solar wind disturbance arrives at the Earth's magnetosphere, there is a sharp increase in geomagnetic field strength by about 10 nanotesla (nT), known as *sudden commencement.* This field is due to a compression of geomagnetic field by the enhanced solar wind, such that the distance of the magnetopause from the Earth in the sunward side is decreased from $10R_E$ to about $6R_E$. A shock wave precedes the wind direction. A magnetic storm is a world–wide phenomenon, unlike crochets which occur only on the sunlit side. After an initial increase in the geomagnetic field, there is a prolonged period when the field remains extremely disturbed, with a variation of several

100 nT. Large geomagnetic storms give rise to intense auroral activity, and induce currents in electric power lines, which lead to voltage and frequency variations and even trip protective relays in power systems. Magnetic field from such currents can induce a voltage potential on the surface of the Earth of up to 6 volts per kilometer in the electrical transmission lines. A great storm on 13 March 1989 associate with a large solar flare of class X4 caused voltage reduction in the distribution system in Sweden and a complete power failure of Hydro-Quebec power in Montreal, destroying transformers, throwing circuit breaker off, and shutting down power grid system for several hours.

Besides the effect of flares on geomagnetic field, the EUV and soft X-rays also enhance the electron density in the various layers of the Ionosphere, which in turn affects short and long distance radio communications. The energetic particles from flares and coronal mass ejections are known to affect and damage delicate satellite electronic equipment and may harm astronauts if they are out side the spacecraft during a severe flare. Thus space-weather studies and forecast of solar activity is becoming a very important field of solar research now.

8. *Aurora.* Unlike geomagnetic storms *substroms* give rise to less magnetic intensity, which occur in a limited region of the Earth. These are also generated by solar flares and mass ejection activities on the Sun, and may last for about half an hour, with burst like initial phase. The geomagnetic effect of substroms is a colorful display of aurorae, near the north and south magnetic pole of the Earth. Aurorae are generally **not** visible at the magnetic poles but are seen in almost circular zones, some 4000 km in diameter and centered on the magnetic pole known as auroral ovals. The visible emission seen in aurorae occurs around 90-130 km above the earth and is due to the excitation of oxygen and nitrogen atoms and molecules in our earth's atmosphere by energetic electrons directed towards the earth's north and south magnetic polar regions. The characteristic strong red (6300Å) and green (5576Å) auroral emission arise due to excitation of neutral oxygen atoms in our earth's atmosphere. The northern aurora is known as *aurora borealis* and the southern as *aurora australis*. Detailed study of aurorae is now possible from space platforms.

5.8.10 *Radio Emission from Flares*

Several different types of radio emissions are observed in association with flares and active prominences, surges, sprays etc. In Figure 5.47 their distribution and duration is shown. At the onset of a flare, the most common radio emission are the Type III bursts, which occur only for a few minutes and extend over frequencies from a few tens of KHz to

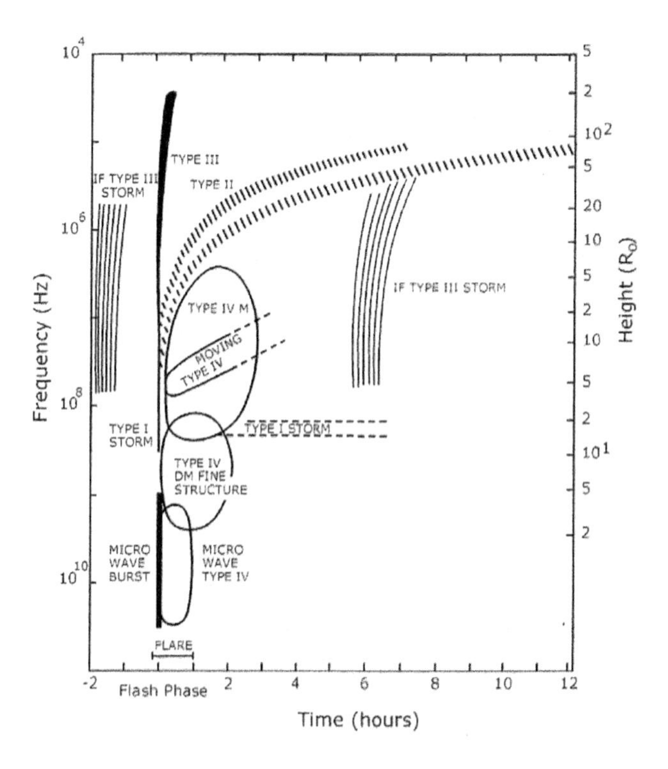

Figure 5.47 Radio spectrum of a large flare indicating the time distribution of various radio emissions. The height-scale on the right hand side corresponds to the plasma level of the frequency scale on the left hand side.

hundreds of MHz. Type III burst are caused by streams of electrons accelerated outward from the flaring active region, which excites plasma oscillations at progressively lower plasma frequency as the electrons pass successively through various coronal layers of lower densities. Type III bursts can be generated by filament eruption also which are often associated with flares, but even without flares. Type III emission can occur in non-flaring regions also.

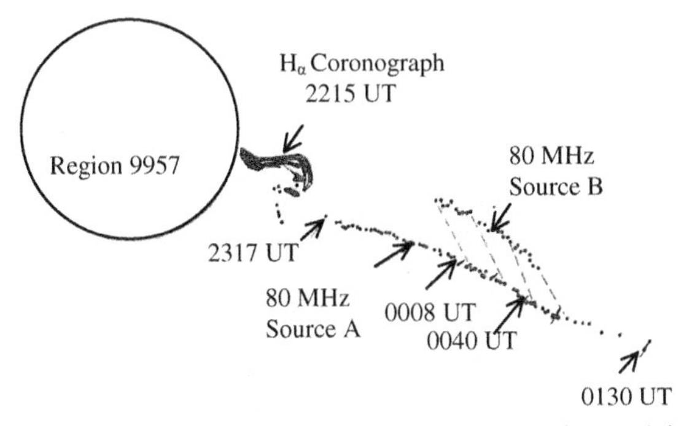

Figure 5.48 An erupting flare generated spray prominence observed from the Mees Observatory in Hawaii followed by a moving Type IV source at 80 MHz recorded with Culgoora Radio heliograph in Australia .

Another type of impulsive broad band continuum radio emission has been observed associated with flares, known as Type IV bursts. These extend in frequency from microwave down to about a hundred MHz. Even moving Type IV emission has been observed at frequencies between 10 and 100 MHz range. The moving Type IV emission lasts for tens of minutes, as the source moves out to several solar radii at speed of few hundred Km/s to 1500 Km/s. It is generally considered that the moving Type IV bursts are radio signatures of Coronal Mass Ejection (CME) produced perhaps by flares and eruptive prominences undergoing disparition brusque. In Figure 5.48 is shown an eruptive spray prominence observed from the Mees Solar Observatory in Hawaii and associated moving Type IV source at 80 MHz, observed from Culgoora radio-heliograph. The Type IV emission is interpreted as due to gyro-synchrotron radiation emanating from energetic electrons, which are magnetically trapped in plasma cloud accelerating outward from the flare region.

The third major meter wave burst is the Type II, which produces intense radiations at frequencies below 100 MHz. The Type II emission consists of two discrete frequency bands, which drift to lower frequencies over time scales of about 5 to 10 minutes. The frequency drift rate corresponds to outward velocities ranging between 500 and 5000 km/s. As the observed Type II velocities of around 1000 km/s are much higher than the local Alfvèn speed, therefore the moving disturbance must be an interplanetary shock wave moving outward from the Sun, triggered by the flare. Interferrometric observations of Type II sources indicate that they are spherical 'shock' fronts, expanding out from the active region.

The fourth type of radio emission associated with active regions and flares is the Type I, or noise storm emissions which are very long lived events emitting radiations between the frequency of 50-400 MHz. They may or may not be associated with flares, and are almost 100% polarized and display high brightness temperature. It is suggested that the Type I emission is due to plasma oscillations excited by energetic electrons.

The highly energetic electrons that produce impulsive hard X-rays also emit microwave (cm and mm wavelengths) radiations. The observed similarity in time profiles of microwave and hard X-rays during the impulsive phase (see Figure 5.40), suggests that the high speed electrons that produce hard X-rays and microwaves are accelerated and originate at the same place with energies of the order of 10 to 100 KeV.

5.8.11 *EUV and X-ray Flare Emissions*

Solar flares are extremely hot features, ranging in temperature from 10,000 to several million degrees K. They emit more than 70% of their energy in the X-ray domain and a small fraction in EUV line emission and about 25% in visible and radio wavelengths. For a short time, a large flare can outshine the whole Sun in soft X-rays, while the same flare in white light may not be even perceptible. According to Wien's displacement law, the hottest gases radiate most intensely in the short wavelengths, hence a 10 million degree gas would emit peak radiation λ_{max} at about 3Å (λ_{max} = hc/kT= 0.0029/T, where λ is in meters). The X-rays in the energy range of $\geq$ 10 Kev, emitted during flares are *thermal bremsstrahlung* radiation by virtue of intense heat and depend on the random thermal motion of very hot electrons. From the measurement of flaring X-ray bremsstrahlung power, one can determine the electron density also. The EUV emission arises from plasma temperature of 1-2 million degree K.

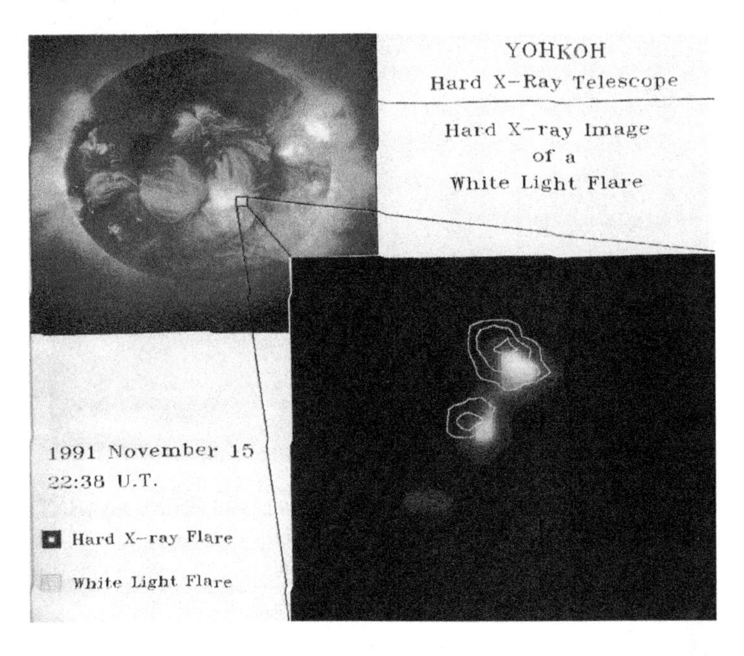

Figure 5.49 Showing the location of hard X-ray and white light flares corresponding to a two foot–points at the end of soft X-ray magnetic loop, seen on the YOHKOH soft X-ray image taken during 15 November 1991 flare. The time profiles of this flare exactly matches with white light and hard X-ray emission regions, the location and simultaneity of the two hard X-ray bright points appearance suggests that the non–thermal electrons transport the impulsive phase energy along the flare loops.

The flare emission of hard X-rays in the energy range $\leq$ 30 Kev is also due to bremsstrahlung, but they are produced by *non-thermal* electrons that have much higher energies than those emitting soft X-ray. The non-thermal electrons are accelerated above the tops of coronal loops and also beam along the legs of the loops at the foot-points, into the denser chromosphere and low corona. Hard X-rays observations of flares made from YOHKOH observations in 1990-2001and earlier ones from Solar Maximum Mission (SMM) in 1980s, clearly show that these arise simultaneously at the double foot points of a flaring magnetic loop visible in soft X-ray wavelengths. On November 15, 1990, YOHKOH had observed a double hard X-ray emission source at the foot points of

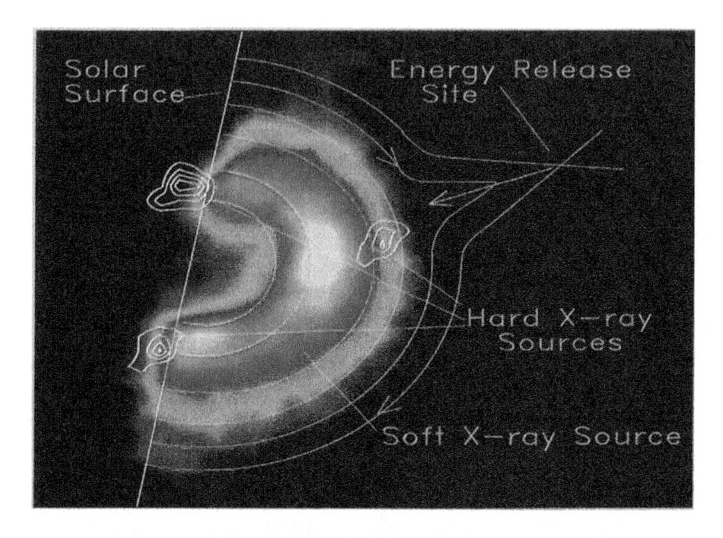

Figure 5.50 Limb flare observed on 13 January 1992, with Hard X-ray (HXT) & Soft X-ray (SXT) Telescopes aboard YOHKOH distinctly shows locations of hard X-ray sources, one at the loop top and two at the foot points of Soft X-ray loops.

soft X-ray loop, and simultaneously also a white light flare at the photospheric level, which was almost co-spatial with the hard X-ray emission sources, as shown in Figure 5.49. From such observations in soft X-ray, hard X-ray and white light, it is well established that the non-thermal electrons transport the *impulsive phase energy* along the loops, and penetrate deep down to the photospheric level to produce white light flares. In Figure 5.50 is shown a limb flare observed by YOHKOH on 13 January 1992, displaying a soft X-ray loop and a double hard X-ray source at the foot points of the loop and another hard X-ray source at the top of the loop.

The time profile of impulsive EUV flare radiations closely resembles that of hard X-rays. This suggests that EUV emission also arises from same downward propagating heating mechanism, resulting in heating of the lower chromosphere emitting Hα and filling the magnetic loops with hotter plasma at million degrees temperature or more, as indicated by emission of soft X-rays. Doppler shift measurements of EUV and visible lines show plasma motion in flares. Profiles of Hα line and other chromospheric lines observed during flares, indicate first an upward and later a downwards motion of up to 100 km/s in the cool plasma. The time profile of impulsive EUV flare radiations closely resembles that of hard X-rays, this suggests that some kind of *evaporation* of the hot solar plasma occurs in the flaring region, which settles down on cooling.

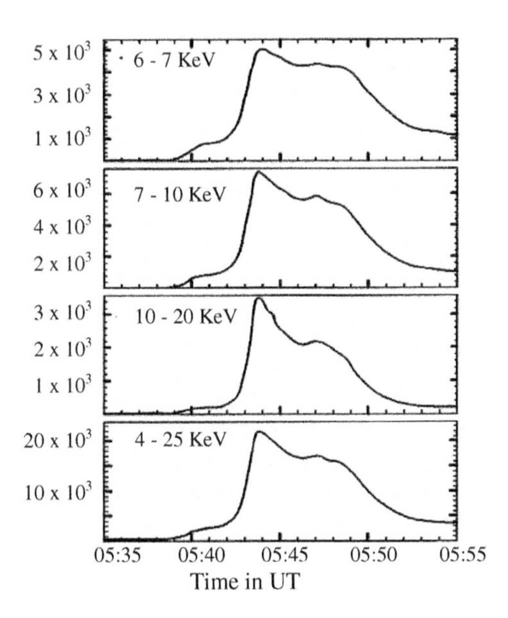

Figure 5.51 Showing time profiles of X-ray emissions from a solar flare observed on 14 August 2004 at 05:43 UT from SOXS in the energy range of 6 to 25 KeV from Si detector.

Geostationary Operational Environmental Satellites (GOES), of the National Oceanic and Administration (NOAA) of USA, provide minute to minute soft X-ray peak flux in the energy range of 3.1-24.6 keV (0.5-4Å) and 1.6-12 keV (1-8Å) from the Sun, along with low resolution solar images in soft X-ray through their Solar X-ray Imager (SXI) instrument. Electron and proton flux measurements are also available for Space-weather studies and solar activity forecasts. These data are extremely useful for monitoring of X-ray radiation from the flares. Recently some more geostationary satellites have come into operation for recording of hard and soft X-ray fluxes on continuous basis. One such satellite called SOXS (SOlar X-ray Satellite) was launch by the Indian Space Research

Organization in 2003, to observe the solar integrated flux in the energy range between 6 and 90 KeV with separate detectors (Si & CZT). A sample data obtained from a flare observed on 14 August 2004 is shown in Figure 5.51.

5.8.12 *Gamma Ray Flares*

Intense solar flares also emit very energetic γ-rays. During the impulsive phase of an intense flare, protons and heavier ions are accelerated to high speed and are beam down to low chromosphere and photospheric levels, where they bombard the hydrogen nuclei and other heavier ions and shatter the atomic nuclei and produce nuclear reactions. Thus on the surface of the Sun, we do see occasionally nuclear reactions associated with flares, which occur all the time in the deep solar interior. The nuclear fragments are initially excited but relax to former state by emitting γ-rays at discrete energies between 0.4 and 7.1 MeV. The strongest γ-ray line at 2.223 MeV is formed by bombardment of atomic nuclei by flare accelerated ions emitting neutrons. When one of the neutrons is captured by a hydrogen nucleus (proton) available in the solar atmosphere, forms deuteron, which decays with release of a γ- ray line at 2.223 MeV $\{{}^{1}H + n \rightarrow {}^{2}H^{*}, {}^{2}H^{*} \rightarrow {}^{2}H + \gamma\ (h\nu = 2.223 MeV)\}$. There is another γ-ray line at 0.51 MeV, which is formed by annihilation of positron-e^{+} and electron e^{-} $\{e^{+} + e^{-} = \gamma + \gamma\ (h\nu = 0.51\ MeV)\}$. Several other narrow γ-ray lines have been observed by Gamma Ray Spectrometer (GRS) on board the Solar Maximum Mission (SMM) satellite in 1980-81. These lines are formed during flares through the process of de-excitation of carbon, nitrogen, oxygen and heavier nuclei.

Short lived fundamental particles called *mesons* are also produced in intense solar flares, when protons with energies above 300Mev interact with hydrogen nuclei in the solar atmosphere. The decay of *neutral* mesons produce a broad γ-ray peak at 70 MeV, but the decay of *charged* mesons gives rise to bremsstrahlung, yielding continuum γ-rays.

Neutrons with energies in excess of 1000 MeV can also be produced in flares. Such high energy neutrons have been detected in space near the Earth in 1980 from SMM and in 1991 from Compton Gamma Ray Observatory, aboard YOHKOH satellite. Ground level neutron monitors have also detected relativistic neutrons produced by meson decay.

5.8.13 *Cosmic Ray and Proton Flares*

Most of the very energetic flares emit protons with energies >500 MeV, and take about 15-20 minutes to reach the Earth, after the on-set of the flash phase of the flare. These particles produce *Ground Level Effects* (GLE). Flares emitting protons with energies between 10 MeV$\leq$ E $\leq$100 MeV are known as *proton flares.* These flares produce anomalous ionization in the D-layer of the polar ionosphere and thereby cause *Polar Cap Absorption* (PCA), which causes a strong absorption of radio waves or cosmic radio noise in 1-50MHz range of galactic or extra galactic origin in high latitude polar regions of the Earth. The effect of a maximum PCA event is observed within a day or two from the onset of flare, but the recovery may take 10 days or more. Cosmic ray and majority of proton flares are two ribbon flares and are associated with Type IV and microwave bust and hard-X-ray bursts. All proton flares are accompanied by non-relativistic electrons with energy > 40 KeV flux up to nearly 5000 electron $cm^{-2}s^{-1}ster^{-1}$. Many proton events may occur without detectable relativistic electron flux, these are always associated with Type III and microwave bursts and sometimes hard X-rays may be also present.

5.8.14 *Flare Theories*

As we have seen in earlier Section that flare energies range from 10^{26} to 10^{32} ergs, the question is how such large energies are build–up, stored and suddenly released. We know that flares always occur in magnetic active regions on the Sun, the more complex magnetic region is, stronger flares are expected to emit. Thus it is well established now that due to interplay of magnetic field with solar plasma, extremely energetic events like flares, mass ejection, eruptive prominences, coronal heating etc., take place. But how exactly the transformation of magnetic energy into kinetic energy of mass motion and heating to several million degrees takes place is a big question. For several decades scientists had been working on theories to explain the flare mechanisms, as to how the enormous flare energy is stored and suddenly released.

It is now well established that the magnetic field and the field configuration are responsible for flare energy build-up and perhaps release, therefore one would expect to see changes in magnetic field strength and orientation connected with flares. Several attempts have been made by a large number of workers to detect changes in magnetic field strength and configuration before, during and after a flare. There are observations which show marked changes in magnetic field strength and there are equal number of observations which show no change. The question whether field changes are associated with flares or not is still open. The problem seems to be, because the magnetic field one measures is at the photospheric level, while the flares are generated high above in the chromosphere and the corona. There is no direct measurement available of the coronal magnetic field. However, some inferences can be derived from the structures seen in soft X-ray and EUV emissions, which delineate the magnetic lines of force. Such structures do show rapid changes in their position and orientations, which could be interpreted as manifestation of magnetic field changes in the coronal region associated with flares. In this section we shall discuss some of the few plausible scenarios explaining the electromagnetic and particle emission from flares.

5.8.15 *Flare Energy Build-up*

It is now believed that the *free* energy needed to power a flare is stored in the low corona in form of non-potential magnetic field components, that is with electrical currents. The excess 'free' magnetic energy is produced in the corona through magnetic loops connecting opposite magnetic polarity regions, at the photospheric level. The free magnetic energy builds up and accumulates in the corona. This energy comes from the dynamo below the solar surface through differential rotation and turbulent churning in the convection zone, which would shuffle around the photospheric foot-points of coronal loops and shear, twist and braid them. Due to such shearing and twisting of the loops, which demark the magnetic field lines, large electric current densities and non-potential magnetic field is created in the coronal gas. But what triggers the instability and releases enormous amount of energy remains the question.

5.8.16 *Flare Energy Release*

In all flare theories the primary energy release is thought to be through re-connection of magnetic field lines. How such reconnection takes place?, two main mechanisms have been proposed. (i) the Petschek model, and (ii) the 'tearing-mode' model.

It is assumed that a volume of coronal plasma with magnetic field lines is brought in close contact with another volume, with magnetic field lines in opposite direction. Such a magnetic field system will be far from being 'potential', that is *without* electric current. Such a field configuration can exist only if there is an associated electrical current located at the boundary and perpendicular to the field lines. The boundary between the two sets of parallel field lines of opposite polarity is known as 'current sheet' or 'neutral sheet'. As the conductivity of the coronal plasma is very high, electric current will flow unhindered along the current sheet. However, there could be localized regions due to atoms, thinly ionized plasma, waves or turbulence, which may impede the flow of electrons and may give rise to some resistance. This would lead to heating and disappearance of magnetic field or diffusion of the field near the current sheet. Thus magnetic reconnection can take place through a small amount of resistivity of coronal plasma, which would result in heating of the plasma and particle acceleration. The rate at which current dissipates, depends on the width of the region where the diffusion occurs. For a plasma temperature of a million degrees, and that the primary energy release takes place within a few seconds, it is estimated that such current sheets would be at most one kilometer thick, which is beyond the observable limit. Therefore the proposed re-connection mechanisms have to remain only a theoretical possibility, but some predicted consequences can be observed.

1. The Petschek model

Based on the above idea Petschek in 1964 proposed a model for re-connection of magnetic field lines. In Figure 5.52 is shown a schematic of the proposed magnetic field configuration. The two oppositely directed slightly convex magnetic field systems are pushed towards each other, such that the central field lines form an X-shape at the center of the diffusion region, indicated by dotted box. Extended away from the diffusion region, a pair of slow mode MHD shock waves remains stationary or standing as the plasma flows through them from either side. The re-connection takes place in the diffusion region with heated plasma, carrying new field lines along the length of the current sheet between the pair of standing waves, as shown in Figure 5.52.

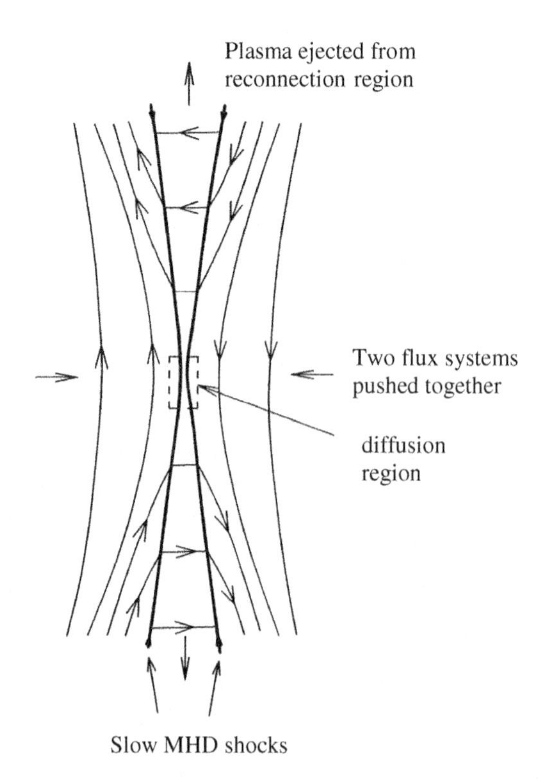

Figure 5.52 Schematic showing magnetic field configuration as per Petschek mechanism.

2. The 'tearing-mode' model

In this model it is proposed that the re-connection takes place through 'tearing-mode' instability, the current sheets which are in the form of a cylinder tears into ribbons, as shown in the schematic diagram in Figure 5.53. Two regions with opposite magnetic polarity come in close

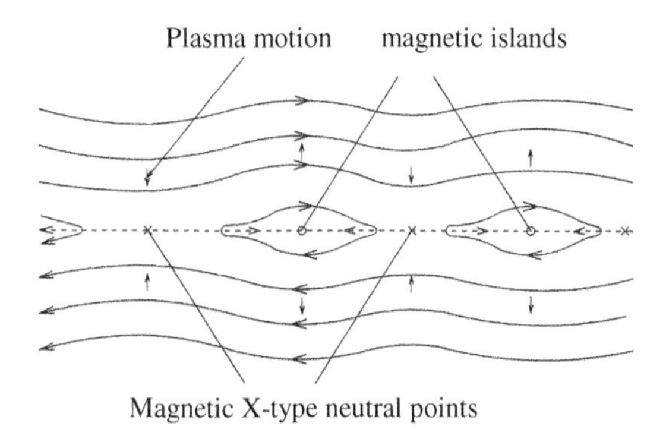

Figure 5.53 Schematic showing *tearing mode* instability.

proximity forming a current sheet, with plasma flowing along the field lines. In this case the magnetic regions are not pushed as in the Petschek's model. At the current sheet itself, the plasma can not move perpendicular to the field lines, but if the magnetic field becomes zero at the same points along the current sheet, the restraining force is much reduced and instability will set-in. The plasma will be driven towards such points in the current sheet due to non-uniformities in the field outside the sheet. An X-shaped neutral point develops and the sheet 'tears'. This could happen repeatedly along the length of the sheet forming 'islands' of magnetic field. This mechanism does not seems to generate enough energy appropriate for a flare, but through coalescing number of 'magnetic islands', much greater energy release can be expected.

5.8.17 *Flare Models*

Several flare models have been proposed to explain in practice how a magnetic field configuration might give rise to re-connection, resulting in flare generation. Peter Sturrock (1980) proposed that the re-connection takes place above a helmet steamer, consisting of mostly closed field lines, but some may be open too, as shown in Figure 5.54 (a). The

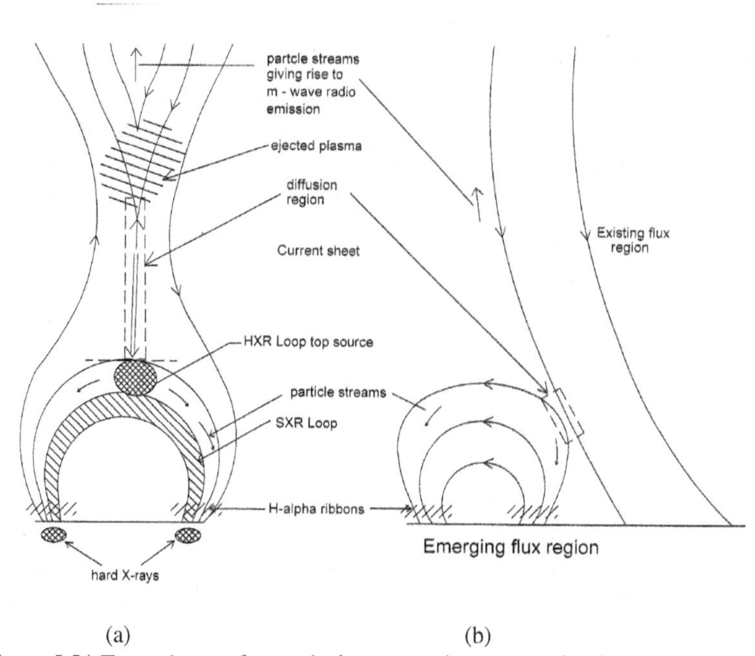

(a) (b)

Figure 5.54 Two schemes for producing magnetic reconnection in current sheet in the solar corona. Field topology (a) as suggested by Sturrock (1980) and (b) as per Heyaerts *et al.,* (1977).

photospheric motion at the foot-points of magnetic field configuration may lead to stressing of the field lines and formation of current sheet, which is subjected to 'tearing-mode' instability, creating reconnection of the field lines in the diffusion region, indicated by a dotted 'box' in the Figure 5.54. The compressed plasma in the region is then ejected in the form of a plasmoid and is driven outwards by the magnetic force. The particles mostly electrons, which are left in the closed loop system below in the diffusion region are accelerated downward, traveling to the foot points of magnetic loops and bombarding the denser low chromosphere and photosphere, giving rise to hard X-rays and even white light flares. These electrons also emit cm and micro wave radio emissions. Heyvaerts *et al.,* (1977) proposed that an emerging magnetic field near an existing active region loop can lead to a current sheet formation, if the emerging flux has opposite polarity as shown in Figure

5.54(b). In this model the current sheet formed between the two magnetic field configurations will have a rapid expansion phase. The impulsive phase of flares is identified with the expansion of the current sheet. The energetic particles travel down the legs of the loop which generate hard X-rays at the foot-points. This scenario is confirmed by hard X-ray images observed from YOHKOH's HXT (Hard X-ray Telescope) as seen in Figures 5.49 and 5.50. The magnetic loops also show expansion which explains the mass motion in flares. Type II radio emission bursts manifests generation of shock waves and ejection of plasmoid.

All these flare theories accept that, much of the energy released goes into accelerating particles. However, the major outstanding theoretical problem is to explain how particles to such vast range of energies are accelerated? For example, the hard X-rays are produced by electrons of a few hundred electron volts energies; while the Gamma ray emission and direct spacecraft observations indicate presence of relativistic electrons. Can such a range of electrons produce in a single process? It is proposed that perhaps it is a two step process; in the first step, non-relativistic energies are produced and in the second step relativistic electrons. In the tearing-mode mechanism, the particles could be accelerated in the first step by electric field to non-relativistic values. In the second stage acceleration, a small fraction of these particles are further accelerated to relativistic energies. This could be achieved if the particles repeatedly collide with moving magnetic field and gaining energy at each collision.

From these proposed flare models, one can understand how various forms of radiations are generated. The hard X-rays are emitted by non-thermal bremsstrahlung, as the accelerated particles are stopped by the dense material at the loop's foot-points and also at the loop tops which are bombarded by the particles from above, as shown in the schematic Figure 5.54(a). The micro wave emission arises due to synchrotron mechanism, as the electrons gyrate in a spiral manner along the field lines and are accelerated to relativistic velocities. The thermal soft X-ray emission results from the hot gas heated in the transition and chromospheric regions by energetic particles in the loop structures. The Hα and other visible radiations arise due to heating of the chromosphere by conduction of hot soft X-ray emitting plasma at million degrees in the flare loops. As mentioned earlier the white light flares are generated by high energy electrons bombarding the photospheric and temperature minimum regions.

The Type III radio emission is caused by electrons those escape the flare loops and Type II radio emission is by generated by shock waves by flares and the ejected plasmoid generate Type IV bursts.

5.9 Coronal Mass Ejection (CME)

Our Sun manifests a spectrum of mass motions, ranging from the smallest scale size motions seen in granulation, supergranulation, sunspots to large scale, such as solar wind, Coronal Mass Ejection (CME), self–contained structures of plasma - plasmoids, and magnetic fields. In this section we shall discuss the interesting phenomenon of CME which has direct bearing on our terrestrial environment.

5.9.1 *Morphology and Development of CMEs*

The CMEs are essentially large magnetized volume of solar plasma moving outward from the Sun at high speed. It is seen as a huge bubble of gas; rapidly expanding in size and hurls billions of tons of solar plasma in the interplanetary space at millions of degree temperature with speeds ranging from 300–2000 km/s (slow moving CMEs with speeds of 10-100 km/s have been also observed). In Figure 4.18 is shown a well developed CME extending to more than 30 solar radii.

The CME material takes on an average about 100 hours to reach the Earth. The largest CMEs may eject masses more than 10^{13} kg, the kinetic energies may exceed the total energy of a large flare (10^{32} ergs), and its volume could be several hundred times that of our Sun. The mass flow rate in CMEs is about $2x10^{8}$ kg per sec. When such large amount of solar plasma hits the Earth, several geomagnetic and terrestrial effects occur. The leading bright edge of CME is followed by a relatively darker region, called the 'cavity' which is depleted in plasma density; this is generally followed by a bright knot, which could be a prominence. As the leading edge of CME hits the interplanetary medium, a 'shock front' is developed ahead of the leading edge. The average frequency of occurrence of CME during solar maximum period is about 3.5 events per day, and during minimum is about 0.2 events per day. Frequency of occurrence of CME follows the 11-year solar cycle. Each time a CME is ejected from the Sun, it carries with it 5×10^{12} to 5×10^{13} kg of coronal material. Assuming on an average 2 CMEs per day, our Sun is loosing more than 36×10^{14} to 36×10^{16} kg of material every year! CMEs usually have curvilinear shapes that resemble a cross-section of loops, shells, or filled bubbles; suggesting close magnetic structure. The upper portion of the magnetic loops is carried to great heights while the remaining tail-ends are attached and rooted in the Sun, as if caught by the solar magnetic field. As the material in CME expands, it stretches field lines until it snaps, taking the magnetic field with it and lifting off in space. Whenever a large closed loop of magnetic field is unable to hold itself, a CME takes off. CMEs are accompanied with shocks and accelerate large quantity of high speed particles ahead of them. A typical large CME may release about 10^{32}ergs of energy, equivalent to a large flare, however it is mostly kinetic. Some CMEs display acceleration as they move out in space, while other may slow deceleration after the initial phase. There is a wide variation of particle density in CMEs; it ranges from 10^5 to 10^8 cm^{-3} around 3 solar radii.

In Figure 5.55 is shown a sequence of evolution of CME observed from LASCO on board SOHO spacecraft. The maximum extend of CME seems to extend to more than 30 $R_{\odot}$ from the Sun. In recent years, the interest in the study of CMEs has tremendously increased because this is related to number of phenomena on the Sun and the Earth, for example; re-structuring of large scale solar magnetic field, emerging new flux, erupting prominences, disparition brusque, flares, geomagnetic effects, aurorae etc. Earlier white light coronagraphic observations were limited to nearly 1.5-2.0$R_{\odot}$, but they did show coronal transits near the Sun, which now we identify as CMEs. Since the introduction of orbiting coronagraphs on spacecrafts, first on OSO 7 from 1971 to 1974, then on Skylab during 1973-74, on P78-1 satellite between 1979-1985, on SMM in 1980 and from 1984 to 1989 and from the most recent LASCO (Large Angle Spectrometric Coronagraph), beginning from 1996, large amount of data on CMEs extending to 30$R_{\odot}$ from the Sun's center is being accumulated. A large number of scientists all over the world are engaged

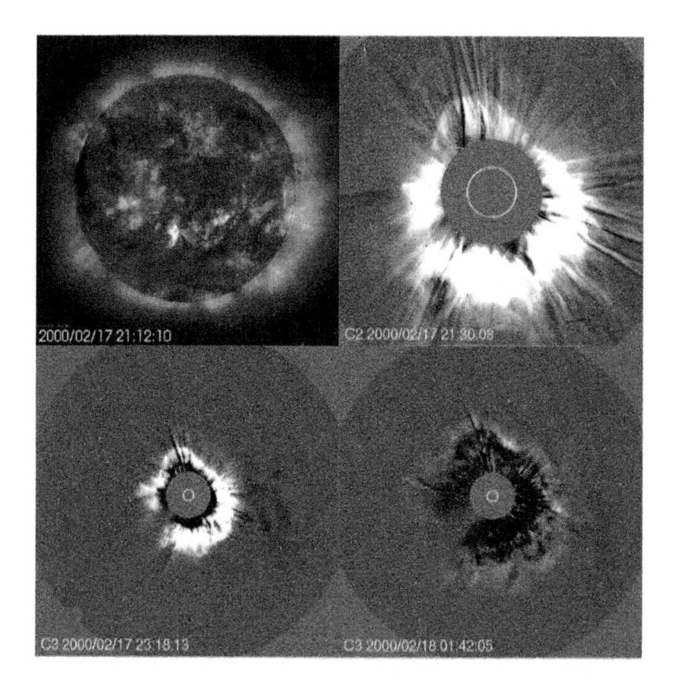

Figure 5.55 Showing development of 'Halo' type CME on February 17, 2000, images taken by LASCO. The 1st panel on the left shows a full disk image taken by EIT showing the location of the active region on the disk, which was responsible for this CME. The upper right panel shows picture taken by C2 coronagraph, the white circles indicate the size of the Sun. The pictures shown in lower 2 panels were taken by C3 coronagraph. The dark region seen around the Sun is due to cancellation of successive pictures taken at different times, to show the movement of the bright leading edge of CME.

in their study. The LASCO aboard SOHO spacecraft has three coronagraphs; C1- extends from 1.1 to 3.0 R_0, C2 - extends from 2.0 to 6.0 R_0, and C3 - extends from 3.7 to 30 R_0, from the Sun. CMEs move out in space with varying speed and distance and their effect is felt even up to the Earth and beyond.

Space observations have shown that CMEs occur in various shapes and sizes and evolve from some kind of disturbance on the Sun, and acquire huge structures, even several hundred times the size of the Sun, but where and how they occur and what triggers them is a subject of intense research now.

5.9.2 *Source Regions of CMEs*

Generally CMEs arise from large–scale, closed structures from pre-existing coronal streamers. Temporal and latitudinal distribution of CME show similar distribution as coronal streamers, prominences and filaments. During solar minimum period they mostly confine to low latitudes, while during maximum are seen all over the Sun. Frequency of occurrence CMEs varies in step with the 11-year solar cycle. Many of the most energetic CMEs involve pre-existing coronal streamers, which show increase in brightness before the erupting CME. Earlier it was thought that CMEs are always related or associated with flares and filament eruptions. Feynman and Martin (1994) made a detail correlation study of Hα filament images from the Big Bear Solar Observatory taken during 1984-86 and the observed CMEs. In several cases they found erupting filaments as well as new emerging region in the same quadrant as the CME, and high degree of correlation with emerging flux near a pre-existing filament. Some flares (1B) of moderate intensity were found 'related' with CMEs.

The ideas about CMEs are that there is no one-to-one correlation with flares or erupting filaments. Only about 40% of all the observed flares may be somehow connected with CMEs. There are number of instances wherein a flare was seen but no CME and vice versa is also true. CMEs may even occur before, during or after a flare, therefore a localized flare event does not seems to be having any connection with global CME phenomenon.

5.9.3 *Mechanism for Generation of CME*

The build–up and ejection of CME raise several theoretical questions, concerning their origin, driving mechanism and long term evolution. Our understanding of why and how the Sun expels huge amount of magnetic flux into space is still inconclusive. Some observational evidence has been obtained from SMM's synoptic coronal maps and soft X-ray images from YOHKOH satellite, it is suggested that magnetic field continuously builds-up in the corona and is released suddenly. Static coronal loops and streamers are observed to expand outwards, sometime in continuous manner, and sometimes in explosive manner displaying a sudden blow-up of a CME. The question is how the magnetic flux builds–up in the corona and why it suddenly disappears in an explosive manner? The answer to the first question can be traced to the generation and transport of magnetic field in the convection zone. As we have seen in Section 5.4.6 that the magnetic field in the form of flux tubes is brought to the surface via the mechanism of magnetic buoyancy. Unless **all** the magnetic field brought to the surface returns back below the photosphere, the emerged magnetic field is bound to build-up in the corona. In practice not all the flux can return back to the solar interior, therefore this is the most likely process through which magnetic flux builds up in the corona. It is now generally agreed that perhaps it is re-structuring of the magnetic field on global scale that triggers a CME.

5.9.4 *Driving Mechanism*

The most widely accepted view about the origin of CMEs is that the principal driving force of CME is the magnetic field. There are two basic types of driving mechanisms proposed for CMEs; one that large-scale motions in the convection zone or in the photosphere impulsively drive a current along a magnetic loop. Another is that magnetic energy is slowly stored in the corona, over period of hours and days and then is released suddenly due to loss of ideal MHD equilibrium. From large number of CME observations it is now generally accepted that it is the global re-structuring of the magnetic field that initiates and drives a CME.

5.9.5 *X-ray Blow Outs (XBO)*

CMEs have been known to be somehow associated with flares, eruptive prominences, filament eruption, active regions, new emerging active regions etc. But there are equally reliable observations which indicate that CMEs have no relation with any of these phenomena. Soft X-ray images obtained from YOHKOH provided a wealth of data on development and ejection of *soft X-ray* CMEs. Normally we refer CME as white light phenomenon emanating out of the Sun, but X-ray images also show CME like blow-offs. Hudson (1994) has found that *some of*

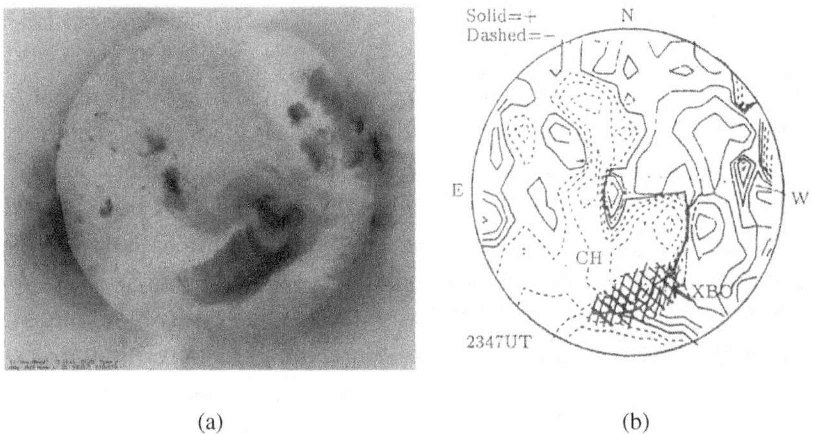

(a) (b)

Figure 5.56 (a) A huge *X-ray Blow Out* observed by YOHKOH on 26 January 1993 at 12:50 UT, (b) showing Stanford low resolution full magnetic map on the same day. Note that the XBO occurred just over the neutral line at the boundary of the coronal hole.

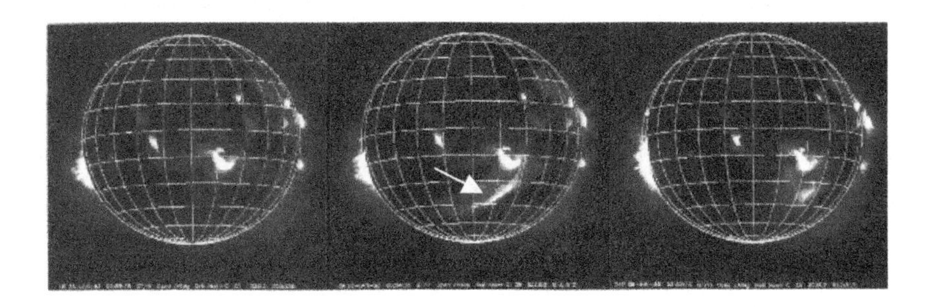

Figure 5.57 Showing evolution of the 26 January 1993 soft X-ray XBO, Yohkoh images taken at 03:50, 08:50 (maximum phase) and at 23:32 UT shown by an arrow.

the X-ray ejections correspond to CMEs. Khan *et al.,* (1994), Kahler (1992) and Tsuneta *et al.,* (1992) have shown that huge soft X-ray structure develop in both active region latitude and polar crown filament zones. Earlier in 1984, Wagner (1984) pointed out that 10-17% CMEs were associated with flares, 30-40% with eruptive prominences or filaments, and a large fraction of CMEs, about 30-48% of the observed CMEs were found unrelated to any optical activity on the solar surface. This motivated Bhatnagar (1996) to look for soft X-ray CMEs which were *unrelated* to any 'optical' activity seen on the Sun. The

YOHKOH's full disk soft X-ray movies taken during the period May 1992 through November 2003 were examined in detail. 15 cases of X-ray CMEs were identified as *X-ray Blow Outs* (XBO), a term similar to coronal streamer *blow out* used by corona observers. At the location of these XBOs there was **no** visible chromospheric or photospheric activity seen before or after the ejection of the X-ray CME. In all the 15 cases XBOs occurred at or along the boundary of a coronal hole and over the magnetic neutral line. A typical example of this phenomenon is shown in Figure 5.56 (a). where in on 26 January 1993, a huge soft X-ray arcade of about 400,000 km length suddenly erupted and vanished within less than 12 hours, as seen in Figure 5.57. This arcade was located at the boundary of a coronal hole, between the two opposite polarity regions, as shown in Figure 5.56(b), on a Stanford low resolution full disk magnetic map. Before the XBO on 26 January, a very small Hα dark filament of only 6 degree long was seen at the location of the eruption. What could have triggered this blow out? It may be that near the boundary of the coronal hole, with open field lines, a new active must have emerged with opposite polarity as shown in the schematic Figure 5.58. As the new flux evolved and extended up into the corona, re-connection of magnetic field lines took place at much higher level with open field lines of the coronal hole. The plasma material from the filament and the coronal hole ejected out in the form of X-ray Blow Out, which is manifested as series of rising X-ray arcades.

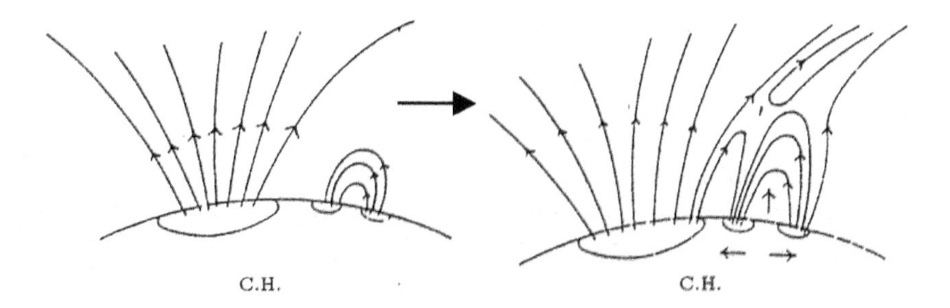

Figure 5.58 A proposed scenario of magnetic field topology responsible for XBO, due to reconnection of open field lines of coronal hole with oppositely directed magnetic field of an emerging active region.

Chapter 6

Observational Techniques

6.1 Evaluating Solar Seeing

All solar observers are familiar with the term '*seeing*', which in some sense gives an idea of the image quality of celestial objects. Image quality is distorted due to the turbulence in our atmosphere, which arises because the Sun heats up the ground and the surrounding air. Hot and lighter air columns rise up and thus thermal currents are built up as the day progresses and the solar rays (wave-front) passing through such an air column experience distortion and tilt resulting in defocusing, movement, and a shifting of the solar image in the focal plane. There are several criteria to estimate solar seeing from white light images. Kiepenheuer (1964) proposed a scheme to evaluate solar seeing from white light images of the Sun. In actual practice it is difficult to measure in arc sec the image motion etc., hence it is best to simply assign seeing depending upon the appearance of the solar image, limb motion, sharpness etc., and grade the seeing in term of 'Poor', 'Fair', 'Good', 'Very good' and 'Excellent'. The estimation of seeing during observations is an important parameter, as it gives some idea about the quality and reliability of the data. Under exceptionally good seeing conditions, visual observations show much more details and fine structure as compared to a photographic record. The reason is that the human eye has the capability to sort out fast variations in seeing, it can follow rapid image motion, and the brain freezes and remembers the best images, while a photograph integrates over a certain time duration, which results in blurred integrated image. Thus visual solar observations by experienced observers are of immense importance even today.

The drawings made visually of solar granulation, sunspots and prominences by A. Secchi, by Langley and by Professor Fernley, show extremely fine details comparable to the best photographs obtained with modern equipment. Over a hundred drawings made by Fernley's in early 1860-70 at Oslo have been 'uncovered' by Jensen, Rustad and Engvold at the University of Oslo. To have an idea what an eye can record of solar features is shown in Figure 6.1. This is a drawing made by Fernley in 1870 of a hedgerow solar prominence. In Figures 4.1 and 5.3 are Father Secchi's high resolution drawings of solar granulation and sunspots. We see fine granulation, sub-arc sec intergranular lanes, sub-arc sec penumbral filaments and even bright umbral dots.

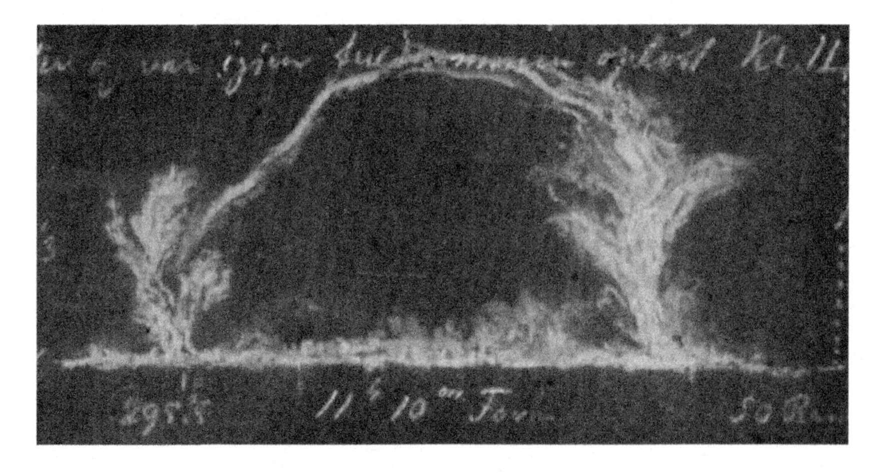

Figure 6.1 Drawing by Professor Fernley of hedgerow solar prominence made in 1870.

To evaluate solar seeing through unbiased human intervention, Seykora (1993) had devised a 'scintillation meter'. This measures the high frequency integrated solar brightness variation which is due to the air turbulence in the path of sunlight. A good correlation between the observed brightness fluctuation and solar seeing has been reported. For many solar site exploration surveys this instrument is being used. However some observers have reservation towards this technique. To evaluate the solar seeing several methods are being used, such as differential image motion technique, measurement of Fried parameter 'r_0'

6.2 Determination of Fundamental Solar Parameters

6.2.1 *Solar Parallax and Distance*

The solar parallax is defined as the angle in seconds of arc subtended by the equatorial radius of the Earth, at the Sun's mean distance. As the distances of objects in the solar system can be determined directly with high accuracy, solar parallax may be obtained from measurements of the parallax of a nearby planet like Venus or an asteroid. The asteroid Eros which occasionally comes as close to Earth as 0.15 times the distance to the Sun, has been extensively used by Jones (1941) to determine the solar parallax. Rabe (1950) had obtained an accurate value of solar parallax from the study of perturbations by the Earth on near-by objects (e.g. Eros).

Besides the triangulation method, more accurate determinations of the solar distance have been made using radar echoes. In this method the Sun is not directly used because the solar corona from which the radio waves are reflected is far from being a good reflector. Rather planets are used to determine the travel times. Now that such travel times can be determined with very high accuracy, in 1976 the International Astronomical Union (IAU) adopted a value of the solar parallax as 8.79418 arc sec. Knowing the Earth's equatorial radius (6378.39 km) to high accuracy, the corresponding mean Sun-Earth distance, the Astronomical Unit (1 AU) turns out to be = 149,597,900 km, and in terms of light travel time, τ_A for unit Astronomical distance A, is given by $\tau_A = 499.009$ seconds.

For all practical purposes a rounded value of 1.496×10^8 km is used as the mean solar distance. As the Earth's orbit is elliptical, the variation of solar distance lies between 1.471×10^8 km at perihelion in January and 1.521×10^8 km at aphelion in July. This variation in distance causes an apparent change in the angular size of the Sun, corresponding to 710 and 734 km per arc second at perihelion and aphelion respectively.

6.2.2 *Solar Mass*

From the study of perturbation of Eros's orbit, Rabe (1950) obtained a value of the solar mass in terms of the (Earth + Moon) system as 328,452±43. He also obtained an improved value of mass ratio of Earth: Moon = 81.375, hence the mass ratio Sun : Earth is 332,488. If the Earth's mass is taken as 5.975×10^{24} kg (Birge 1942) then the mass of the Sun will be 1.9866×10^{30} kg. Recently high precision time and distance measurements yield equally high precision product of G, the gravitational constant and the solar mass $m_\odot$, taking the laboratory determination of G as :-

$$G = 6.672 \times 10^{-11} \mathrm{m}^3\,\mathrm{kg}^{-1}\mathrm{s}^{-2} \quad \text{or} \tag{6.1}$$
$$G = 6.672 \times 10^{-8} \text{ dynes cm}^2 \text{ gram}^{-2}.$$

Thus the solar mass turns out to be:-

$$m_\odot = (1.989) \times 10^{30} \text{ kg},$$

a figure adopted by the IAU.

6.2.3 *Solar Diameter, Density and Surface Gravity*

To determine the linear diameter of the Sun, we must know its distance and the angular diameter of the visible disk. Earlier the angular diameter was visually measured with a meridian circle. These measures gave larger values than the true diameter, depending on the Sun's altitude and the 'seeing' conditions. In recent times, using photoelectric photometry and the inflection points of the intensity profiles at the two opposite limbs of the Sun, the angular diameter has been more precisely measured. The value given by Cox (1999), in the Fourth edition of Allen's Astrophysical Quantities, is 959.63 arc sec. Depending on the wavelength used for observation, the inflection point of the limb profile corresponds to an *optical depth* between 0.001 and 0.004. Taking the latest value of the mean Sun-Earth distance and the Astronomical Unit as 149,597,900 km the linear solar semi-diameter, $R_\odot$ comes to

6.95508×10^5 km, at an *optical depth* of 1. This is considered as the 'true' solar surface or the photosphere.

Beginning from Father Secchi and Rosa's time in mid nineteenth century, several workers have measured the solar diameter and have indicated that the diameter of the Sun undergoes periodic variation that perhaps correlate with the solar cycle. This has proved to be difficult to confirm and is a controversial topic. From theoretical calculations of the Sun's evolution as a main sequence star, it is known that presently the solar radius is changing at a rate of 2.4 cm per year. This long-term change will be apparent only in hundred million years or so. It is likely that superimposed on this extremely slow change there could be faster variations. These have been proposed by early investigators, as well as by recent helioseismology techniques, which will be discussed in a later Chapter.

Knowing the mass and the radius of the Sun, the mean density ρ is given as:-

$$\rho = 1.409 \text{ g cm}^{-3}, \qquad (6.2)$$

and the surface gravity $g_o = Gm_0/r^2 = 274 \text{m s}^{-2}$.

The gravitational acceleration or the surface gravity is an important parameter and influences the structure of the solar atmosphere and the solar model.

6.2.4 *Solar Luminosity $L_\odot$*

The luminosity $L_\odot$ is defined as the total energy output per unit time in the form of electromagnetic radiation. It is related to the total radiation received at the mean distance of Earth, known as the Total Solar Irradiance (TSI) and is given by:-

$$\text{TSI} = L_\odot / 4\pi R_{es}^2, \qquad (6.3)$$

where R_{es} is the mean Earth–Sun distance equal to 1 AU. TSI and $L_\odot$ measurements should be made over the whole electromagnetic spectrum, but our atmosphere attenuates the infrared and completely blocks solar radiation short-ward of 3000Å. Therefore the most accurate measurements of TSI have been from spacecraft above the atmosphere.

The ACRIM (Active Cavity Radiometer Irradiation Monitor) and several other instruments were used to precisely measure TSI from space. ACRIM-I was launched in 1980, on board the Solar Maximum Mission (SMM) spacecraft. This instrument gave a mean value of TSI as:-

$$\text{TSI} = 1365 \text{ to } 1369 \text{ w/m}^2, \tag{6.4}$$

and considering $R_{es} = 1.496 \times 10^{11}$m, the solar luminosity turns out to be $L_\odot = (3.845) \times 10^{26}$ watts.

According to Kuiper (1938) the apparent visual magnitude of the Sun is – 26.75 on the International scale. Adopting Kuiper's value, the Sun's absolute photo-visual magnitude is + 4.82, which is consistent with Kuiper's spectral classification of the Sun as GV2.

6.2.5 *Temperature of the Sun*

The Sun's outer layers are **not** in thermodynamic equilibrium. Hence, a single unique temperature cannot be assigned to the Sun. The solar temperature derived by different methods of observation yield different values, especially when observations refer to different atmospheric layers. For example, the radiation temperature of the outer boundary of the photosphere, as determined from the limb darkening measurement is about 5500° K, while the kinetic temperature of the corona is a million degrees! Therefore, astrophysicists define different '*kinds*' of temperatures, depending on the type of observational data used. When referring to temperature one must be careful to mention, which '*kind*' of temperature one is referring to. For example, there is the '*effective*' temperature, the '*brightness*' temperature, the '*color*' temperature, the '*excitation*' temperature, the '*ionization*' temperature, and the '*kinetic*' temperature. Deviation from thermodynamic equilibrium further complicates the matter to assign a unique temperature.

6.2.5.1 *Effective Temperature*

The most important quantity describing the surface temperature of a star is the *effective temperature* T_{eff}. The effective temperature T_{eff} of a star is defined as that temperature which a blackbody sphere of the same radius must possess in order that the total energy output L, equals that of the star. Since the effective temperature depends on the total radiating power, integrated over all frequencies, it is well defined for all energy distributions even if it deviates from the Planck's law. Using the Stefan-Boltzmann's law:-

$$F_{out} = \sigma_R T^4_{eff}, \tag{6.5}$$

where σ_R is the Stefan-Boltzmann constant $\sigma_R = 5.67 \times 10^{-5}$ erg cm^{-2} deg^{-4} s^{-1} and ($\sigma_R . T^4_{eff}$) is the black body energy output per cm^2. Thus $L_\odot$, the total energy output of the Sun, in ergs per second is given as $L_\odot = 4\pi R^2_\odot F_{out}$, where $R_\odot$ is the radius of the Sun. Thus the flux density F', at a distance r is given by:-

$$F' = L_0 / 4\pi r^2 = R_o^2 / r^2 F_{out} = (\alpha/2)^2 x \sigma_R T^4_{eff}, \tag{6.6}$$

where $\alpha = 2\, R_\odot / r$, is the observed angular diameter of the Sun. Therefore, for determining the effective temperature T_{eff} we have to measure the total flux density and the angular diameter α of the Sun. In the case of the Sun, α is very precisely known. The T_{eff} of the Sun may be determined from the recent measures of the Total Solar Irradiance. The total amount of energy passing through a sphere or through the mean radius of the Earth's orbit at 1.495×10^8 km, must be equal to the total energy output of the Sun = 3.79×10^{33} ergs s^{-1} or 6.25×10^{10}ergs cm^{-2} s^{-1}. Thus the corresponding T_{eff}, of the Sun is obtained using the Stefan-Boltzmann's law is given by:-

$$6.25 \times 10^{10} = (5.672 \text{ X } 10^{-5})\, T^4_{eff}, \tag{6.7}$$

hence T_{eff} = 5760° K.

6.2.5.2 *Brightness Temperature*

The brightness temperature T_b is defined as that temperature of a black body which will give the same energy output per Ångstrom as the star, at wavelength λ. A quantity that can be measured for number of stars is the radiation $cm^{-2}sec^{-1}Å^{-1}$ for selected points in the continuum radiation. Actually, the brightness temperature is *only* a parameter that expresses the rate of energy radiated at certain wavelengths. It can not be readily converted into physical temperature of the emitting stellar surface. For example, E. Pettit measured the radiation over the whole solar disk and found that the brightness temperature T_b = 6200 K at λ 4500, and T_b = 6000 K at λ 6500. Labs (1957) at λ 5263 found T_b = 6470 K. Higher solar brightness temperatures are found in the infrared. In the rocket ultraviolet T_b falls to about 4500 K. Generally in radio astronomy the brightness temperature T_b is used to express the intensity (or the surface brightness) of the source.

6.2.5.3 *Color Temperature*

A stellar *color temperature* T_c is a parameter employed in the Planck's black body energy curve to represent the slope of the energy distribution. It does not necessarily represent any physical temperature of the photospheric layer. The color temperature can be determined even if the angular diameter of the star is not known. We have to only know the relative energy distribution in some wavelength range [λ_1, λ_2]; the absolute value of the flux is not needed. The observed flux density as a function of wavelength is compared with the Planck's curve at different temperatures. The temperature that gives the best fit is the color temperature in the interval [λ_1- λ_2]. The color temperature is usually different for different wavelength intervals, since the shape of the observed distribution may be quite different from the black body spectrum, because of spectral line absorption. Generally the '*color index*' of stars is measured through broadband UBV filters which gives a crude estimate of color temperature.

6.2.5.4 *Kinetic Temperature*

The *Kinetic temperature* $T_{k,}$ is related to the average speed of gas molecules or atoms or ions. The kinetic energy of an ideal gas molecule as a function of temperature follows from the kinetic gas theory.

$$\text{Kinetic energy} = \tfrac{1}{2}\,\text{mv}^2 = \tfrac{3}{4}\,\text{k}\,T_\text{k}, \tag{6.8}$$

$$\text{or}\quad T_\text{k} = \text{mv}^2/\,3\text{k},$$

where m is the mass of the molecule, v its average velocity and k the Boltzmann constant. For ideal gases, the pressure P is directly proportional to the kinetic temperature,

$$\text{P} = \text{n}\,\text{k}\,T_\text{k}, \tag{6.9}$$

where n is the number density of the molecules.

6.2.5.5 *Excitation Temperature*

The *excitation temperature* T_exc is defined as a temperature which if substituted into the Boltzmann distribution gives the observed population numbers of an atomic or molecular species. The excitation temperature can be found from a comparison of the number of atoms in different energy levels, using the Boltzmann's equation:-

$$\frac{N_r}{N_{r'}} = \frac{b_r g_r}{b_{r'} g_{r'}} e^{-\frac{\chi_{rr'}}{kT_{exc}}}, \tag{6.10}$$

where N_r and $N_{r'}$ are the relative number of atoms in the r and r'^{th} levels, factors b_r and $b_{r'}$ are corrections required for Boltzmann's equation to allow for possible deviations from local thermodynamic equilibrium (LTE). The quantity $\chi_{r\ r'}$ is the difference in excitation energy of the two levels r and r', and g_r and $g_{\ r'}$, are the statistical weights, k the Boltzmann constant, and T_{exc} the excitation temperature. The number of absorbing atoms per gram of stellar material can be determined from the measurement of the equivalent widths of spectral lines. In the above Boltzmann's equation, if we arbitrarily put $b_r = b_{\ r'} = 1$, and determine T_{exc} from the ratio $N_r\,/\,N_{\ r'}$. It may *not* necessarily agree with the ionization temperature which is defined in another way. Determination of the excitation temperature from spectrum line intensity (equivalent widths) have led sometimes to contradictory results, partly because of the line transition probabilities values are poorly known, and that the lines are formed at different depths in solar atmosphere, and also due to deviations from the Local Thermodynamic Equilibrium (LTE) condition. If the distribution of the atoms in different levels is a result of collisions only, the excitation temperature would equal the kinetic temperature.

6.2.5.6 *Ionization Temperature*

The *Ionization temperature* T_I is found by comparing the number of atoms in different states of ionization. It can be determined from the level of ionization in an atmosphere, if the electron pressure is known. Using the Saha ionization equation, one can determine the ionization temperature as follows:-

$$\frac{N_1 P_e}{N_o} = \frac{(2\pi m)^{3/2}(kT)^{5/2}}{h^3} \frac{2u_1(T)}{u_o(T)} e^{-\chi_o/kT} . \qquad (6.11)$$

For numerical calculations, following logarithmic form is more useful:-

$$\log \frac{N_1}{N_o} p_e = -\frac{5040}{T} I + 2.5 \log T - 0.48 + \frac{\log 2u_1(T)}{u_o(T)}, \qquad (6.12)$$

where T is the ionization temperature, I denotes the ionization potential in electron volts, p_e is the electron pressure in dynes cm^{-2}, N_I is the number of ionized atoms cm^{-3}, N_o is the number of neutral atoms cm^{-3}, and $u_I(T)$ and $u_o(T)$ are partition functions of the ionized and neutral atoms respectively. The u's can be calculated with the aid of a term table for the atom or ion in question as a function of the temperature and electron pressure.

6.2.6 *Position Determination of Solar Features*

The heliographic position of solar features on the disk is an important and essential parameter for many investigations, such as the study of sunspot development, movement of spots, solar rotation, proper motion of sunspots, active regions, flares, and other transient phenomena. We may wish to correlate and identify solar features observed in various wavelengths. Several positioning methods are in use. In this section we present some simple methods to determine the heliographic coordinates on a full disk photograph or from a drawing of the solar image obtained through an equatorial or alt-azimuth mounted telescope.

Two terms are used in the following paragraph: 1) 'location' and 2) 'position', to mark a certain solar feature. By 'location' we mean where the feature is on the *solar image*. By 'position' we mean where on the Sun itself. The location on the image can be given in either Cartesian (x, y) or polar (r,θ) co-ordinates. Converting from Cartesian to polar system is done by:-

$$r = (x^2 + y^2)^{½}\ ,\ \ \theta = \text{arc tan } y / x, \tag{6.13}$$

and from polar to cartesian by:-

$$x = r\cos\theta, \qquad y = r\sin\theta \tag{6.14}$$

In Figure 6.2 various heliographic co-ordinates such as L, L_0, B, B_0, and the P angle are explained. The heliographic latitude, L of any solar feature is measured from 0^0 to + 90^0 from the solar equator to north, and from 0^0 to - 90^0 to the south, similar as the Earth's latitude is measured. For measuring the longitude of a feature on the Sun there is no fixed point, or meridian, on the Sun, from which one can reckon longitude, as in the case of the Earth where we have the zero meridian passing through Greenwich, England.

For this purpose, an internationally agreed *'zero'* meridian L_0= 0 is defined as that central meridian which passed through the apparent center of the disk on 1 January 1854 at Greenwich local noon, called the *Carrington meridian.*

From this so called Carrington central meridian, we count the heliographic longitude (positive) towards the West from 0^0 to 360^0. The change in the position of the zero meridian is expressed by the heliographic longitude L_0 of the apparent center of the Sun's disk, which lies between 360^0 and 0^0. L_0 is given in the Astronomical Almanac for each day at 0 hour UT.

A convenient method had been devised to keep track of the synodic solar rotation (synodic rotation period is 27.2753 days, which includes

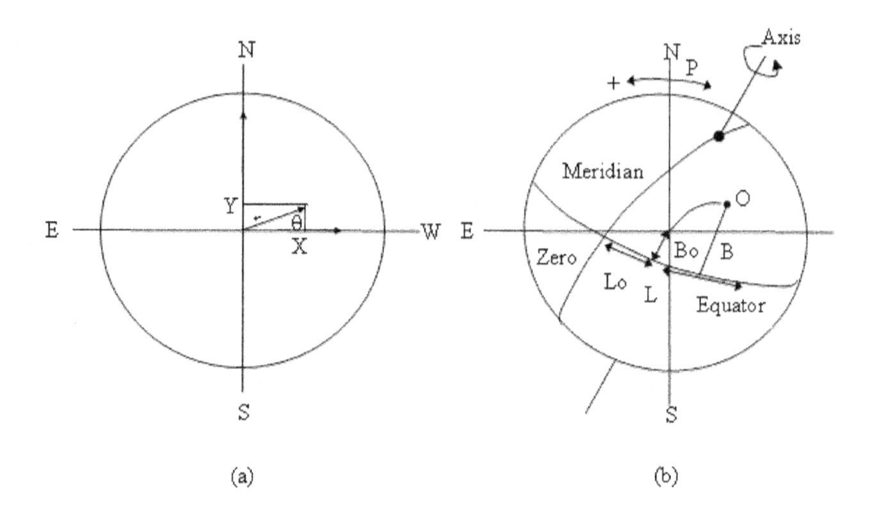

Figure 6.2 Explains the various heliographic co-ordinates. (a) Cartesian and polar co-ordinate system, (b) showing heliographic latitude B, longitude L of any feature O on the solar disk. Position angle P of solar axis is measured from the Earth's north point on the disk, and is + eastward. B_0 and L_0 are respectively the heliographic latitude and longitude of the center of Sun's disk.

both the true solar rotation on its axis, that is the equatorial sidereal period (25.38 days) and that due to the Earth's orbital motion around the Sun), this is called the *Carrington rotation number*. The beginning of each synodic rotation is that instant at which the solar meridian L_o passes through the central point of the apparent solar disk, i.e., when the heliographic longitude L_o of the central point is zero. The beginning of Carrington rotation number 1, is assigned as the meridian which crossed the disk center on 9 November 1853. The Carrington rotation number (CRN) for any epoch can be derived by the following relation:-

$$CRN = \text{int}\left[R_o + \frac{(JD - JD_o)}{27.2753}\right], \tag{6.15}$$

where int(x) is given in the nearest integer $\leq$ x. Here R_o is the Carrington number for any known day and can be obtained from an almanac, JD_o is the Julian day number for that day, also available from an almanac, JD is the Julian day number for a particular day for which Carrington rotation number is required. Although the Sun does not rotate as a rigid body, the value of a 'mean' period is assumed for the purpose of defining heliographic longitude and the Carrington rotation numbers.

The Sun's rotational axis in inclined to the ecliptic plane by an angle of $7^{\circ}.25$. This angle was determined by Carrington and since then this value has been used. The tilt of the Sun's axis was discovered from the observations of sunspots traversing the solar disk. At times during the year, sunspots appear to move in straight lines across the disk, while at other times in semi-elliptical paths. This suggested to early observers that the Sun's N-S axis is inclined at an angle to the ecliptic plane. During the course of the year, the Sun's North and South hemispheres are alternately more inclined towards us. This angle, or the heliographic latitude B_0 of the center of the solar disk varies between $\pm 7^{\circ}.25$. In June and December when the $B_0 = 0$, sunspots appear to traverse in straight line.

The position angle P of the Sun's N-S rotational axis in the sky, also varies during the year. This angle is determined by superimposing the Earth's equator on to the ecliptic inclined at an angle of $23^{\circ}.37$. Due to the combined effect of both these tilts (B_0 and P), the P angle varies between $\pm 26^{\circ}.37$. For positive P angles, the solar axis is inclined towards the East while for negative P, to the West. Daily values of P, B_0 and L_0 are given in the section for Ephemeris for Physical observations of the Sun in the Astronomical Almanac.

6.2.6.1 *Determining Solar E-W*

To accurately determine the heliographic co-ordinates of solar features (sunspots, faculae, filaments, flares etc.) on the Sun with the help of small refracting telescope, it is essential to know as accurately as possible the orientation of the solar equator (E_S - W_S) or solar (N_S - S_S) axis. To achieve this, the following simple procedure is generally used:-

A solar image is formed by an equatorial mounted telescope with a diurnal drive. The solar image is enlarged to about 150 to 180 mm in diameter by an un-cemented Huygens or Ramsden eyepiece and projected on a white screen. A circle of the same diameter as the solar image is drawn on a paper and placed on the projection screen. The solar image is centered on the circle and the telescope drive is switched off. The image will drift from East to West due to the Earth's rotation. As the limb of the Sun intersects the circle at two points, these two points are quickly marked by a soft pencil and the line joining the two points P_1 & P_2 mark the Earth's North – South axis, as shown in Figure 6.3, while a perpendicular line to this defines the Earth's E-W in the sky plane. After making the N-S markings, the telescope drive is again switched on, care should be taken that the telescope remains sharply focused and that the diameter of the solar image fits the circle accurately. This is important because during the year, the Sun's apparent semi-angular diameter changes from 16'15".93 arc (around 5th January) to 15'43".83 arc (around 5th July), and thus the size of the solar image also changes.

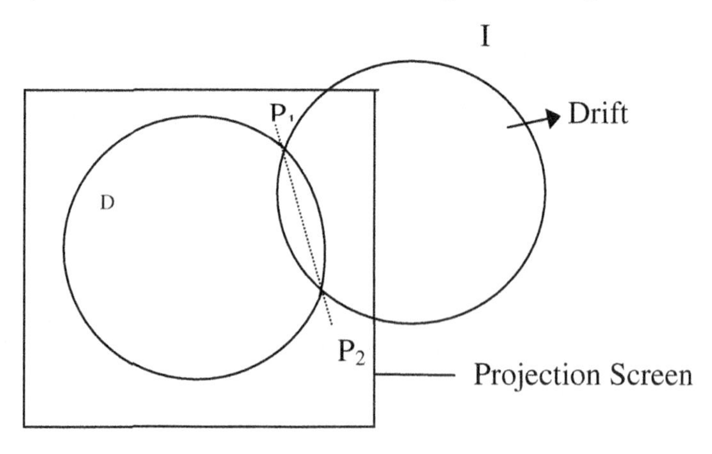

Figure 6.3 Showing the method to determine the projection of the Earth's North – South (N_E - S_E) axis by the 'drift method'. D indicates the circle drawn to the same size as the solar image on the projection screen. I is the drifted solar image, with P_1 – P_2 the intersection points of the image and circle marking N_E - S_E .

To know which is the North and South hemisphere the Sun, tip the telescope slightly towards higher declination, the North point in the sky will disappear last from the field of view. For better accuracy this procedure is repeated several times before and after taking the observations. The same procedure can be followed for fixed focus telescopes like heliostats or coelostats for each setting of the instrument every day. But in the case of equatorial telescopes, it may not be necessary to determine the North-South axis every day. Generally two fiducial marks are permanently made near the focal plane and recorded on the photographic plate to mark the N-S direction of the Earth's axis. At some observatories, a ring with fiducial marks is placed just before the photographic plate, which is turned by the appropriate P angle for the particular day and time of observation.

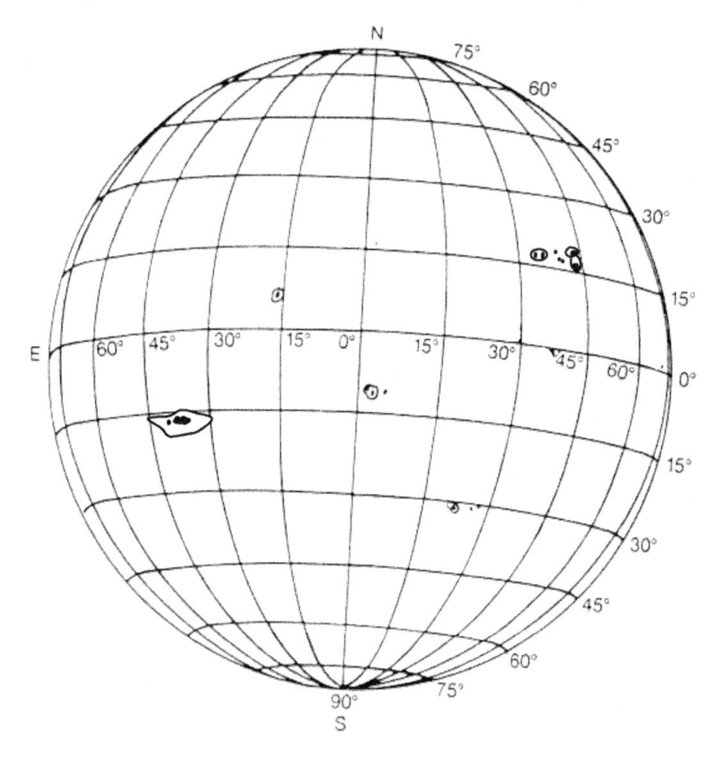

Figure 6.4. Illustrating the use of a Stonyhurst disk on a drawing. The heliographic co-ordinates, e.g., heliographic latitude and longitude of any solar features are directly read off from this grid

Thus a permanent record is imprinted on a full disk picture with the solar N-S axis marked. Sunspots visible on the Sun can also be used to determine the sun's N-S axis on the solar image by the same drift method.

Photographic techniques give higher accuracy to determine the solar axis. The photographic technique is as follows:-

Double exposures of two full disk solar images are made on a single plate, the first at one instant and then a second exposure after drifting the image for a little while. Thus one obtains two solar images on the same plate, intersecting near the solar limb, and the line joining the two intersecting images mark the geocentric N – S axis projected on the sky.

Depending on the accuracy required, one could determine the heliographic co-ordinates either by using overlay grids known as *Stonyhurst disks,* as shown in Figure 6.4, or calculate mathematically by measuring the Cartesian or polar co-ordinates of solar features and converting them into heliographic co-ordinates, using the formulae given in this section.

6.2.6.2 *Grid Overlay Template Method*

To determine the heliographic co-ordinates of solar features to fair accuracy, the easiest, quickest and widely used method is that of the overlay grid. A Stonyhurst disk with heliographic latitude and longitude printed on a transparent sheet is used as an overlay template. A set of 8 Stonyhurst disks corresponding to B_0 values from 0 to 7 degrees are shown in Figure 6.5. Copies of these Stonyhurst disks can be made on transparent sheets of appropriate size, corresponding to the solar image used, and with the help of these grids the coordinates of the solar features can be directly read off. As the value of B_0 varies over a small range of $\pm$ $7^{\circ}.25$, generally 8 grids are sufficient ($B_0 = 0^{\circ}$, 1°, 2°, ... 7°) to yield an accuracy of about $0^{\circ}.5$ in latitude and longitude determination. An appropriate Stonyhurst grid for a particular day's B_0 value is overlaid on the full disk drawing or on the full disk photograph of the Sun, so that the equator coincides with the Earth's projected equatorial plane on the solar disk (which is determined earlier by the drift method). The central meridian line is turned by the P angle for the particular day, P being the position angle of Sun's N – S axis with respect to the Earth's axis. For a positive P angle, the solar N point will be towards the East, and for negative P angle towards the West. For negative B_0 values the Sun's South pole is pointed towards the Earth and the Stonyhurst disk is so orientated that the South pole faces the Earth. The same disk is turned upside down for positive B_0 values, and positions of the solar features are read off from the grid. Once the solar N-S or E-W line is correctly determined, B_o and the P angles are aligned on the drawing or photograph with the Stonyhurst disk template and the heliographic coordinates B' and L' (angular distance in longitude from the central meridian) can be directly read off to better than $\pm 0^o.5$ accuracy.

To convert further these approximate measured co-ordinates into more accurate B and L heliographic co-ordinates, the following equations may be used:-

$$\sin B = \cos B_o \sin B' + \sin B_o \cos B' \cos L', \quad (6.16)$$

$$\cot L = \frac{\cos B_o}{\tan L'} - \frac{\sin B_o tanB'}{\sin L'}. \quad (6.17)$$

6.2.6.3 *Mathematical Method*

This method is based on the measurement of the polar co-ordinates (r, θ) which can be determined from the Cartesian co-ordinates x and y of a solar feature on a projected drawing or a full disk photograph. Refer to Figure 6.2, where θ is measured from the solar E-W line, along the north-east-south-west direction. The angular distance ρ of a solar feature from the solar disk center is obtained from the equation:-

$$\sin \rho = r / R, \quad (6.18)$$

where R is the radius of the projected image and r is the distance of the feature from the disk centre. To calculate the heliographic latitude B and heliographic distance l from the central meridian, the following equations are used:-

$$\mathrm{sinB} = \cos\rho \mathrm{sinB_o} + \sin \rho \cos B_o \sin \theta, \quad (6.19)$$

$$\sin l = \frac{\cos\theta \sin \rho}{\cos B}. \quad (6.20)$$

Any of the above two methods provide directly the heliographic latitude B of a solar feature but not the heliographic longitude L. To determine the 'true' heliographic longitude L, from the measured heliographic distance l, the following relation between the heliographic longitude L_0 of the central meridian distance l is used. L_0 for each day is given in the Astronomical almanac:-

$$L = L_0 + l. \qquad (6.21)$$

Details of how to calculate the solar heliographic latitude and longitude are described by Peter Duffett-Smith in "Practical Astronomy with your Calculator", page 70, Publ. Cambridge University Press.

Degrees from Central Meridian

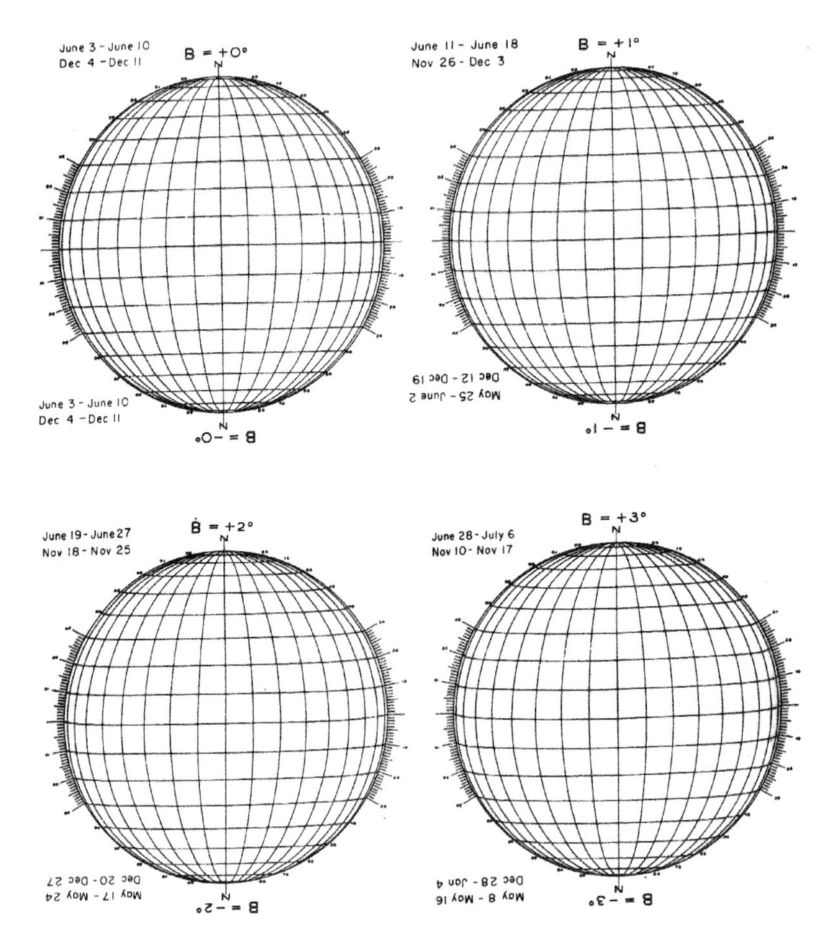

Figure 6.5(a) Stonyhurst disks for B = 0 to ± 3°, copies of these grids can be made of appropriate size on transparent sheets for measurement of heliographic coordinates of solar features for particular dates.

Degrees from Central Meridian

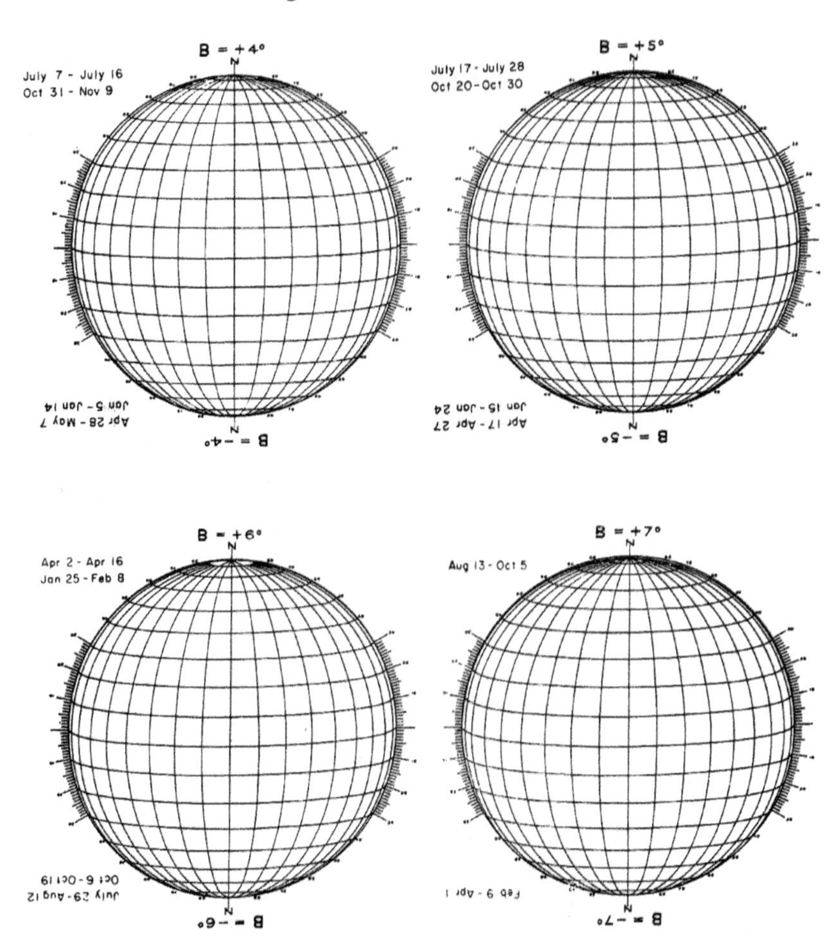

Figure 6.5(b) Stonyhurst disks for B = $\pm4^0$ to $\pm7^0$, copies of these grids can be made of appropriate size on transparent sheets for measurement of heliographic coordinates of solar features for particular dates.

Chapter 7

Solar Optical Instrumentation

7.1 Solar Optical Telescopes

Amateur observations of the Sun typically involve a refractor of modest size and a narrow band filter centered on H-alpha (6563 Å) for chromospheric observations, or more simply a yellow filter for the white light image. Or, there may be no filter and the solar image is projected onto a card for viewing as shown in Figure 7.1

Figure 7.1 Simple and safe method to examine the solar image is by projecting the image on a white card. The image is formed by a small 50-150 mm aperture size refracting telescope with an eyepiece, or even by a binocular.

Professional installations often consist of a solar feed, such as a coelostat or heliostat, which directs the sunlight to a fixed telescope, followed by a spectrograph or other instruments. A refractor on an equatorial mount has the advantage of simplicity and freedom from spurious instrumental polarization (certain types of magnetic field observations are compromised by telescope polarization). Depending on the transmission of the objective it will have limited wavelength coverage and also different wavelengths may have slightly different focus positions. If the accompanying telescope is a reflector the image focus is independent of wavelength. Various types of feeds include the coelostat, heliostat, polar siderostat, uranostat, and the altazimuth mount. The size of the mirrors and the focal length of the telescope determines the degree of vignetting at the solar image, if any.

7.1.1 *Coelostat*

A coelostat is a two-mirror system. A flat mirror, whose plane coincides with the polar axis and rotates at half the diurnal rate, directs sunlight to a movable second flat. This in turn sends the light along the optical axis to a fixed refracting or reflecting telescope objective. To accommodate the seasonal change in the Sun's declination, and to minimize the angle of incidence for reasons of induced polarization, one or both mirrors may need to be moved between observations. For eclipse observations at a temporary site such movement must be planned for before the eclipse. In the case of a coelostat, for any given setup the image does not rotate with time. But image orientation differs each time the 2nd mirror is translated. Pointing to an arbitrary point in the sky can be complex, although pointing to the Sun is of course simple. A photograph of the Mount Wilson 150-foot tower telescope along with the optical schematic of its 12-inch aperture telescope, and the 75-foot underground spectrograph is shown in Figure 7.2a. The two mirror coelostat arrangement is the most convenient means of attaining a fixed focus telescope. The major advantage of this set-up is that the image does not rotate as in the case single mirror heliostat. Major solar observatories around the world have coelostats as light fed. Even amateur astronomers prefer coelostats due to their simplicity and easy to build.

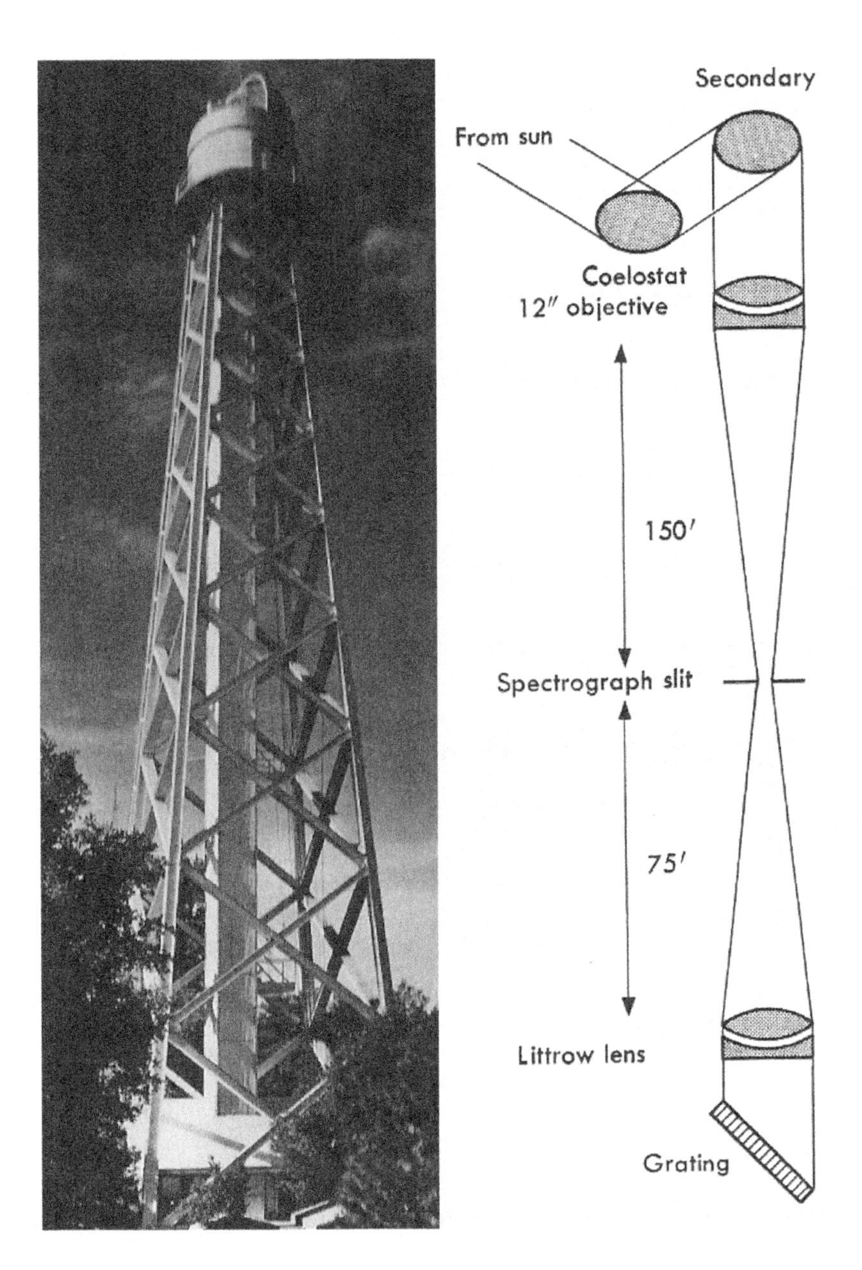

Figure 7.2 (a) The 150-foot solar tower telescope at Mount Wilson Observatory, along with a schematic of the telescope–spectrograph. It was here in 1908, that Hale discovered magnetic fields in sunspots.

Figure 7.2 (b) The 60-cm 2-mirror coelostat of Mount Wilson observatory. Observer Pam Gilman adjusts the 2nd flat mirror to direct the sunlight to a 30-cm objective triplet lens, forming a 38-cm diameter solar image at the base of the tower in the observing room.

7.1.2 *Heliostat and Siderostat*

In the case of heliostat, only one mirror is required to direct the sunlight, provided that the telescope and polar axis coincide, as in the case of McMath-Pierce facility of the National Solar Observatory on Kitt Peak in Arizona, USA, otherwise a second flat is called for. This system has the advantage that no seasonal mirror translation is required, but the disadvantage is that the image rotates at the diurnal rate. This may be a problem for high resolution eclipse observations where provision must be made for instrument rotation. Pointing is by right ascension (RA) and declination (DEC) coordinates. Reflection angles may not be optimum in terms of polarization. Because of RA and DEC pointing, non-solar objects can be observed (except near the pole of the sky).

In Figure 7.3a is shown a photograph of the 2.1m heliostat of the McMath-Pierce solar telescope, and the optical path of the telescope–spectrograph arrangement is indicated in Figure 7.3b.

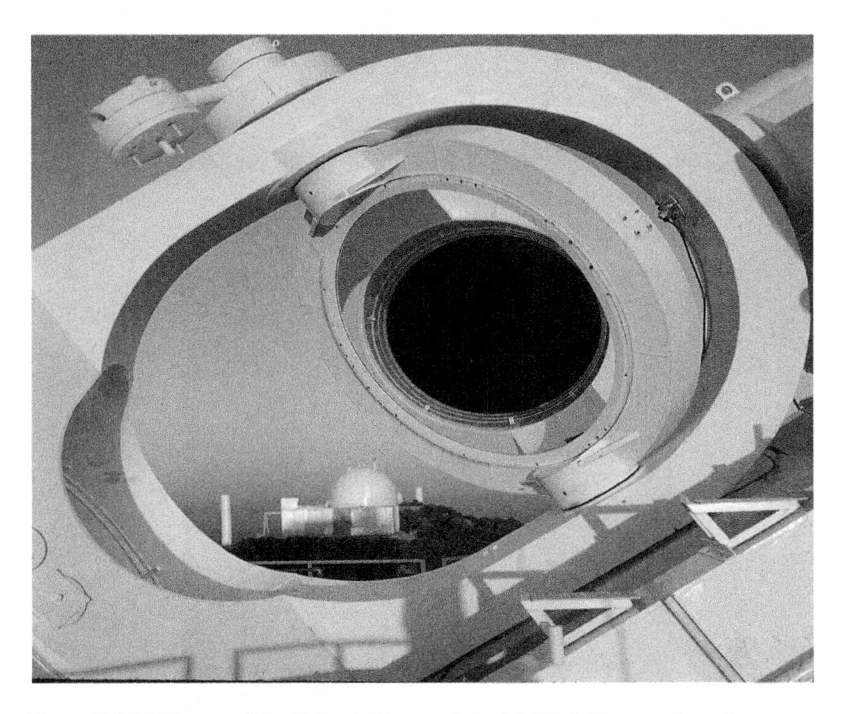

Figure 7.3 (a) Picture of the 2.1 m heliostat of the McMath-Pierce solar telescope. In the background is the 2.1-m stellar dome.

A polar siderostat is identical to a heliostat except the tracking mirror directs the light pole-ward. An uranostat is an off-axis siderostat which only tracks correctly when the Sun is near the meridian.

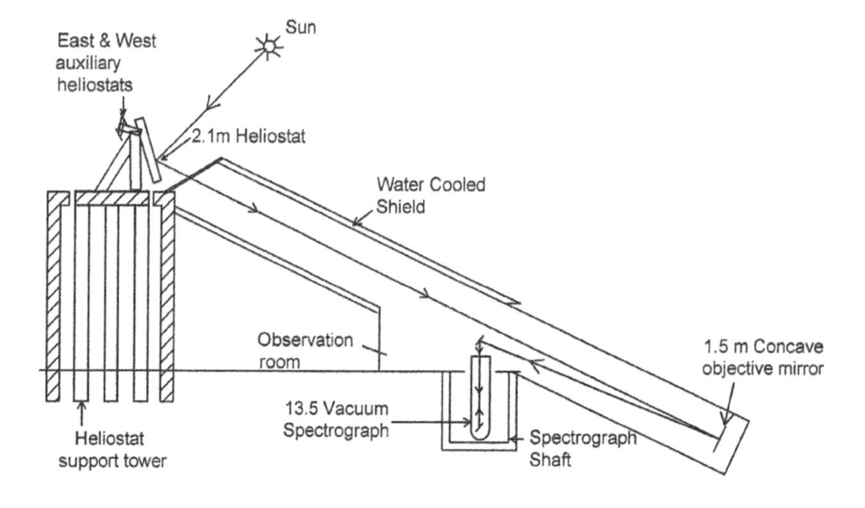

Figure 7.3 (b) Schematic of the optical path of the McMath-Pierce solar telescope and the 13.5-m spectrograph.

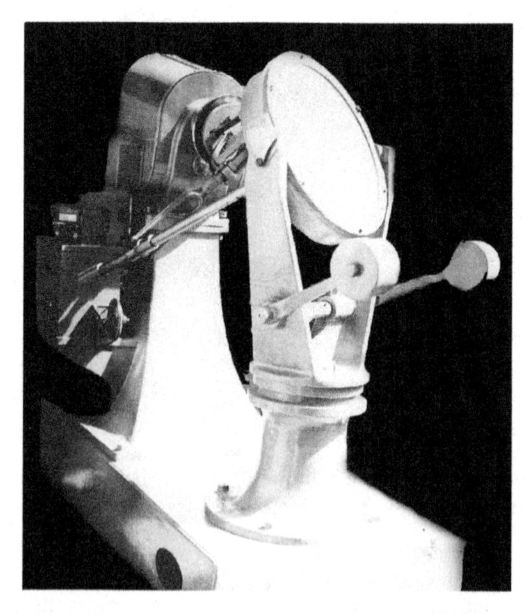

Figure 7.4 30-cm aperture siderostat of the Kodaikanal Observatory, India.

A hybrid siderostat can be an alt-azimuth single mirror heliostat which directs the sunlight along the north-south direction. Because of alt-azimuth mount of siderostat, polar axis tracking is only possible through a series of mechanical gears. In Figure 7.4 is shown a typical siderostat of the Kodaikanal Observatory.

Very large telescopes and systems where it is essential that all reflections be normal to the mirrors often employ an alt-azimuth mounting. Computer control of tracking is necessary and also compensation must be provided for the non-uniform image rotation. Observations become awkward near the zenith. The planned 4-meter aperture Advanced Technology Solar Telescope (ATST) will certainly be an alt-azimuth system. Pointing to non-solar objects may also be achieved by a computer.

7.1.3 *Coronagraph*

Outside of a total solar eclipse the corona is too faint to be seen with an ordinary telescope. Scattered light from the optics and atmosphere is several hundred times brighter than the white light corona. Bernard Lyot (1930, 1939), a French astronomer solved this problem by inventing the coronagraph.

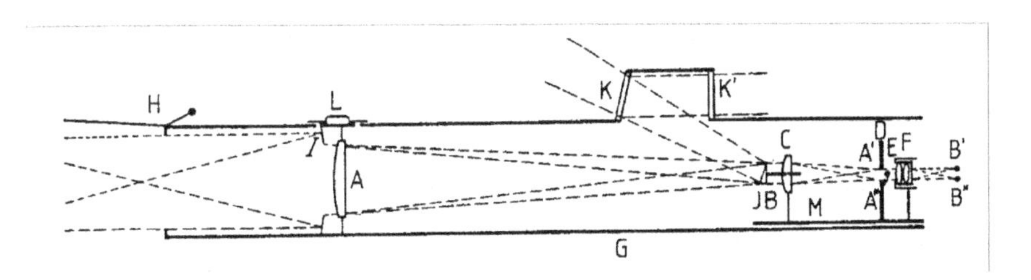

Figure 7.5 Schematic of a Lyot coronagraph. H is a long tube to protect the objective singlet lens-A from dust and stray light. The primary image is formed by this highly polished and scatter-free lens–A, at B, there is an inclined mirror which reflects the central portion of the solar image to a window K and out of the telescope. Any residual scattered light in the window is further reflected out through window–K′. Following the occulting disk J, a field lens C images the objective on a diaphragm D, this further reduces any stray light entering the focal plane. The final image of the inner corona is formed by an achromatic relay lens F onto the focal plane BB′.

First, the telescope objective must be a single lens, made of extremely homogeneous glass that is finely polished. In front of this lens is an extended baffle tube with a shutter opens only when in actual use. This is to prevent dust accumulation. Second, at the focus is a circular reflecting disk that reflects the photospheric light out from the telescope. Behind this `Lyot stop' is a field lens designed to correct the color error introduced by the singlet objective. Third, this telescope must be placed on a high mountain, preferably over 3000 meters, selected for its clear air. During times when scattered skylight is a few times 10^{-6} of the disk, the inner corona may be recorded. In Figure 7.5 is shown an optical diagram of a Lyot coronagraph. For coronal studies it is useful to observe in the light of emission lines formed only in the corona. These are the highly ionized lines of Fe XIV at 5303 Å, Ca XV yellow line at 5694 Å, and the red line of Fe X at 6374 Å. Observations in red or infrared wavelengths have an advantage since Rayleigh scattering, which makes the sky blue, has an intensity proportional to λ^{-4}. The sky in the infrared at 2 microns has an intensity of 0.004 compared to that at 5000Å. Scattering by aerosols goes down more like λ^{-1}, scattering by even larger particles is independent of wavelength. Flying insects can be a problem. A promising high altitude site in Hawaii had to be abandoned for coronal work because of insects.

To reach further into the IR, the next generation coronagraphs may be all reflective. Low scatter mirrors can now be made by means of 'super polishing' techniques. A 20-cm aperture mirror coronagraph is in use at El Leoncito, Sonjun, Argentina. The ATST hopes to operate also as a coronagraph. The aim is to measure magnetic fields in the corona, using the Zeeman sensitive line of Si IX at 3.9 microns.

7.2 Solar Image Guiders

A requirement for a solar telescope is good pointing stability and guiding of the image. We must keep our telescope pointed at the same place as long as our observations require. For any serious and long period observations, it is most important to minimize image jitter due to seeing and inaccuracy of the telescope drive. This is achieved through the use of photoelectric guiders. There are two kinds of guiders for image stabilizing: limb guiders and correlation trackers. Details of a simple photoelectric solar limb guider are given by Bhatnagar (2003). The guider should not be too sensitive to light level and seeing or focus. These conditions can be met with a vibrating mirror guider, which scans the limb and seeks the inflection point in limb intensity. Such image guiders are operational at the solar telescopes at Kitt Peak. Correlation trackers work on sunspots or on the granulation pattern. A problem with the correlation tracker is the low contrast of granulation in poor seeing. With the advent of fast data processors, now it has become possible to work on larger scale out-of-focus patterns. Such a technique has been developed by C. Keller of the National Solar Observatory which is almost seeing insensitive.

7.2.1 *Active Mirrors and Adaptive Optics*

The spatial resolution in arc sec of a telescope is set by the diffraction limit $\delta\theta$ of the objective:

$$\delta\theta = \frac{0.25(\lambda)}{d}, \tag{7.1}$$

where d is in meters and λ in microns. For example, the Sac Peak's Dunn Solar Telescope (DST), with an aperture of 0.76 metres, in the yellow at 0.5 microns, has a theoretical resolution of

$$\delta\vartheta = \frac{0.25(0.5)}{0.76} = 0.16 \text{ arc sec.} \tag{7.2}$$

More often the spatial resolution is about 1 arc sec (or worse). That means the telescope performs as though its aperture is only 13 cm. This value is sometimes called the 'Fried parameter'- r_o. What has happened is that the light from a point on the Sun got distorted from a flat wavefront into a distorted one when it arrives at the telescope. This happens because of temperature inhomogeneities in our atmosphere. By sensing this distortion with a *wavefront sensor*, and then appropriately *distorting* the optical system with a fast acting *`rubber mirror'*, the full diffraction limit can be restored from 0.13 to 0.76 m effective aperture. This technique of restoring the image is called *adaptive optics*. The active mirror consists of a mosaic of small mirrors each positioned by some sort of fast servo-mechanism. These mirrors may be discrete, or the surface may be continuous but flexible. Such an adaptive optical system has been incorporated into the DST on an experimental basis, and has achieved great success at the Swedish Solar Vacuum Telescope at La Palma.

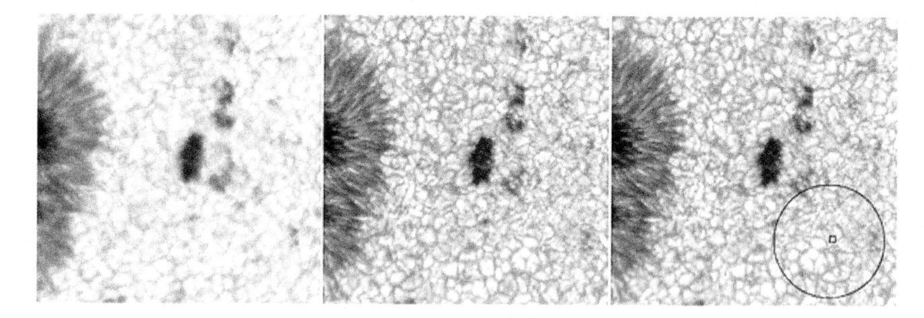

Figure 7.6 Left panel shows an image of an active region without AO, other two pictures show images with AO, displaying marked improvement in image quality.

In Equation (7.1) notice that the spatial resolution is dependent on λ/d. By observing in longer wavelengths, the same resolution is achieved with larger apertures. The Fried parameter, r_0 increases in size with wavelength. In the visible region of the spectrum, where the Fried parameter is small, a rubber mirror may require a hundred or more actuators to correct the wavefront. At a wavelength of 4 microns, the McMath-Pierce telescope at Kitt Peak with its 1.5-meter aperture objective, has a theoretical resolution identical to the Fried parameter. Diffraction limited 0.67 arc sec resolution is achieved without adaptive optics. The image will still move around somewhat but this movement can be restored with a fast acting guider or a `tip-tilt' corrector. Atmospheric compensation is thus much easier in the IR. Another important aspect of adaptive optics is that static optical aberrations of the telescope can also be corrected by a rubber mirror. Using the adaptive optics (AO) technique, very impressive images of the Sun have been obtained by the Adaptive Optics team at the Sac peak Observatory. In Figure 7.6 is shown a photograph of an active region taken with and without adaptive optics.

7.3 Spectrographs

A general purpose solar spectrograph is called on to transduce the solar spectrum from nearly 2900 Å to 200,000 Å, with a spectral resolution appropriate to the Doppler width of solar line (about 0.1Å in the visible). Many interloping terrestrial (telluric) lines have half of this width. A spectral resolution of about 0.01Å or 5000/0.01 = 500,000 around 5800Å is required for solar lines, not to be appreciably widened by the instrument. The design of a solar spectrograph could be of many types. As an example we describe the 13.5 meter focal length spectrometer of the McMath-Pierce facility which has the above resolution capability.

The f-ratio of this telescope is f/54, to which the spectrograph is *matched*. By 'matched', we mean that the telescope and its collimator have the same f-ratio, so that no light is lost by over-filling the mirrors. The design is a Czerny-Turner (C-T) type, which means the optical train consists of a collimator (fl=13.5m), a grating followed by a camera mirror of same focal length. Optical aberrations are minimized in the C-T design. Generally, two gratings are necessary to cover the above range of wavelength; for the visible region from 2900 to 20,000Å the grating has 632 grooves/mm over an area of 42x32 cm; for the IR, from 10,000 to 200,000 Å, the second grating has 120 grooves/mm over an area of 47x37cm. The two gratings are mounted on a spindle which can be rotated into one of the two positions. As the theoretical resolving power (RP) of a grating is given by mN, where m is the order and N the total number of grooves, the visible grating in the 5th order has a RP = 5(470)(620) =1,327,200. In practice RP is set by slit widths. For a slit width of 0.1 mm the RP will be 350,000. Often to increase the photon flux, and where spectral resolution is of secondary importance, a slit width of 0.5 mm is used and the RP drops to 71,000. In the IR region, a slit width of 0.5 mm or even wider may be optimum for diffraction reasons. For precise work in the visible and UV, the spectrograph can be operated in the `double pass' mode. This means that, instead of exiting the spectrograph, the dispersed light after the camera mirror is sent through an intermediate slit and passes the grating a second time, before finally exiting the spectrograph. It is as though we have two spectrographs in tandem. The result is that scattered light in the spectrograph can be measured by shuttering the intermediate slit and subtracted off. This may be important at short wavelengths where precise line depth information is desired. An optical layout of the National Solar Observatory's double pass spectrograph, designed by Keith Pierce (Pierce 1963) is given in Figure 7.7.

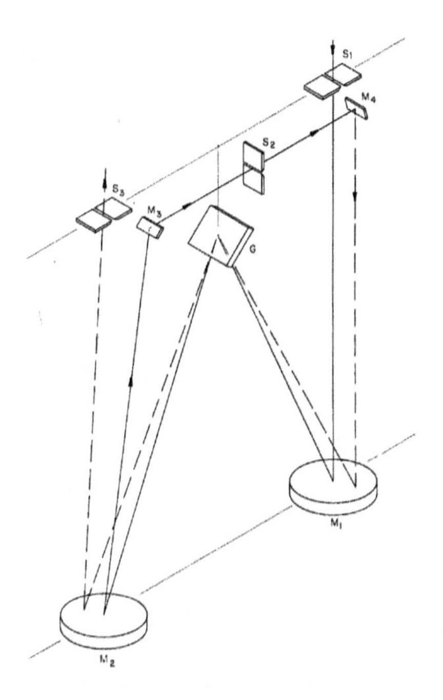

Figure 7.7 Optical layout of a double pass spectrograph. S1 is the entrance slit; M1 and M2 are the collimator and camera spherical mirrors. The rays shown by continuous line form the first spectrum and is picked up by mirror M3 and sent through an intermediate slit S2 and returned back to mirror M4, and to the same spectrograph consisting of M1, G-grating and M2. The rays forming the double pass spectrum are shown in dotted lines. As the major portion of the scattered light is eliminated at slit S2, the double pass spectrum formed at exit slit S3 is free to a large extent from scattered light and grating ghosts.

Another spectrum 'mapper' is the Fourier Transform Spectrometer (FTS), which is based on the Michelson's interferometer principle. The FTS offers several advantages; wide spectral coverage in a single observation, resolution dependent on the path difference of its scanning mirrors and as high as 1,000,000 in the visible, and wavelength stability equal to the stability of its control laser. The spectrum range is determined by mirror coatings and detectors. The 1-m FTS at the McMath-Pierce facility is capable of mapping from the atmospheric ozone cutoff at 2900 Å in the violet to 200,000 Å in the infrared. Spectrum maps covering this range, with solar line identifications, are available from the National Solar Observatory in Tucson, Arizona. Compared to the spectrograph, it is less suited for high spatial resolution and high signal/noise scans of limited spectral coverage. Detectors for the above equipment are typically silicon devices in the visible and InSb or arsenic doped silicon out to 20 microns.

Ancillary to the spectrograph are prism pre-dispersers, which separate over-lapping grating orders, particularly in the violet and UV. 'Image slicers' are useful for converting a square image area on the Sun into a slit-like array to match the spectrometer input. For example, an area 1x1 mm on the image can be transformed to 0.1x1 cm to study details of a granulation field or a sunspot.

7.4 Imaging the Sun

White light or broad band pictures are useful for granulation and sunspot studies. Recent observations obtained with the 1-meter aperture New Swedish Solar Telescope in the Canary Island at La Palma have resolved solar structure to better than 0.1 arc sec, see Figure 4.2. These high resolution pictures show the existence of solar structures smaller than 70-100 km in size. Such high spatial resolution images have been obtained using real-time frame selection, adaptive optical system and image restoration based on '*phase-diversity*' techniques, over a small field of view either in broad band or through G-band filters. Atmospheric distortions may be thought of as phase errors, where image points have multiple phase delays. Phase retrieval is possible if two images are recorded, one an in-focus 'conventional' image, and other one by a known amount *out-of-focus* image called the 'diversity' image. This technique requires massive amount of post processing computations to retrieve high resolution images, but this has recently become feasible with the advent of fast and large computers.

For chromospheric and special purpose observations, such as the velocity and magnetic field measurements, narrow passband solar images are required. This is achieved by either of the two techniques; 1) photographs taken through very narrow pass-band (~0.12-0.5Å) filters, or 2) pictures made through a spectroheliograph. Most monochromatic pictures in the chromospheric lines are taken in Hα, Ca K, or He 10830Å, and are obtained using birefringent filters or other types of narrow passband filters. Birefringent filters use calcite crystals which have become difficult to obtain of required optical quality. These filters are sometimes called Lyot filters. In the 1950-1990 era, Messers B. Halle Company in Germany turned-out perhaps more than 200 such filters. The bandwidth of a Lyot filter made by Halle was 0.5Å at Hα. The Carl Zeiss Company in Germany also made high quality Lyot filters having a passband as narrow as 0.25Å. These firms have now discontinued making narrow passband birefringent filters. Locating and rebuilding a used birefringent filter is one method to acquire a filter. Good quality Lyot filters are also available from the Nanjing Optical Instrument Factory in China. Lesser quality filters, as compared to birefringent filters, are also available in the market from Messers DayStar and Coronado Filters instrument companies, in USA. Both of these outfits use the Fabry-Perot principle for making filters of passband about 0.5-0.7Å centered at the H-α and Ca K chromospheric lines. The DayStar filters use mica and Coronado uses glass as substrates. Even the Interference filters with bandwidths of 1 to 4 Å can be used for broad-band solar prominence observations. A spectroheliograph can also be used to obtain a 2-dimensional monochromatic photographs of the solar chromosphere. George Hale in America, H. Deslandres in France, and John Evershed in England independently invented the spectroheliograph in the late nineteenth century, and obtained remarkable monochromatic images of the solar chromosphere. Detail description of spectroheliograph is given in the subsequent section.

7.4.1 *Spectroheliograph*

In 1891 George Ellery Hale (1868-1938) at his private Kenwood Observatory in Chicago invented an instrument to photographically record monochromatic images of the solar chromosphere. It combines the principle of a spectroscope with the photographic plate to yield a permanent record. Around the same time, in 1891-92, H. Deslandres (1853-1948) at the Meudon Observatory in France and John Evershed in England independently devised similar instruments. The spectroscope breaks up the Sun's light into its spectrum of colors crossed by thousands of dark Fraunhofer absorption lines. These dark lines are essentially images of the spectroscope slit and convey spectral information about the Sun in those lines. To observe another portion of the Sun, one must either move the slit or the image. A series of slit images placed side by side records a portion of the Sun or the whole Sun in that line. In Figure 7.8 is shown an optical diagram of a spectroheliograph.

To observe the Sun in a particular line, say in the H-α line of hydrogen at 6563Å, we move the image of the Sun across the slit, we will see in the focal plane the spectrum of solar features falling on the slit. To isolate a particular spectral line, a second slit is placed at the exit focal plane of the spectrograph and a photographic plate is placed in the focal plane behind the slit.

As the solar image and the plate are moved at the same speed one builds a composite picture of the Sun as seen in that line. Or the solar image and the photographic plate can be kept fixed and the spectroheliograph is moved. In this way one obtains monochromatic pictures of the Sun in any desired wavelength. Spectroheliograph has the advantage that the bandpass can be varied and made very narrow to achieve high spectral purity. However it has the disadvantage that to make a picture of the Sun, the spectroheliograph takes considerable time exposing the plate. Scanning the image through a narrow slit is relatively slow and errors due to seeing and guiding may degrade image quality.

To make a monochromatic image the spectrograph is set to the desired wavelength, usually in the core or wing of a spectrum line. This wavelength interval is isolated by the exit slit. A photographic plate or an array detector is moved across the exit slit synchronous to the passage of

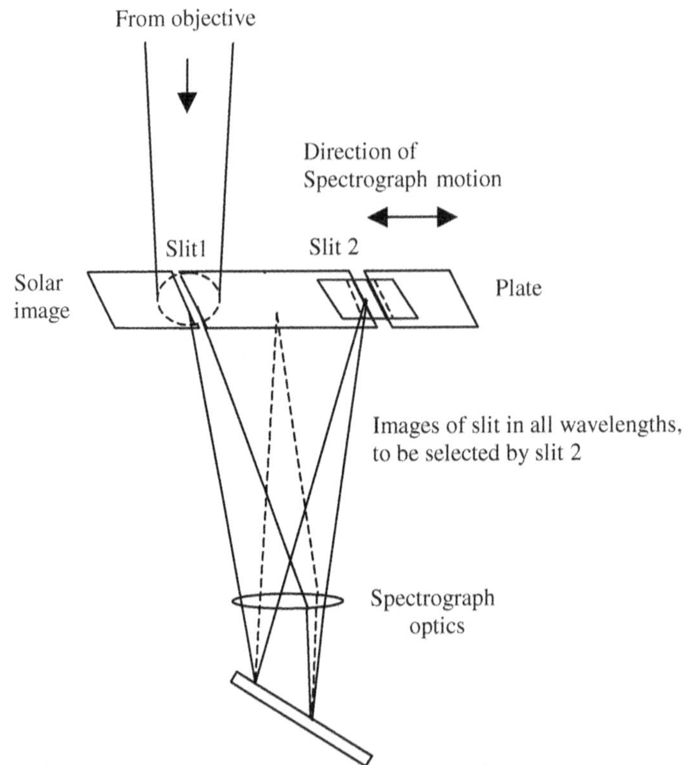

Figure 7.8 Optical layout of a typical spectroheliograph.

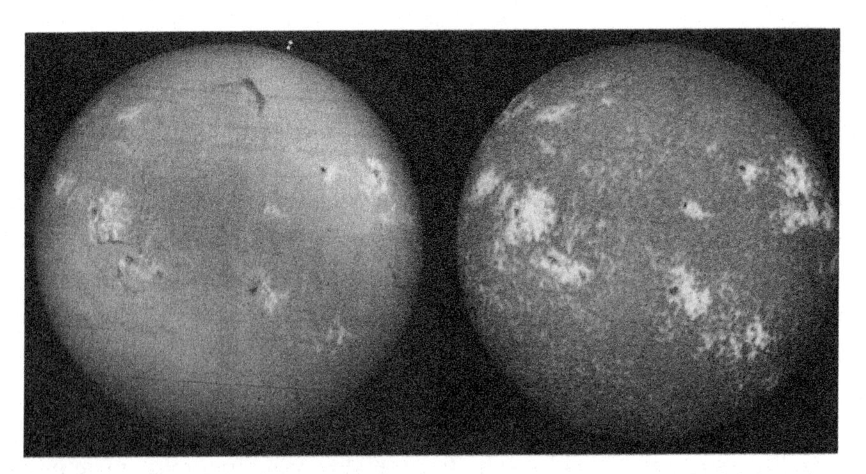

Figure 7.9. Spectroheliograms made in H-α (left) and K-line of CaII (right) on 1 April 1958 at the Kodaikanal Observatory.

of the image across the entrance slit. This results in a monochromatic image or spectroheliogram in that particular line. In Figure 7.9 is shown spectroheliograms made in H-α and K-line of CaII on the same day at the Kodaikanal Observatory. We see differences in the chromospheric structures.

7.4.2 *Narrow Band Filters*

To achieve a narrow passband filter there are several available methods; (i) by using the principle of interference in polarized light, (ii) by using a solid or air space Fabry-Perot (F-P) etalon filter, such as mica or electrically tunable Lithium Niobate F-P filter, or just an air-spaced F-P, (iii) employing the Macaluso-Corbino magneto-optical effect in the sodium or potassium resonance lines, (iv) using Michelson Interferometer, or (v) simply using narrow pass-band interference filters (3-4 Å pass-band filters have been used for solar prominence observations). Each of these systems has advantages and disadvantages.

7.4.2.1 *Principle of Lyot Type Birefringent Filters*

Among the most important contributions to instrumentation in solar astronomy was the invention of the birefringent filter. This was done independently by Y. Ohman (1938) in 1937 in Sweden, and by B. Lyot (1944) in 1933 in France.

The principle of the birefringent filter is to make 2 rays of polarized light interfere with itself, as a function of wavelength. This is achieved by decomposing a linearly polarized ray of light into two polarized components; the ordinary (o) and the extraordinary (e) rays, traveling in the same direction. A path difference is created between the two decomposed waves and then by recombining the two components, interference results. To understand the principle, let us allow white light to pass through a polaroid sheet and then through a birefringent plate, as shown in Figure 7.10(a). Looking at the exit end of the birefringent plate, nothing seems to happen to the light beam, but once another polaroid sheet is placed with its plane of polarization either parallel or perpendicular to the first polarizer some colors will appear excluded and some are transmitted. By placing the entrance polarizer at 45° to the fast and slow axes of the birefringent plate, two components of equal amplitude are decomposed by this plate.

A relative path (phase) delay is introduced between these components due to the difference in the propagation speeds of the ordinary (o) and extra-ordinary (e) rays traveling through the crystal plate. In the case of calcite, the o-ray travels faster than the e-ray (n_e-n_o = -0.17), while in quartz it travels slower (n_e-n_o = +0.009), where n_e and n_o are the refractive indices of the extra-ordinary and ordinary rays respectively. The phase difference (retardation) upon emergence from the birefringent crystal element depends on the thickness d of the crystal element, the birefringence of the crystal (n_e-n_o) and the wavelength λ. The retardation n is given by the following expression:-

$$n = \frac{d(n_e - n_o)}{\lambda} . \tag{7.3}$$

There is a fundamental thickness d_o for completion of this "in phase" and "out of phase" cycle which is determined from the formula:-

$$d_o = \frac{\lambda}{(n_e - n_o)} . \tag{7.4}$$

Since the fundamental thickness d_o depends directly on the wavelength, it will be different for other wavelengths, and this relationship is the principle of a birefringent filter. The exit polarizer forces the two waves, the ordinary and extraordinary to recombine, which results in interference. Now if the emergent light is analyzed with a spectroscope, the continuous spectrum is seen broken up into series of broad alternate dark and bright bands. This is called a *channel spectrum*. The spacing and number of bands depends on the wavelength λ and the thickness of the birefringent element. For some wavelengths, retardation n will be a whole integer number, and in this case the plane of polarization is rotated back to the same orientation. For other n-numbers, circularly or elliptically polarized light is observed. For ½ integers, the plane of polarization is rotated by 90° and the emerging light will be absorbed by the second polaroid. Thus a birefringent crystal rotates the plane of polarization by an amount proportional to its thickness d,

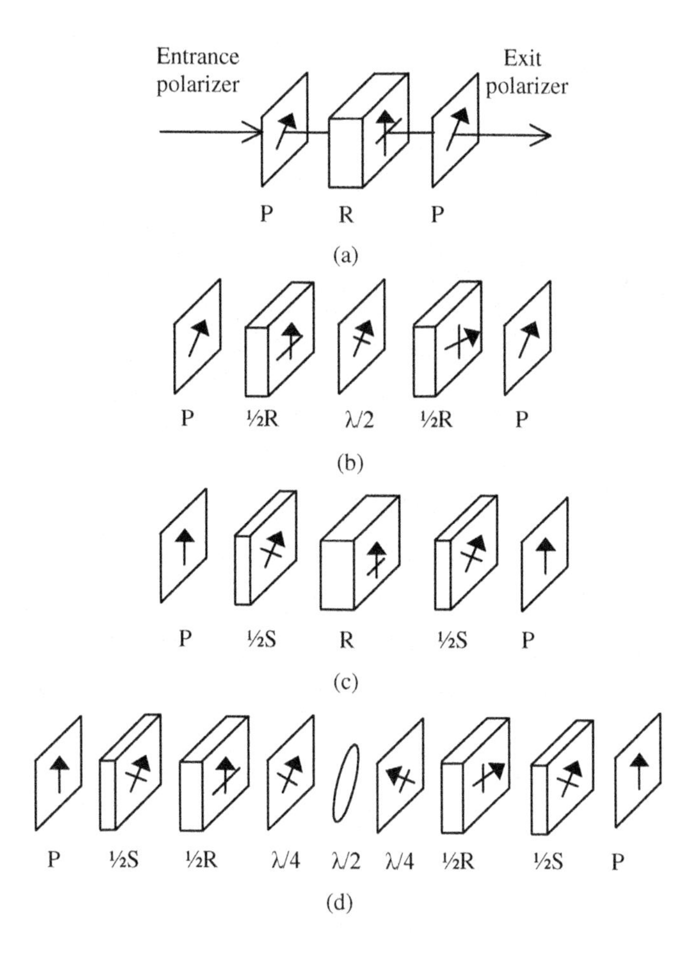

Figure 7.10 (a) Single birefringent element, with a retardation plate R between 2 polaroids P, with their axis at 45° to the axes of R. (b) Lyot's type wide field element retardation plate R is split into two halves, with axes crossed and separated by a λ/2 wave plate. (c) Evan's scheme of split elements for a wide field configuration. One retardation plate R lies between the two halves of another retardation plate S. (d) Evan's scheme of wide field element with tuning of the inner element. The inner retardation plate R, as in (c), is split and separated by a pair of λ/4 wave plates, and in between is put a rotatable λ/2 wave plate. The retardation plate R is tunable but not S.

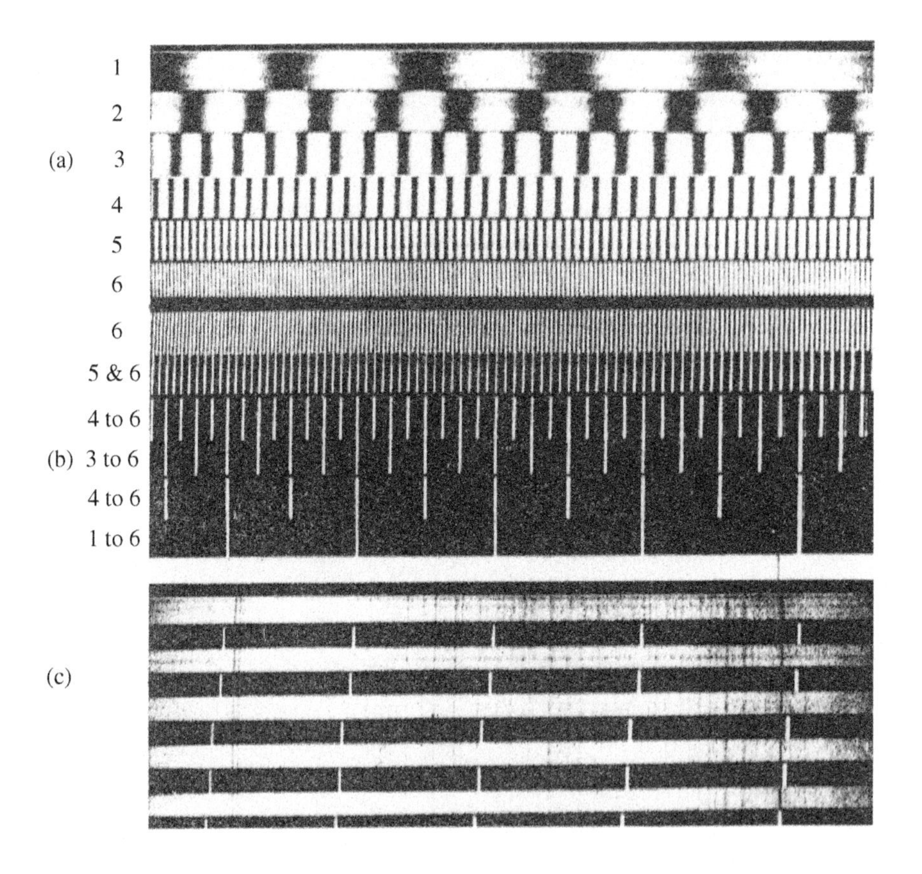

Figure 7.11 (a) Showing 6 channel spectra formed by birefringent elements of increasing thickness in ratio of d, 2d, 4d, 8d, 16d, 32d. (b) 5 strips of channel spectra formed due to the combination of elements 5 & 6; 4,5 & 6; 3, 4,5 & 6; 2,3,4,5 & 6; and 1,2,3,4,5 & 6. (c) The passband as the result of all 6 elements, along with the solar spectrum around the H-α line. With increasing temperature the transmission band shifts towards short-wave lengths for quartz elements.

divided by the wavelength λ. If the birefringent plate is thin, few interference bands are seen in the spectroscope, as shown in Figure 7.11 (a-1). As the plates get thicker and thicker, the number of maximum and minimum bands increases, as seen in Figure 7.11(a-6). A typical modern day Lyot filter consists of series of birefringent plates of calcite and quartz, along with linear polarizers, half and quarter wave plates.

Generally the clear aperture of such Lyot filters is about 30-35 mm and it is housed in a temperature-controlled oven. In Figure 7.12 is shown the layout of a typical Lyot filter of 0.5Å pass band. A broad band blocking interference filter of about 40Å pass band, centered on the line of interest is placed in front of the birefringent elements to isolate the adjoining maxima in the 'free spectral range' of the filter, which is decided by the thinnest element. Practical techniques for making birefringent filters out of quartz for observing solar prominences are given in three very detailed articles by Dunn (1951), by Paul (1953), and by Petit (1953).

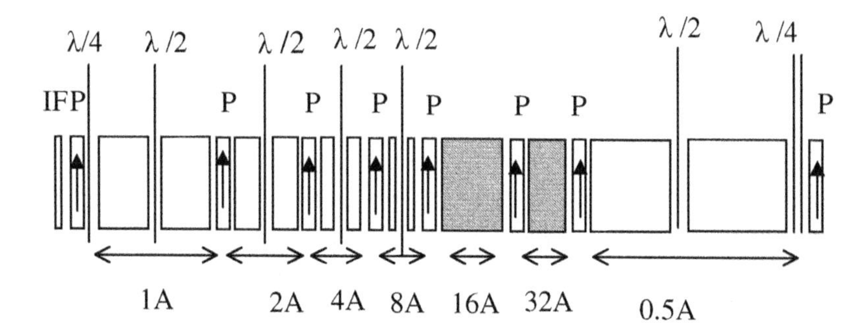

Figure 7.12 The layout of a typical wide field Lyot filter of 0.5 Å passband. It consists of series of birefringent elements; some are split elements to make the filter wide field, polaroids, quarter and half wave plates. Just after the 0.5 Å element, a λ/4 plate and a rotatable λ/2 plate are put before the exit polaroid to tune the filter across the 1 Å free spectral range of the filter.

7.4.2.2 *Principle of Birefringent Šolc Filter*

Ivan Šolc (1965) showed that it is possible to construct a birefringent filter *without* any intermediate polarizers. This has a great advantage as the major attenuation of light takes place in the large number of polaroids used in a Lyot type filter. Fewer the number of polaroids, the greater the throughput. Šolc filters are of two designs: '*fan*' and '*folded*' types. N identical birefringent plates of same thickness are stacked, and the orientations of the optical axis of the plates are arranged as α, 3α, $5\alpha, \ldots (2\alpha\text{-}1)\alpha$, where $\alpha = 45^\circ/N$. Thus a four-plate *fan filter* would have its plates with their optic axes oriented at $11^\circ.25$, $33^\circ.75$, $56^\circ.25$ and $78^\circ.75$ and the entrance and exit polarizers are oriented parallel at 0°. In the case of a *folded filter*, the birefringent plates of identical thickness are oriented with their axes alternately at $+\alpha$ and $-\alpha$. A four-plate folded filter would have its plates oriented at $11^\circ.25$, $-11^\circ.25$ and $11^\circ.25$ and $-11^\circ.25$. The entrance polarizer is placed at 0° and the exit at 90°.The spectral transmissions of the fan and folded Šolc filters are similar. Evans (1958) has shown that the transmission profile of an N plate Šolc filter is given by:-

$$T(\lambda) = [\frac{\sin N\chi}{\sin \chi}\cos\chi \tan\alpha]^2, \tag{7.5}$$

where $\cos\chi = \cos\alpha \cos \pi c \Delta t / \lambda$ and α is $45^\circ/N$.

Comparing the transmission profiles of Šolc and Lyot filters, for the same number of total crystal thickness, it is noticed that Lyot filter has slightly lower secondary maxima (side lobes) for similar bandwidths.

7.4.2.3 *Transmitted Intensity through Birefringent Filter*

Let us determine the intensity of the light transmitted by a birefringent filter as a function of the wavelength λ, assuming that the incident beam is normal to the plane of the crystal plate and neglecting the light loss in the plates and polaroids. Let A be the amplitude of the light vibration emerging from the first polaroid, λ its wavelength, N the number of plates, d the thickness of the thickest element and $\mu = (n_o - n_e)$, the difference between the ordinary and the extra-ordinary indices of the plates. The vibration A is decomposed by the thinnest plate into two vibrations of amplitude $A/\sqrt{2}$ with a phase difference of $\varphi = 2\pi\mu d/\lambda$. The second polarizer renders them parallel and reduces their amplitude by A/2. The two rays then combine into parallel vibrations as they emerge from the second polarizer and the resulting amplitude is given by $A\cos(\phi/2)$ or $A\cos(\pi\mu d/\lambda)$. The resulting intensity is given by $A^2\cos^2(\pi\mu d/\lambda)$. Now we do the same for each plate of the filter with plate thickness twice the previous one. Say we have N plates, and then the intensity transmitted by combination of all the plates in the filter is equal to the product of the intensities transmitted by each plate:-

$$A'^2 = A^2\cos^2(\varphi/2)\cos^2(\varphi)\cos^2(2\varphi)\ldots\cos^2(2^{(N-1)}\varphi/2), \qquad (7.6)$$

where $\varphi = 2\pi\mu d/\lambda$.

The numerical calculation of the intensity of the transmitted light by the filter with N number of plates can be obtained by Equation (7.6). However, it is easier to obtain this by another expression, which can be obtained directly by combining the vibrations at the end of the filter. After the first plate and before the 2nd polarizer, we have two parallel vibrations of amplitude A/2 and with a phase difference of φ. Similarly after the 2nd plate and the 3rd polarizer, we get four parallel vibrations of A/4 and with phase differences of 0, φ, 2φ and 3φ. At the end of the filter we get 2N parallel vibrations of amplitude A/2n and phases 0, φ, 2φ, ...$(2N-1)\varphi$. Hence, their resultant amplitude is given by:-

$$A' = \frac{A\sin 2^N \varphi/2}{2^N \sin\varphi/2} = \frac{A\sin 2^N \pi\mu d/\lambda}{2^N \sin\pi\mu d/\lambda}, \qquad (7.7)$$

and the intensity is given by:-

$$A'^2 = \frac{A^2\sin^2(2^N \pi\mu d/\lambda)}{2^{2N}\sin^2(\pi\mu d/\lambda)}. \qquad (7.8)$$

According to Equation (7.6) or (7.8) the transmission of the filter passes through a maximum equal to 1 for every value of λ which makes the fraction $\mu d/\lambda$ a whole number, while it will be zero for other values of λ, which make the fraction $(2^N\mu d/\lambda)$ a whole number. For intermediate wavelengths the transmission intensity passes through a series of secondary maxima. This results in (2^n-2) secondary maxima between the two principal maxima. But if N is large, greater than 4 for example, their intensities tend to decrease rapidly towards zero, as one goes farther from the principal maximum. In Figure 7.13 is shown transmission versus frequency for components of a 4 element (A, B, C, D) Lyot filter and the

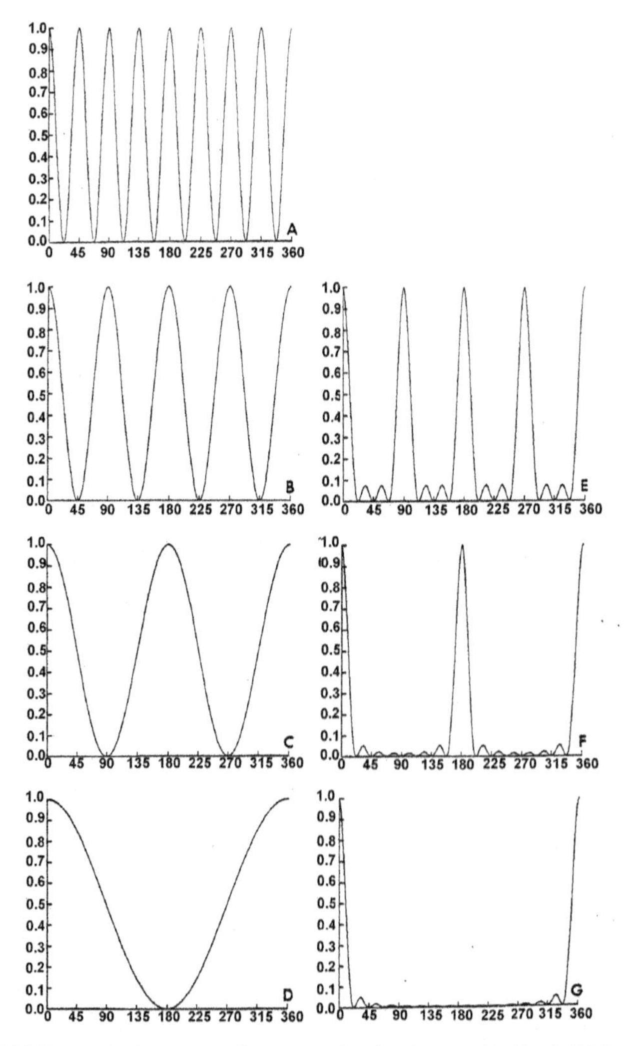

Figure 7.13 Transmission versus frequency for 4 - element (A, B, C, D) Lyot filter and transmission through successive stages of combinations of elements (E, F, and G).

transmission through successive stages E (A+B elements), F (A+B+C elements) and G (A+B+C+D elements), besides the series of principal maxima and secondary maxima. For transmission maximum at λ_o, the *full width at half maximum* (FWHM) or the bandwidth of the filter, is approximately given by:-

$$FWHW = 0.88 \frac{\lambda^2{}_o}{2^N \mu d} . \tag{7.9}$$

The *free spectral range* (FSR) is given by:-

$$FSR = \frac{\lambda^2{}_o}{\mu d} . \tag{7.10}$$

Hence, the finesse F, which is the ratio of the free spectral range to bandwidth of a Lyot type birefringent filter, is given by:-

$$F = 1.13 \times 2^N . \tag{7.11}$$

Birefringent filters have been built with FWHM as narrow as of 0.12A at 5000Å and a finesse of 290. The highest intensity of the first side lobe, or the first secondary maximum relative to the peak transmission of Lyot filter is approximately 0.045.

7.4.2.4 *Contrast Element*

Small amplitude side lobes appear in the transmission of birefringent filters. Due to which a small amount of continuum leaks through such filters, when a strong absorption line like H-α is centered on the filter peak transmission. To suppress the side lobes, a 'contrast element' is introduced which reduces the height of the side lobes and thereby increases the signal-to-noise, or the contrast in the filter transmission. This element, instead of being twice the thickness of the largest element, is approximately equal to the second longest element. For example, the commercially available Lyot filters of passband of 0.5Å at H-α made by B. Halle Company has a contrast element of 0.7 Å pass band.

7.4.2.5 *Tuning of Birefringent Filters*

One of the major advantages of birefringent filters over multilayer dielectric interference filters is the convenience of tuning the passband of the filter over a limited spectral range. There are 3 basic tuning techniques used in birefringent filters. One of the direct methods is to change the optical path length by varying the temperature, which affects both the birefringence and thickness of the crystal elements. For quartz elements, it is approximately –0.5Å shift per degree C, while for calcite elements it is +0.3 Å shift per degree C in the visible range. But this method takes considerable time to attain the desired temperature to shift the pass band.

The best tuning method is through the use of quarter or half wave plate analyzers. A quarter wave plate for the particular wavelength of interest is inserted immediately following the retardation plate (birefringent filter element), see Figure 7.14. The tuning of the filter is then done by rotation of the exit polaroid. Briefly the principle of operation is as follows; the output of the retardation plate is elliptically polarized, whose ellipticity ε is a function of wavelength.

$$\varepsilon = \text{Ellipticity} = \text{Horizontal axis/Vertical axis} = \tan \pi d/\lambda\ \mu. \quad (7.12)$$

The quarter wave plate acts as a polarization analyzer and transforms the elliptically polarized light into a linear polarized wave, whose orientation is a function of the ellipticity,

$$\theta = \tan^{-1} \varepsilon. \quad (7.13)$$

The orientation of the quarter wave plate as a function of wavelength is given by :-

$$\theta = \pi d/\lambda\ (\mu)^{-n\pi}. \quad (7.14)$$

At $\theta = 0^{\circ}$, or vertical, the standard wavelength λ_0 is selected, and then the tuning relationship is given by:-

$$\theta = \pi d\ \mu\ (1/\lambda - 1/\lambda_0). \quad (7.15)$$

Rotation of the polaroid by half a revolution (180°) tunes through a free spectral range of the element. With this technique the tuning of the

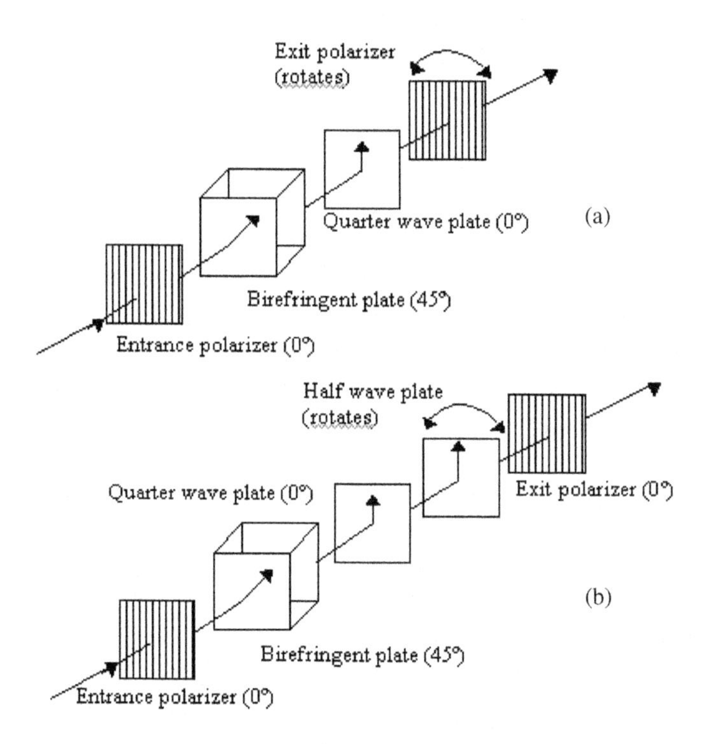

Figure 7.14 (a) Simple quarter-wave tuning element by rotating the exit polarizer, (b) A more practical arrangement for tuning a filter is to rotate a half-wave plate introduced between the exit polarizer and quarter wave plates.

filter can be achieved to within a small fraction of the filter bandpass. For practical purposes, it is desirable to avoid rotating the exit polarizer, since that would require rotating the following retardation plate also and all the subsequent elements which becomes complicated. This can be avoided by introducing a half wave plate just before the exit polaroid, which can then be rotated as shown in Figure 7.14(b). In this arrangement the polarizers and retardation plates remain fixed, only the half wave plate rotates, this also improves the imaging performance. In the case of Šolc filters, the tuning can be achieved by introducing a rotating half wave plate between the two fixed quarter wave plates, which is placed following the birefringent element. The extra quarter wave is just the retardation needed to correct the phase difference between the orthogonal polarization components.

Recently, electro-optical tuning of birefringent filters has been also tried. The advantage of this type of tuning is that there are no moving parts in the optical path. Materials such as ADP and KDP are used for this purpose. A few thousand volts are required to introduce quarter wave retardation. A constant DC high voltage, however, generally damages the ADP or KDP crystals. With the introduction of liquid crystal technology, it is now possible to obtain half wave and quarter wave retardation by applying small voltage of only about 5 volts. Fast response liquid crystals are now available in the market.

7.4.2.6 *Field of View of Filters*

Besides the advantage of easy tuning, birefringent filters also provide a larger field of view compared to interference filters, or Fabry-Perot etalons. Unlike the interference filters in which the off-axis behavior is isotropic, uniaxial birefringent crystals have a strong azimuthal dependence. The fringe pattern of a uniaxial birefringent element is a series of hyperbolic isochromes, while an interference filter or Fabry-Perot etalon gives a circular pattern. The change in retardation actually changes sign between the consecutive quadrants of the fringe pattern. For rays at 0° azimuth with respect to the optic axis, the retardation decreases with incidence angle as:-

$$\Delta(i,0^o) = \Delta_o[1-\frac{\sin^2 i}{2n_o^2}], \tag{7.16}$$

while in the next quadrant, at an angle 90°, the retardation increases as:-

$$\Delta(i,90^o) = \Delta_o[1+\frac{\sin^2 i}{2n_o n_e}]. \tag{7.17}$$

At an arbitrary azimuth angle θ, the retardation variation is given by:-

$$\Delta = \Delta_o [1 - \frac{\sin^2 i}{2n_o^2} (\cos^2 \theta - \frac{n_o}{n_e} \sin^2 \theta)] . \qquad (7.18)$$

Azimuthal behavior is the key for making wide-field elements. If two plates are combined with their optic axes crossed, then the positive increase of retardation in one plate will be offset by the negative change in the retardation of the other plate. Lyot (1933, 1944) has shown that this could be achieved by two methods. First, by using two identical plates of the same uniaxial material with their fast and slow axis at 90° to each other. To avoid having the plates cancel and produce a wide field with zero retardation, a half wave plate is placed between the two crossed elements. The second method that Lyot proposed was by crossing two retardation plates, one being of uniaxial positive and other of uniaxial negative crystal. A wide field element may be made from a combination of quartz and calcite crystals.

To achieve a wide field, Evans proposed a scheme of 'split elements' as shown in Figure 7.10(c) and (d). One retardation plate R, lies between the halves of another retardation plate S. To obtain both wide field and also tuning of the filter, Evans proposed that the inner retardation plates in Figure 7.10(d) be split into two halves and separated by a pair of ¼ wave plates, placed between a rotatable half-wave plate. In this arrangement retardation plate R is tunable, but not the S element.
The variation of transmitted wavelength as a function of incident angle is given by:-

$$\frac{\delta \lambda}{\lambda_o} = -\frac{\sin^2 i}{4n_o^2} (n_e - n_o) , \qquad (7.19)$$

where i is the angle of incidence, λ_0 the nominal wavelength, and n_e and n_0 are the refractive indices for extra ordinary and ordinary rays respectively.

In the case of F-P etalons, the variation with incident angle is given by:-

$$\frac{\delta \lambda}{\lambda_o} = -\frac{\sin^2 i}{2n_s^2} , \qquad (7.20)$$

where n_s is the refractive index of the spacer layer. Thus the field advantage of the wide field birefringent element results from the factor $(n_e-n_O)/2n_e$. For quartz this factor is 0.003 and for calcite is –0.058, resulting in large field advantage for birefringent elements as compared to the F-P etalon. In Figure 7.15 is shown the 'off axis' wavelength shift for wide field birefringent quartz and calcite elements and F-P etalon.

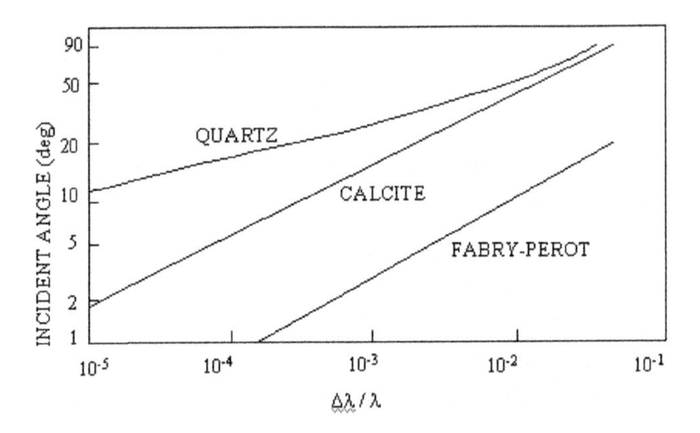

Figure 7.15 Off-axis wavelength shift for wide field quartz and calcite birefringent elements and a Fabry-Perot etalon.

It is better to use a telecentric beam (parallel concentric beam) of light through the F-P filters. In the case of Lyot filters, as a general rule, beams *faster* than f/15 should not be used even with wide field birefringent filters, otherwise the field of view will exhibit non-uniform transmission.

7.4.2.7 *Throughput or Filter Transmission*

For any filter system we can ask: what is the transmittance of the filter? This depends on number of factors, for example, the aperture of the filter, solid acceptance angle and loses in the filter itself. The birefringent filters with wide field elements can operate over a larger solid angle compared to a F-P. In Figure 7.16 are shown plots displaying the advantage in the throughput of birefringent filters made of quartz and calcite over F-P. In practice a gain of about 50 to 200 can be achieved by using birefringent filters over F-Ps.

The throughput advantage of birefringent systems is somewhat

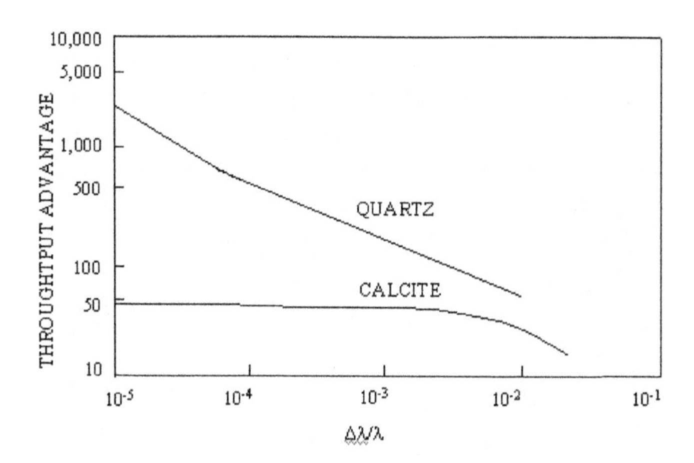

Figure 7.16 Throughput advantage of wide field birefringent filters compared to a Fabry- Perot for the same passband.

decreased by internal attenuation due to the number of polarizers in such filters. An unavoidable loss of about 50% occurs at the entrance polarizer, unless the incident light is polarized. A Lyot filter with a finesse of 250 requires 9 polaroids. If 95% is the transmission in polarized light, then the overall attenuation is a factor of 63% in polarized light, and the overall transmission in unpolarized light of such a filter would be 32%. The overall throughput advantage of birefringent filters relative to an ideal F-P, is a factor of 15 to 80 in unpolarized light and 30 to 160 in polarized light.

7.4.3 *Principle of Fabry-Perot (F-P) Filter*

A Fabry-Perot (F-P) interferometer consists of two parallel flat semi-transparent mirrors separated by a fixed (or variable) distance. This arrangement is called an *etalon*, a name give by its inventors, Charles Fabry (1867-1945) and A. Perot (1863-1925). If the monochromatic light is incident upon an etalon at an arbitrary angle to the normal of the mirror surface, it will undergo multiple reflections between the mirrors. The intensity distribution of the reflected and transmitted beams by the etalon interfere and are brought to focus by a converging lens to yield a series of interference fringes. Because of the circular symmetry of the device, a set of bright concentric rings or fringes results on a dark back ground in the case of transmitted light, while a complementary set of dark fringes results in a light background for the reflection case. The angular diameter of these fringes depends on the spacing between the etalon mirrors and the inverse of the wavelength (wave number) of the radiation. The basic function of a Fabry-Perot device is to transform wavelengths into an angular distance.

The wavelengths of maximum transmission are determined by the well known formula:-

$$m\lambda = 2\,n\,d\cos\theta, \tag{7.21}$$

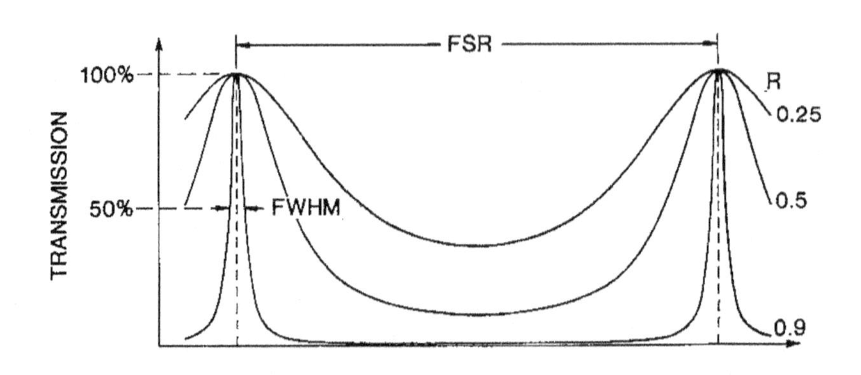

Figure 7.17 Transmission spectra of a Fabry-Perot etalon as a function of the coating reflectance-R.

where m is the order of interference, λ is the wavelength, n is the refractive index of the transparent medium, d is the thickness of the medium and θ is the angle of incidence within the medium. The reflectance or the coating on the F-P mirrors is an important parameter to decide the width of the fringe system. In Figure 7.17 is shown transmission spectra of a F-P filter for several different coating reflectivities. It will be noted that the narrowest fringe width (FWHM) occurs when the coating reflectance R is 0.9.

There are several important parameters that characterize the performance an F-P. When F-P etalons are used as a narrow band filter, three important parameters characterize the filter; these are (i) the effective *finesse* N_e, (ii) transmission band width $\Delta\lambda$ (FWHM), and (iii) the *free spectral range* (FSR). These parameters are given by the expressions:-

$$N = FSR / \Delta\lambda. \tag{7.22}$$

The free spectral range of a F-P etalon is the wavelength interval between the two consecutive transmission bands as shown in Figure 7.17 and is given by:-

$$FSR = \lambda^2 / 2\, n\, d. \tag{7.23}$$

The effective finesse N_e is determined by contributions from the reflective coatings N_c, from the thickness non-uniformities N_t, and surface roughness N_r, according to the expression given by Atherton *et al.* (1981):-

$$N_e^{-2} = N_c^{-2} + N_t^{-2} + N_r^{-2}. \tag{7.24}$$

Here the reflectance finesse N_c is given by:-

$$N_c = -\pi / \ln R, \tag{7.25}$$

where R is the reflectance of the etalon coatings, N_t and N_r are the defects in finesses due to spherical defect δt_s and the rms surface roughness as δt_r, respectively, and are related by following expression:-

$$N_t = \lambda / 2\, \delta t_s, \tag{7.26}$$

$$N_r = \lambda / 4.7\, \delta t_r \,. \tag{7.27}$$

When F-P is used as an interference filter, the central or the operating order m_o of interference for transmitted light is defined as that occurring perpendicular to the surfaces. Putting in Equation (7.21), $\theta = 0$,

$$m_o = 2\, nd / \lambda. \tag{7.28}$$

One other parameter of interest is the maximum transmissivity T_m of the etalon. It is largely dependent on the reflectance R of the coatings as well as their absorption A and is given by:-

$$T_m = \{1 - A/(1\text{-}R)\}^2. \tag{7.29}$$

Tuning of F-P etalon can be done either by tilting with respect the incident beam (but this broadens the filter profile and shifts the wavelength of the filter transmission towards the short wavelengths), or by changing the spacing in the case of air spaced etalons, or by temperature change, or by changing the refractive index of the substrate of the medium by electrical means, in case of electrically tunable etalons.

7.4.3.1 *Mica Solid F-P Etalon Filter*

Tolanski in the early 1900s investigated the properties of silvered mica and wrote an interesting paper on the subject. In 1962 Dobrowolski (1962) of the Canadian Research Council obtained a patent for silver coated mica sheet. Doug Martin while working with Spectrolab Company in California, USA realized in early sixties the potential of this device for narrow band solar observations, and developed mica etalon filters. Such mica filters of about 25 to 30 mm aperture were made by a company named as DayStar Inc., in California, USA. The DayStar filter is essentially a solid Fabry-Perot etalon. The great advantage of using mica as substrate is that it has the remarkable property of *parting* (capable of being separated into thin sheets of uniform thickness) along cleavage planes, with no lattice steps over areas of a few square centimeters. Sheets as thin as few microns can be cleaved, producing an ideal Fabry-Perot spacer at nominal cost. As mica is transparent over a wide range of wavelengths, such filters can be used from the violet to the near infrared. Dielectric coatings of alternating high and low refractive indices are deposited on the two surfaces of the mica sheet to produce semi-transmitting etalon mirrors. A semi-transparent silver coating can also be used, but this has reduced transmission efficiency. The spacer layer does not have to be a quarter-wave multiple; any thickness can be used and adjusted so that a transmission band is located at the wavelength of interest. As mica is a bi-axial crystal, two sets of interference fringes are formed by the etalon. Both sets of fringes are orthogonally polarized and a linear polarizer is required to eliminate one set if the spacer's optical thickness is not an exact half-wave multiple of the wavelength of interest. But if the spacer is exactly a multiple of $\lambda/2$, this will utilize both fringe systems, superimposed on each other and centered at λ, thus eliminating the need for a polarizer and resulting in twice the transmission. As the mica filter is a standard F-P, the angle of acceptance or the field of view for such filters is limited (see Figure 7.16), thus it would be better to use a telecentric, or parallel beam of light through the filter. The narrowest passband of a DayStar mica filter is about 0.4 Å. These filters are temperature sensitive and should be housed in a temperature-controlled oven to maintain a temperature within ± 0.1 C. The transmission profiles of the Lyot and mica F-P filters are quite different. The transmission profile of mica F-P is exponential with wide *skirts* or wings, while birefringent filters with contrast element have sharp cutoff and the intensity drops to almost zero, just beyond the peak transmission. Due to wide *skirts,* the filter transmission profile of mica F-P filters transmits a significant amount of solar continuum radiation. Thus the chromospheric Hα structures appear somewhat different compared to the Lyot filter. Further wavelength tuning **can not** be easily achieved in the case of mica filters, except by temperature change. Mica DayStar filters are being extensively used by professional and amateur solar astronomers for taking chromospheric observations in Ca II and H-α lines, because they are relatively inexpensive and are small in size, and does not need any high precision oven for thermal control. Smartt (1982) has also used mica etalon filters for radiance measurement in three coronal lines 5303, 5694, and 6374 Å.

7.4.3.2 *Lithium Niobate Solid F-P Filter*

Lithium Niobate ($LiNiO_3$) crystal is a transparent birefringent material whose refractive index can be changed by applying an electrical potential (Pockel effect) across the two faces of the plate. This property has been recently exploited by many solar astronomers (Rust 1985, Mathew *et al.* 1998) and they have used it as a solid Fabry-Perot etalon filter for chromospheric, photospheric velocity and magnetic field observations. Li-Niobate F-Ps were made by the CSIRO, Australia. These F-Ps were available in fairly large clear aperture sizes up to 75 mm in diameter. A large throughput can be achieved, with passband of about 0.175Å, a free spectral range of 3 to 3.5Å, and finesse between 18 and 22 for the central 5 mm portion of the etalon. But for 60% of the full aperture, it drops to about 14 at λ5800. The voltage-tuning characteristic is fairly linear varies at a rate of 4.5 kV/Å, and the wavelength shift is reversed by reversing the voltage. A hysteresis effect is observed, if the voltage change is made faster that 10 Hz. A maximum voltage of ±3000 volts is permissible with a 200 micron thin $LiNiO_3$ wafer.

7.4.4 ***Special Purpose Narrow Band Filters***

Recently milliangstrom narrow passband filters have been developed for special applications in solar physics such as the measurement of velocity and magnetic fields. These filters use either the principle of Macaluso-Corbino magneto-optical effect in sodium and potassium resonance atomic lines (known as Magneto-Optical filter), or the Michelson Interferometer.

7.4.4.1 *Magneto-Optical Filter*

Cacciani and colleagues (Agnellii et al., 1975)*)* utilized the atomic resonance property of lines of sodium and potassium for the construction of narrow passband filters. Light passing through a sodium or potassium vapor cell, if placed in a strong longitudinal magnetic field undergoes resonant scattering, and the σ-transitions are circularly polarized. This is called as Macaluso-Corbino effect. If the vapor cell is placed between crossed polarizers, only the light absorbed and re-emitted in the σ-transitions will have its plane of polarization rotated and pass through the second polaroid. Thus such a sodium vapor cell filter can isolate the Na D line at 5890Å or in case of potassium cell, the 7699Å line. A second cell placed in tandem is used to select alternately the blue or the red wing of the line.

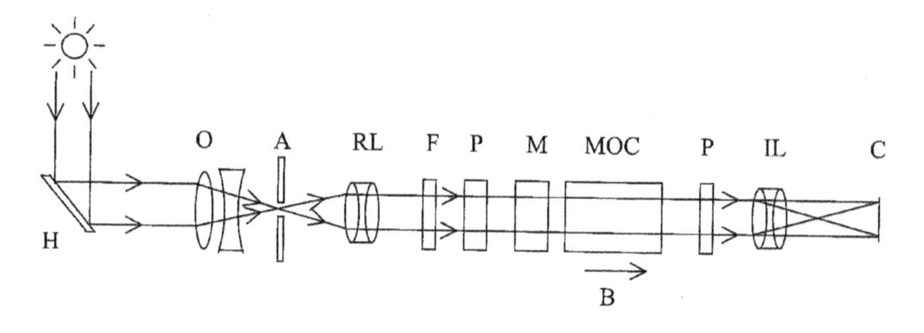

H – Heliostat
O – Objective lens
A – Aperture stop
RL – Relay lens
F – Broadband 10A Filter
P – Linear Polarizer

M – Electro optical modulator
MOC – Magneto-optical sodium
IL – Imaging lens
C – CCD Camera
$\vec{B}$ – Magnetic field of about 1500 gauss parallel to the light beam

Figure 7.18 Schematic of imaging video magnetograph based on Magneto-Optical filter.

To measure magnetic fields, one requires only one cell, switching between right and left circular polarization. To obtain velocity observations one has to use two cells in tandem; the second selecting one wing at a time. The cell provides a very stable bandpass because the atoms always absorb at their fixed wavelength. This also means that we can only see that part of the Sun that is stationary relative to the observer. Corrections due to the Earth's diurnal and orbital motions have to be made. Considering the over all functioning, the Cacciani cell (as it is generally known) is the most stable, clean and rather inexpensive narrow passband filter available for velocity and magnetic field observations.

The main disadvantage of such vapor cells is that they can be used only with the two resonance lines. In Figure 7.18 is shown a schematic layout of Magneto-optical filter (MOF) for magnetic field observations.

7.4.4.2 *Polarizing Michelson Interferometer*

The principle of using Michelson Interferometer as a narrow band filter is essentially the same as in the case of birefringent filters. One splits the incoming sunlight into two orthogonally polarized components by a polarizing beam splitter, sandwiched between the two halves of a Michelson cube. Such filters are used in the GONG and MDI (Michelson Doppler Imager) instruments for observing solar oscillations. A general layout of a Michelson interferometer used as a narrow band filter for velocity measurements is shown in Figure 7.19. A monochromatic beam of light enters the Michelson interferometer cube and is split by a polarizing beam splitter. Light of one polarization is reflected by the beam splitter and enters the air arm. The other beam, which is orthogonally polarized, is transmitted and enters the glass arm. The two beams pass through quarter-wave plates placed in each arm, and are reflected back to pass through the quarter wave plate again. The reflected light has a polarization orthogonal to the incoming beam. The beams combine at the beam splitter, forming interference fringes and exit through the quarter wave plate. The beam leaving the Michelson cube has elliptical polarization, which depends on the wavelength, as in the case of a birefringent filter. The light from the interferometer emerges and then passes through a rotating half wave retarding plate and a fixed linear polarizer. These two elements allow the orthogonally polarized light emerging from the interferometer to interfere but with a phase that depends on the angle of the rotating half wave plate. The presence of a solar spectrum line in the sunlight, passing through the interferometer causes modulation at the rate of 4 cycles per rotation. The signal is integrated for 120° intervals of the modulation cycle to produce three signals I_1, I_2 and I_3. These three signals can be combined to find the phase φ:-

$$\tan\phi = \sqrt{3}\frac{I_2 - I_3}{I_2 + I_3 - I_1} \tag{7.30}$$

and the modulation amplitude is given by:-

$$M = \sqrt{(\tfrac{2}{3})\frac{I}{I_o}}\sqrt{[\sum_{i=1}^{3}(I_i - I_o)^2]}\,. \tag{7.31}$$

The phase of the modulation is related to the Doppler shift of the solar spectrum line but one can only measure its *modulus* within 2π and with no absolute reference. The relation between the Doppler velocity V and the phase $\delta\varphi$ is given by:-

$$V = c\ \delta\varphi/\varphi. \tag{7.32}$$

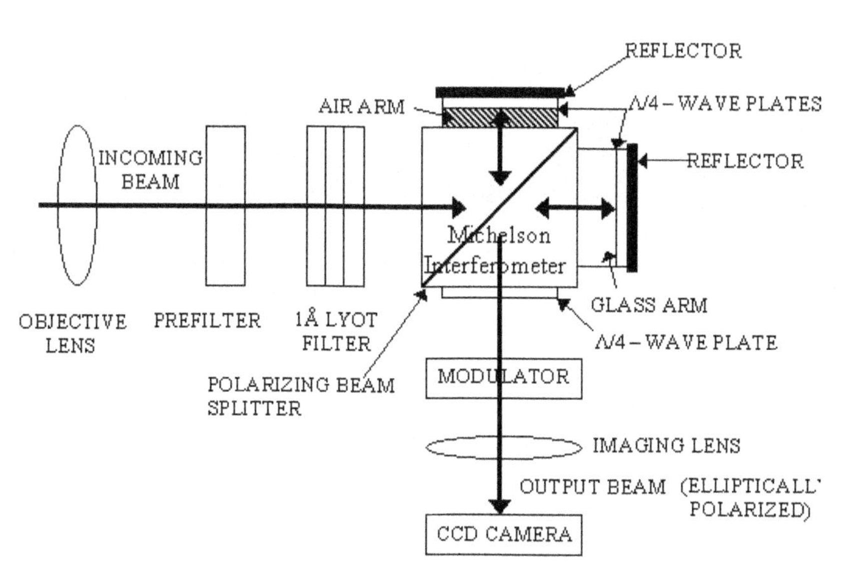

Figure 7.19 The Michelson interferometer used as narrow band filter in GONG and MDI instruments for the measurement of solar velocity oscillations.

and a Doppler solar image is generated after some computation. Using this principle, two major instruments for solar velocity oscillations have been built. One is for the Global Oscillation Group Network (GONG) and another as the Michelson Doppler Imager (MDI) on board the SOHO spacecraft. Both these instruments use the Ni I line at 6768Å. The major advantage of such a system is that Michelson interferometer offers higher throughput, frequency stability, and a very narrow pass-band of the order of 50 mÅ.

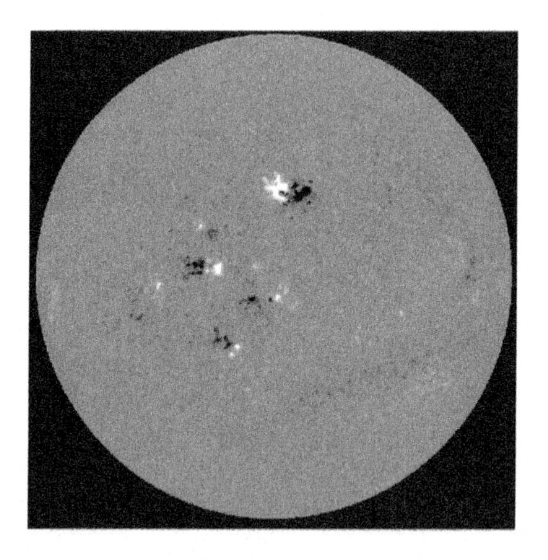

Figure 7.20 Full disk magnetogram obtained on February 25, 2004 from the GONG instrument at Udaipur using a polarizing Michelson interferometer.

Figure 7.20 shows a full disk magnetogram taken through the GONG instrument using the polarizing Michelson interferometer as a filter. Magnetic flux as small as 3-5 gauss can be detected by this technique. At disk center the field is purely longitudinal, that is the field lines are directed towards the observer, but at the limb the field is transverse and the field lines are normal to the line-of-sight.

7.4.5 *Filter-based Solar Magnetograph*

There are several devices for mapping magnetic fields over an active region or over the full disk. The basic principle is to use the Zeeman effect. In the presence of a magnetic field certain spectrum lines split into linearly and circularly polarized components, depending on the line of sight with respect to the magnetic field direction. When we look along the magnetic field (longitudinal field), we see only two σ -components with oppositely circular polarization, displaced by a certain amount depending on the field strength. When we look perpendicular to the field (transverse field), we see one π-linearly polarized component, parallel to the field at the undisplaced position of the line and two σ-components linearly polarized in the perpendicular direction. The total radiation in any direction is unpolarized. The amount of displacement or Zeeman splitting in wavelength is given by:

$$\Delta\lambda\,(\text{Å}) = 4.7\text{x}10^{-13}\, g\, \lambda^2\, B, \qquad (7.33)$$

where g is the spectroscopic parameter called the Landé factor, which ranges between 0 and 3, λ is the wavelength in Å, and B the field strength in Gauss. Since the Zeeman splitting of solar lines increases with λ, because $\Delta\lambda$ is proportional to $g\lambda^2$. Most atomic lines have g values around 1. Molecular species are much less sensitive (but generally not zero). As mentioned earlier, in the case of longitudinal magnetic fields, a line splits into two σ-components which are oppositely circularly polarized. To make a longitudinal field map, one needs two narrow band images, each displaced by $\pm\,\lambda$ (in line width) from the nominal line position, and these images are modulated with some kind of electro-optic crystal to pass one image at a time, in either the right or left circular polarized light. The two images are subtracted synchronously to obtain the magnetic field signal. To avoid seeing induced noise, the modulation frequency should exceed 100 Hz.

At number of solar observatories around the globe use narrowband birefringent filters with a pass-band of some 0.25Å for observing longitudinal and transverse magnetic fields. A typical arrangement for measuring the magnetic and velocity signals is shown in Figures 7.21. In this figure we use the term RHC for right hand circular and LHC for left hand circular polarized light. Compared to the longitudinal fields, it is much more difficult to measure the small transverse magnetic field. This is because the split components are linearly polarized, with the central π-component being opposite to the outer split σ-components (unlike longitude Zeeman effect). Except near spots, where the magnetic field strength is very large on the order of 200 gauss, transverse fields can not be measured with sufficient accuracy and sensitivity.

In Table 7.1 useful lines for magnetic and velocity field measurements are given.

Table 7.1 List of lines used for magnetic and velocity observations. λ is in microns.

Species	λ	Lande	$g\lambda$	Origin	Problem
Fe	0.5250	3.0	1.57	photosphere	temperature sensitive
Fe	0.6303	2.5	1.57	photosphere	
H α	0.6563	1.0	0.65	chromosphere	broad line
Fe	0.6733	0.0	0.0	photosphere	Zeeman insensitive
Ni	0.6768	1.5	1.02	photosphere	
Fe	0.8688	1.66	1.44	photosphere	
Fe XIII	1.0747	1.0	1.07	corona	broad line
Fe	1.5648	3.00	4.69		
Ti	2.2310	2.5	5.6	photosphere	blend, only in umbra
Si IX	3.9343	1.0	3.9	corona	blend, broad
Fe	4.1364	1.7	7.0	photosphere	blend
Mg	12.32	1.0	12.32	upper photo.	not in quiet Sun

Any oblique reflections within the telescope and mirror surfaces also introduce false linear instrumental polarization, which is difficult to account for. That is why image systems for measuring vector fields (i.e. both longitudinal and transverse fields along with the field direction) must avoid oblique reflections (which linearly polarize the light). Further, there is also an ambiguity in defining the direction of the field lines in vector field measurements that must be resolved. This is usually done by theoretical modeling. Except in active regions, the magnetic fields are mostly normal to the solar surface. This is seen in full disk magnetograms by the fall off of longitudinal fields toward the limb. But in flare producing active regions transverse fields play an important role. For this reason there is considerable interest in vector magnetographs. But one should keep in mind their limitations in sensitivity.

Another aspect of magnetographs to be noted is that they measure *flux* (i.e. field strength divided by the area on the Sun) and not directly *field strength*. Outside of sunspots, fields are concentrated in sub-arc sec elements whose true field strength ranges between 500 and 1500 gauss. But because we record flux, the observed signals are more like 5 to 15, may be up to 50 Gauss. This averaging process smears the fields spatially. Thus magnetic areas may appear to cover a fair fraction of the Sun's surface, but in fact the 'filling factor' is far less than 1 percent.

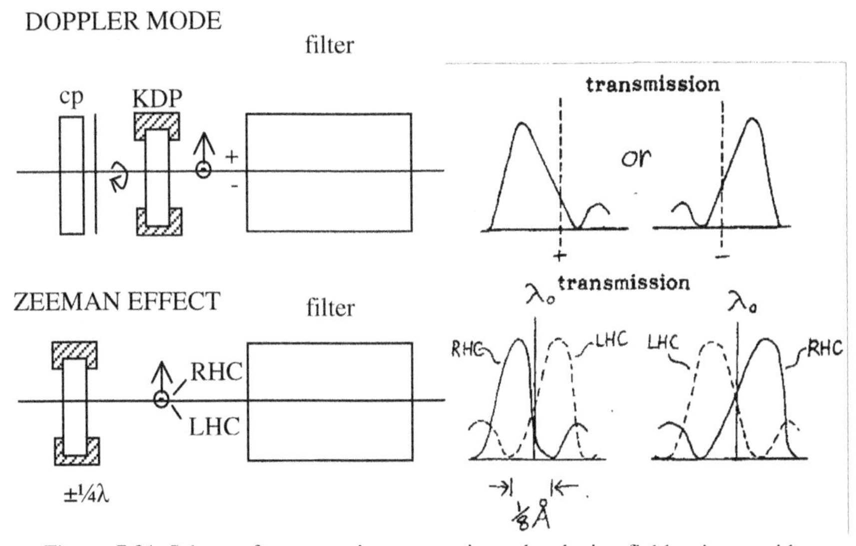

Figure 7.21 Scheme for measuring magnetic and velocity field using a video magnetograph. An electro-optical modulator, a K*DP crystal, is used to provide $\pm\lambda/4$ wave retardation. CP is the circular polarizer and the filter has a 0.25 Å passband at 6103 Fe I line.

A magnetograph becomes a velocity mapper by inserting a circular polarizer in front of the modulator, as shown in Figure 7.22c. Doppler shifts then convert to magnetic signals. Doppler shift measures based on solar rotation can be used to calibrate magnetograms. Since rotation is well known, it can set the scale for other velocity measurements. In Figure 7.22 is shown several standard set-ups used for observing velocity and longitudinal and transverse magnetic fields, using narrow band filters of passband 0.25Å or narrower.

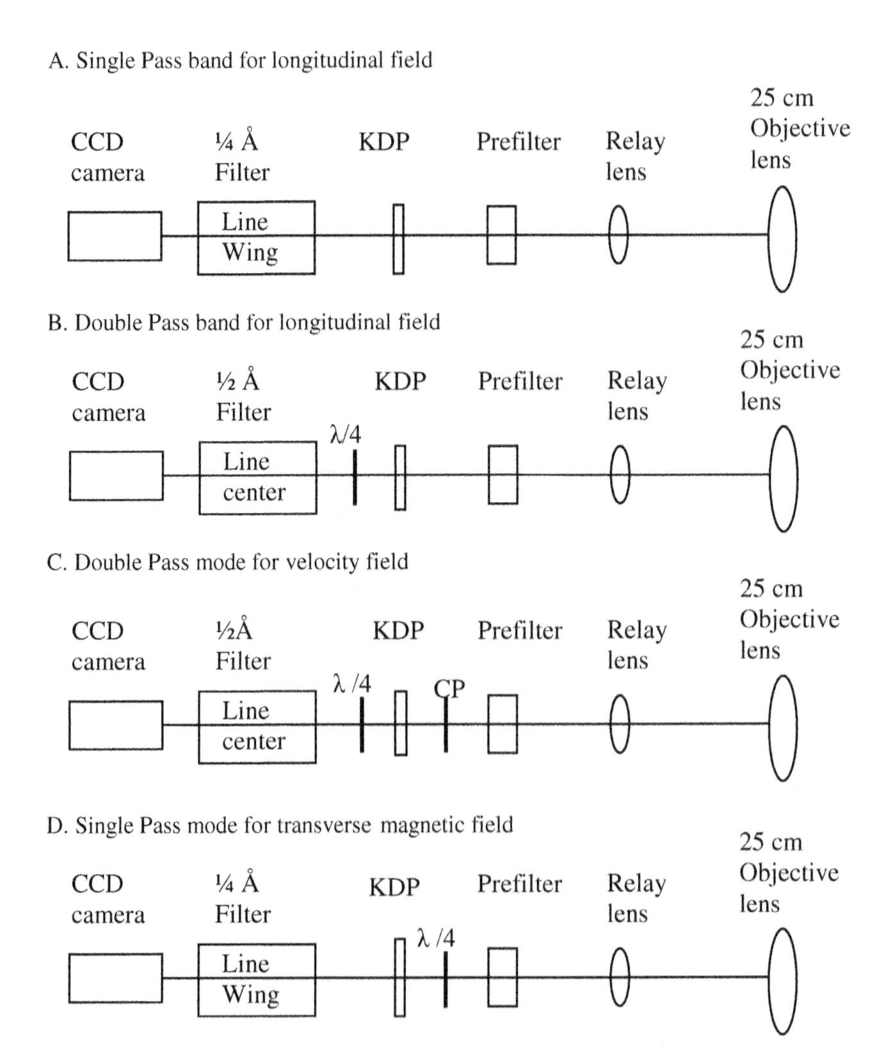

Figure 7.22 Arrangements for observing magnetic and velocity fields using narrow band filters.

Chapter 8

Solar Eclipses

8.1 Eclipse Geometry

By remarkable coincidence the apparent diameters of the Sun and the Moon are nearly the same. Moon is approximately 400 times nearer to the Earth as compared to the Sun and nearly 400 times smaller than the Sun. In Figure 8.1 is shown the relative sizes of various objects, for example, a round Dime, a Rupee coin, a tree, a hill, Moon and the Sun, when seen at various distances from an observer, all subtend the same angle of about half a degree at the eye.

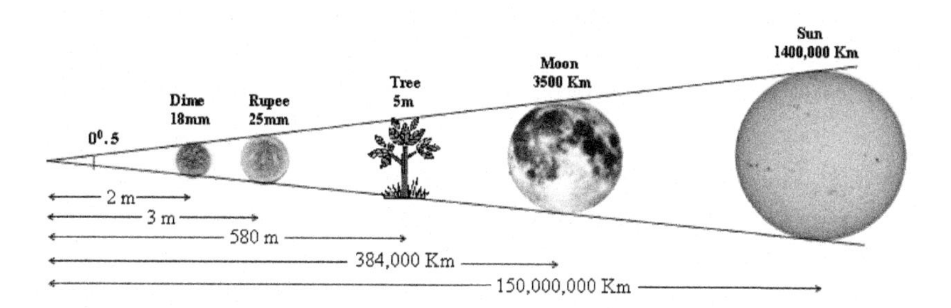

Figure 8.1 Showing relative sizes of the various objects at different distances, subtending same angular angle at the observer. A coin at arms distance and the Sun's disk both subtend the same 0.5 degree at the eye.

For an easy understanding of the circumstances of eclipse phenomenon, let us consider a geocentric system. The planes of the Sun and Moon's orbit intersect along a line. The orbits intersect at two points along this line, these are referred as the ascending and descending nodes, that is when the Moon travels from South to North and from North to South respectively. Solar and lunar eclipses occur when the two objects are near the nodes and all the three; Earth, Moon and the Sun are in a straight line. As the Moon's orbit is inclined to the equator at an angle of about 5 degrees, eclipses do not occur at every new Moon day (solar eclipse) or full Moon night (lunar eclipse). There is an 'eclipse season', lasting for 38 days. This occurs each time when the Sun is near a node and an eclipse will occur when the Moon enters this region. As the Moon's sidereal revolution period (rotation around it's own axis) is 27.3 days and the synodic period of 29.5 solar days is shorter than this interval of 38 days, there can be at least one solar eclipse during each 'eclipse season'. Due to the gravitational attraction of the Sun and Earth, the orientation of the orbit of the Moon changes and the nodes slide with a period of 18.6 years. The interval between the two eclipse seasons is 173.3 days, thus two eclipse seasons make one 'eclipse year' of 346.6 days. From this logic a maximum number of seven solar and lunar eclipses can occur in a calendar year. This happens rarely; that two eclipses may occur spaced four weeks apart in the consecutive two 'eclipse seasons' and one in the next season during a calendar year. This happens because the 'eclipse year' is shorter than the calendar year by about 20 days. In such years, only two lunar eclipses can take place, they occur on full Moon night, halfway between the two solar eclipses in an eclipse season. In 1973 and 1982, we had three eclipse seasons, centered in a calendar year, with two seasons in November and December, giving four solar eclipses and three lunar, thus yielding a total of seven eclipses during a calendar year.

Depending on the position in their respective elliptical orbits, the size (angular diameter) of the Moon varies between 90.2% and 106% of the Sun's angular diameter. When the Moon passes in front of the Sun, as viewed from the Earth, we have a solar eclipse. If the angular size of the Moon is larger than the Sun in the sky and the Moon's umbral shadow intercepts the Earth's surface, within this shadow we see a total solar eclipse. The Sun's faint corona then becomes visible, as shown in the Figure 8.2 drawing by Captain Tupman of the solar corona during the 12 December 1871 total eclipse. It's amazing that naked eye observations made by Captain Tupman in 1871, could record extended fine details of the solar corona, almost similar to modern eclipse pictures.

Figure 8.2 Drawing of the total solar eclipse of 12 December 1871 by Captain Tupman, showing the fine coronal streamers. Compare this with modern eclipse pictures given in this chapter.

If the angular size of Moon is smaller than the angular size of Sun, we have an annular, or ring eclipse, as shown in Figure 8.3. In the case of annular solar eclipse, the bright solar photosphere which is about a million times brighter than the corona remains visible and drowns out the faint corona. Within the penumbral shadow, it is a partial eclipse. On an average the width of totality is only about 100 km, while the area where it is seen partial may extend over 6400 km, covering an appreciable fraction of the Earth's surface. There can be up to five solar eclipses per year, although two is the likely number. At any one place the average frequency for a total solar eclipse is 375 years, and the maximum duration between eclipses is calculated to be 4500 years (Meeus 2002). This is one of reasons that the visibility of a total solar eclipse is a rare phenomenon from a given location but a total lunar eclipse is visible over a large area.

Figure 8.3 Photograph of annular eclipse around sunset on January 04, 1992 from San Diego, California by Livingston. Clouds are seen in the foreground of the solar image.

8.1.1 *Saros Cycle*

A particular eclipse *recurs* when the same geometrical configuration of the Sun, Moon and the Earth prevails. This recurrence interval of about 18-years was discovered by Chadean astronomers as early as 400 BC and perhaps to some extent it was also known even before 3000 BC to Babylonians, Greek, Chinese and Indian astronomers. With this knowledge of the periodicity of eclipses, the astronomer-priests in ancient civilizations could predict the occurrence of solar eclipses to a fair degree of accuracy. There is the famous story that two Chinese Court astronomers, Hsi and Ho, were beheaded because they did not predict the solar eclipse of October 22, 2134 BC. However, some scientists discredit this story, and believe that Hsi-Ho was the name of the Sun god, who was responsible for preventing eclipses. Anyway, the knowledge of eclipse prediction was a well developed science in ancient civilizations. This repetitive cycle of eclipses was named *Saros cycle* by Edmond Halley, as mentioned by Cook (1996), and provides a simple method for approximate determination of the dates of past and future eclipses. To understand how this 18-year *Saros cycle* occurs, we have to consider the motion of Moon around the Earth, and Earth around the Sun, which have the following periods:-

1. **Synodic period of Moon**: called the Lunation, is the interval of time between two consecutive New Moons, or two consecutive full Moons, and is of 29.5306 mean solar days.

2. **Eclipse year**: is the interval between two consecutive passages of the Sun through the lunar node. It is of 346.6201 mean solar days.

3. **Tropical year**: It is the time taken by the Earth to go once round the Sun, starting from one equinox (point of intersection of the ecliptic and the Earth's equator) and coming back to the same equinox. It is 365.2422 mean solar days.

The above three periods give us the following approximate relations:-

223 Lunations	6585.3238 days
19 Eclipse years	6585.7819 days
18 Tropical years and 11 1/3 days	6585.3596 days

Thus the same geometrical configuration recurs at an interval of 18 years 11 days and 8 hours and similar eclipse will recur after 223 Lunations. However, the position of the eclipse path on the Earth shifts by 120 degrees in longitude, due to the Earth's rotation in 8 hours. A Saros series begins with an eclipse that occurs near the Earth's pole. Each successive eclipse of the series shifts towards the west and towards the other pole. When the eclipse reaches the opposite pole, the series is said to end. A Saros series may last from about 900 to 1400 years. The

Saros series are numbered and the system of numbering is so devised that eclipses of an odd Saros number take place at the ascending node, while those with even numbers take place at the descending node. Many Saros series run concurrently. A Saros series may have partial, annular and total solar eclipses. For example, the total solar eclipse of August 11, 1999 belongs to the Saros series no.145. It was the 21st eclipse of the series. This Saros began on January 4, 1639 AD, with a partial solar eclipse near the North pole, and will continue until April 17, 3009 AD. Its duration is 1370.3 years. A total of 77 eclipses belong to this Saros series, out of which 34 are partial, 1 annular, 1 both annular and total, and 44 total solar eclipses have taken place. In Figure 8.4 is shown the paths of the last seven total eclipses of saros 120, each eclipse occurring after an interval of 18 years, 11days and 8 hours.

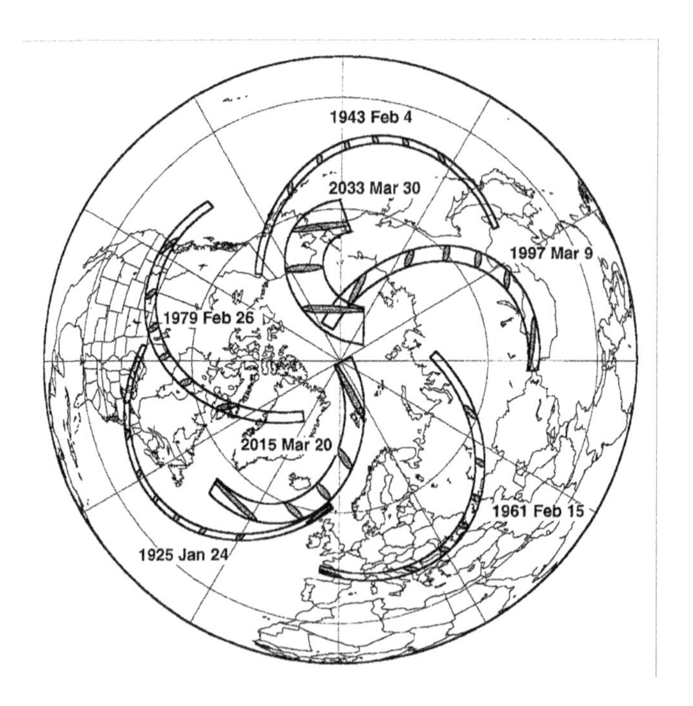

Figure 8.4 Showing eclipse paths of the final seven umbral eclipses of Saros series 120. Note that series ends near the North Polar Region with the last eclipse of the series on 2033 March 30.

Espenak and Anderson of the U.S. Naval Observatory have been calculating and publishing booklets giving the circumstances of solar and lunar eclipses. Several computer programs are also available to calculate the circumstances and parameters of eclipses with reasonable accuracy.

Eclipses have been chronicled for over 4000 years, but the corona escaped notice until relatively recently. The single exception; Leonis Deaconis in AD 968 mentions that, "...a certain dull and feeble glow, like a narrow headband, shinning around the extreme parts of the disk..." (Stephenson 1997). Kepler made a clear comment on the corona seen during the eclipse of 1609 (Zirker1984), but the first real description came from Halley in 1715. He noted its structure and saw red prominences around the rim of the Moon. Kepler and Halley both thought the phenomenon was lunar in origin. Later it was assumed that coronal rays were produced by diffraction at the edge of the Moon, many false drawings were made on this basis.

Why has the world population been blind to what today we consider as one of the most spectacular displays of nature? Education is helpful for one thing; we see what we have been told is there. In olden times it was also unlikely that any single person would see more than one total solar eclipse in a lifetime. Scant attention was paid to the corona, since fear of the darkness and a concern about what the event portended, probably dominated the experience. Ancient man was undoubtedly frightened by the sight of such an unusual scene. Many superstitions about eclipses still prevail. People are advised not to look at, or even go out, during a solar eclipse. This absence of reports about the corona from ancient prompts the question: what else are we missing?

8.2 Eclipses as Time Keepers

Historic records of eclipses, which extend back four millennia, can be used to investigate the rotation of the Earth and hence the constancy of time, which is based on its rotation (Stephenson 1997, Crump 1999). Many ancient eclipse records are questionable. The exact date and even the place have been `adjusted' to fit important events like battles, or the death of kings. Stephenson discusses this and concludes that the

Babylonian eclipse of 15 April 136 BC is probably reported correctly because it is clearly described on cuneiform tablets in the British Museum, without reference to any particular occasion. Assuming the Earth's rotation to be constant and the same as today, a computer simulation places the 136 BC eclipse about 5000 km west of Babylon. Taking into account the tidal forces which slow the rotation of the Earth, and make time variable, the prediction is for 2000 km east of Babylon. Geophysicists propose a number of factors to account for this error. These include changes in the polar ice cap from post-glacial melting, or sea level differences arising from the evolution of glaciers. But just noting the date and place of an eclipse, yields information on this most useful of all physical parameters: time.

8.3 Solar Corona and Cosmic Magnetism

There is a strong suggestion that coronal structures mimic the iron filing pattern of a bar magnet. In 1891 Arthur Schuster, speaking before the Royal Institution on the question "Is every rotating body a magnet?" remarked about the solar corona. He said that "the form of the corona suggests a further hypothesis which, extravagant as it may appear at present, may yet prove to be true. Is the Sun a magnet?" It is interesting that appearances can lead to great discoveries. From the structure of the corona seen during a total solar eclipses, the Sun was suspected to have a general magnetic field like a bar magnet as Schuster had mentioned. To detect the general magnetic field, several attempts were made at the Mount Wilson Observatory by George Hale. He used the Zeeman Effect discovered in 1886 by Peter Zeeman, which showed that spectral lines split in magnetic fields. All attempts by Hale to detect the magnetic field failed, because the fields were much smaller than the sensitivity permitted by the photographic techniques then in use. Although Hale could not detect general magnetic field on the Sun, but he discovered magnetic fields in sunspots (see Chap 5). This was the first concrete detection of *cosmic magnetism*. Eclipse pictures also indicated a complex surface magnetism outside sunspots. Years later the father and the son team of H. D. Babcock and H. W. Babcock (1953) invented an ingenious photoelectric device to measure the small magnitude solar magnetic fields. Consider the corona on 11 August 1999 (Figure 8.5), there we see helmet-like patterns at about 7 o'clock, coronal rays and very fine loop structures, all of which appear to be dominated by magnetic field.

Figure 8.5 Picture taken during the 11 August 1999 total solar eclipse in Iran, by S. Koutchmy, using a radial gradient filter. This picture shows bright prominence, fine coronal structures; helmets, plumes coronal rays apparently directed by solar magnetic fields.

In 1860 photography was introduced as a way to objectively record the corona at an eclipse. Pictures taken from sites 500 km apart produced similar shapes of the coronal. This confirmed that corona was of solar origin. This meant that it was not simply a terrestrial or lunar phenomenon. In 1843 Schwabe discovered the sunspot 11-year cycle. Subsequent eclipses from 1874 to 1882 revealed that the coronal structures varied with the solar activity cycle. Figure 8.6 shows the variation of the shape of the solar corona with solar activity.

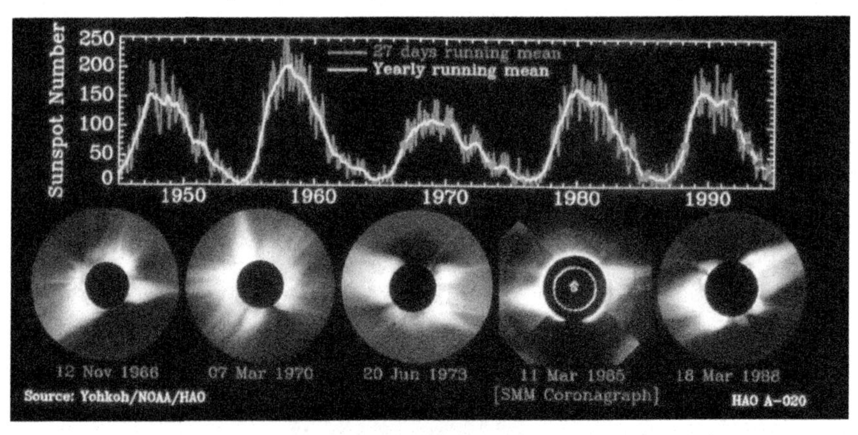

Figure 8.6 Variation of coronal structure with the 11- year sunspot cycle from 1945 to 1995. Note that during solar maximum (7th Mar 1970, 11 Mar. 1985) the outer white light corona appears roundish and rather symmetrical, while during minimum (12th Nov. 1966, 20th Jun. 1973 and 18th Mar. 1988), it appears asymmetrical with extended coronal plumes.

8.4 Scientific Results from Eclipse Observations

We mentioned in the earlier section about determining the constancy of the Earth's rotation and the general nature of the corona. The big coup of the 1920s was the verification of Einstein's *General Theory of Relativity* by the bending of starlight near the Sun. This was revealed by the apparent shift in a star's position near the limb of the Sun on a photograph taken during an eclipse, compared to that taken without the Sun. This experiment made during the 1919 total solar eclipse by Eddington, confirmed the predictions of Einstein's General Theory of Relativity.

The discovery that the temperature of the corona is several million degrees and that the emission lines in it's spectrum arise from extreme ionization states, culminated in Edlèn's identification of the 5303 Å line, as arising from a thirteen times ionized state of Iron, i.e., Fe XIV. This opened-up an entirely new field of study and the concept of non-Local Thermodynamic Equilibrium (non-LTE). Spectra taken of the chromosphere just before and just after totality, called the 'flash spectra' proved difficult to interpret until non-LTE methods of analysis were developed by Pecker, Michard, Athay, Jefferies, and Thomas, (Athay 1976).

8.5 Observing a Total Solar Eclipse

Every person who can possibly travel should at least once ***experience*** the beauty of totality. The unaided human eye excels in its ability to capture the scene. Our eyes have the ability to detect and accommodate to the large range in intensity seen during a total solar eclipse. Your authors suggest that photography is a waste of time for the amateur and even for most professionals. The best photographs are not the equal of what we can see with our eyes. The most thrilling total solar eclipse phenomena occur in a few seconds to minutes, so why fool around with cameras? Instead look at the sky. Sensations of temperature (the noticeable cooling), sounds (the nesting of birds, the exclamations of your companions, in some places the beating of drums), and most of all the astounding light dynamics of the corona demand your full attention. Probably one should use a pair of binoculars to inspect prominences and fine details of the corona.

Dark adaptability is key to see the faintest and most extended parts of the solar corona. An experiment was carried out by one of the authors (AB) during the total solar eclipse of October 24, 1995 in India. A good artist, Ms. Deepti Sunder, who had no previous knowledge of how the corona would look like, was selected and her eyes were covered about 45 minutes before the totality. At the time of totality her eyes were opened and was asked to draw a sketch of the corona, especially the extent of the streamers with respect to the solar disk. In Figure 8.7 is shown the sketch she drew displaying very much extended coronal streamers on the East and West sides of the Sun, with jagged structure towards the end. From this unbiased drawing it is clearly seen that if one dark adapts before the

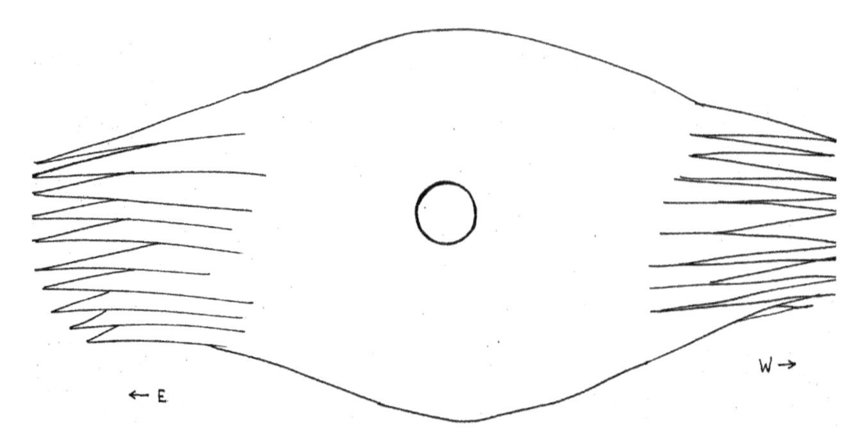

Figure 8.7 A drawing by Ms. Deepti Sundar of the extended coronal structure seen after making her 'dark adapted' before the October 24, 1995 total eclipse. Note the extended streamers to more than 15-16 solar radii, as compared to the solar disk and their jagged ends.

eclipse, it would helps one to see the corona out to more than 15-16 solar radii. No photographic film or CCD camera can record this. Our advice is, to leave the photography to the dedicated eclipse goers with their long focal length lenses and radially graded filters. Rely on your eyes for the wonderful experience.

8.5.1 *Logistic, Site & Weather Conditions etc.*

Logistic planning is amply covered these days on the Internet. Advice on cloud cover, site conditions, political situations, and photography (if you must), is all there. The only suggestion we can offer is to take a comfortable folding chair, some type of shade umbrella and extra water. Because we tend to see what we have been taught, preparation is important. The eclipse sequence consists of 1st contact (the first bite out of the solar disk by the Moon verifies that the Ephemeris is correct. It is exciting to note how precisely such calculations can be made), 2nd contact (beginning of totality, wildly exciting), 3rd contact (end of totality, much verbal to do, congratulations all around), 4th contact (seldom noticed, packing has begun!). After *experiencing* a total solar eclipse many people are inspired to witness another such event.

8.5.2 *Eye Protection*

Because our natural protection against staring at the Sun does not work during the solar eclipse, our eye's retina may be damaged. One must be careful to use special dark filters during the **partial phases**. *Often the retina is burnt during the partial eclipse, if due care is not taken.* However, no protection is needed during the totality. This is because, as the Moon covers the Sun the general light decreases and naturally the pupil of the eye dilates to allow more light to the eye. But the pupil is not able to contract fast enough when the intense photospheric light suddenly emerges, soon after the third contact. Further, the retina does not the have the temperature sensing nerves to warn the human brain to remove the eye from being heated and getting damaged. Due to the intense light many people loose their eyes. *So **never ever** look at the Sun without protective dark filters during partial phases.* If `eclipse filters' are not available one may employ suitably very dense black and white, exposed and developed negative film, but not color negative. Again, no filters are necessary during totality.

8.5.3 *What to Look for*

There are large number of phenomena that can be observed and studied during a total solar eclipse. Some of the most interesting and unusual do not require special equipment.

* The first bite out of the solar disk, it is going to happen.
* Approaching Moon's shadow darkening the horizon.
* Shadow bands on the ground seconds before 2nd contact (caused by temperature in-homogeneities in the air, and are visible when the Sun becomes a slit source, near 2nd and 3rd contact). (See Figure 8.8 of shadow bands)
* Brighter stars and planets seen during totally?
* Edge of Moon opposite the Sun visible? If so, is it coronal light?
* The Diamond Ring (last bit of photospheric light)
* Bailey's beads (photospheric light peaking through lunar craters)
* Inner parts of corona before totality?

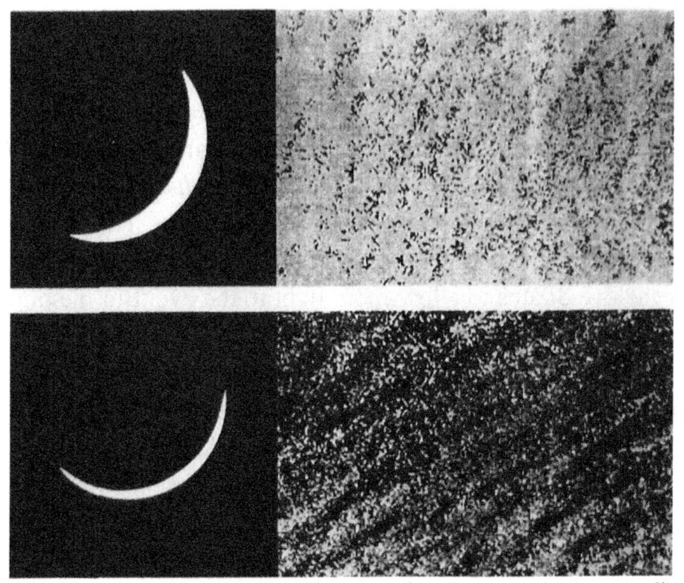

Figure 8.8 Approximate drawings of shadow bands and corresponding crescent solar image on the left.

Figure 8.9 Picture taken of a Coronal Mass Ejection (CME) from LASCO coronagraph aboard SOHO spacecraft and a Sun grazing comet is also seen. Such events occurring during a total eclipse would be very interesting to observe.

* General shape of the corona (is it maximum or minimum type? (see Figure 8.6).
* Prominences and their color (use binoculars).
* Coronal transient in progress? (see Figure 8.10 showing CME taken from LASCO).
* Any comets? (see Figure 8.10 taken from LASCO).
* Moon darker than surrounding sky?
* Earthshine on Moon?
* Departing Moon's shadow darkening the horizon.

Chapter 9

Solar Interior and Helioseismology

9.0 Introduction

Just three decades ago all what we knew about the Sun was from the observations of its outer layers. Except for the neutrinos, no direct radiations from the interior of the Sun can reach us, which could tell us about the physical conditions inside the Sun. Scientists had to depend on theoretical models, as there was no way to check the theory with actual observations. With the advent of a powerful technique of helioseismology in mid seventies, which made use of the Sun's oscillations, it has become possible to indeed *see* inside the Sun. As seismologists use the p- and the q-waves generated by earthquakes to probe the Earth's interior, astronomers make use of the solar vibrations or oscillations to probe the solar interior. Using the technique of helioseismology for the first time astronomers are able to determine very precisely its physical properties like, temperature, density, pressure, rotation, sound speed and chemical composition, which were earlier derived theoretically based on certain assumptions.

9.1 Solar Oscillations

In Chapter 4 we had discussed the pioneering work of Robert Leighton and his students at Caltech, USA. They had discovered large velocity cells called the 'supergranulation' and large regions on the Sun that oscillate vertically, with speed of a few hundred meters per second and period of about 5 minutes. These regions are smaller than the supergranulation but much larger than granulation. At any point on the photosphere, the small amplitude oscillating components reinforce each other, producing these 5-minute oscillations, which grow and decay, and go in and out of phase to combine and disperse again. In Figure 9.1 is shown a 2 dimensional velocity map with time, this diagram reveals existence of *wave packets* or *patches,* both in space and time. When observed at a given site on the Sun, such quasi-periodic oscillations attain peak velocity of 1000 m/s with period of about 5 minutes. These solar oscillations have been observed with periods ranging between 2 to

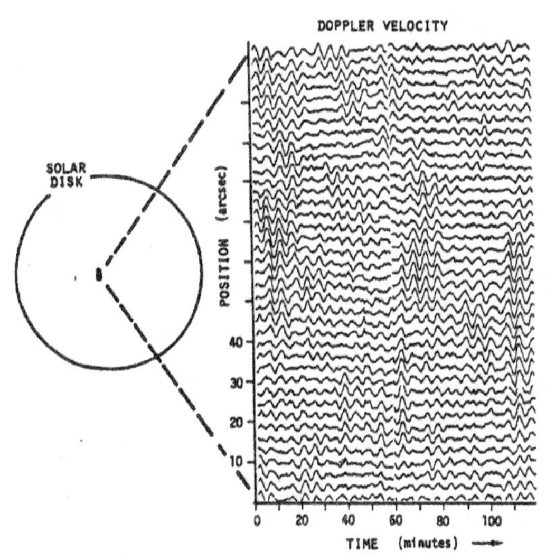

Figure 9.1 Plots of velocity as a function of time, made at number of adjacent points separated by about 2200km on the solar surface along a 60,000 km long strip at the center of the solar disk, over a period of 100 minutes. The periodic 5-minute oscillations are distinctly seen together with clustering of ***wave packets or patches*** over large areas where the phase remains more or less constant. The velocity scale is such that the adjacent curve corresponds to 0.4 km/s.

15 minutes. At first it was thought that these oscillating patches were nothing more than ripples that granulation motion would make as they reach the solar surface.

In early seventies, Roger Ulrich (1970) and independently Robert Strein and John Leibacher (1970) correctly interpreted these 5-minute oscillations as surface vibrations arising from standing sound waves in the deep convection zone. Sound waves within a narrow range of frequencies are trapped in the convection zone and they interfere with one another to form a *standing wave.* The oscillating patches, observed on the solar surface are manifestation of complicated interference pattern of many waves of slightly differing frequencies. The solar surface vibrates like a drum, with nodes (where amplitude of the wave is zero) and antinodes distributed on the surface. These sound waves are essentially pressure waves or the p-modes, generated in the Sun's convection zone by turbulent motions. The human ear can not detect these low frequencies, which are of the order of 3 mHz, but with suitable equipment one finds that they are low pitch noise consisting of many tones playing simultaneously. These sound waves propagate outwards through the convection zone and disturb the gases in the medium and cause them to rise and fall, producing wide spread throbbing motion, which manifests itself as oscillating patches. The amplitude of these motions could be tens of thousand of meters high and speed of few hundred meters per second. These throbbing motions can not be detected by eye, but can be measured using the Doppler Effect. Very sensitive instruments have recorded velocities of these small scale motions of a few meters per second and correspondingly small variation in the total energy output of the Sun.

At the top of the convection zone, near the solar photosphere where steep density gradient is present, this surface acts like a mirror and reflects the sound waves, back towards the solar interior. Near the bottom of the zone, the sound waves gradually *turn back* or are *refracted* towards the surface due to the temperature gradient, as shown in Figure 9.2. The refraction of the wave takes place because the speed of sound waves depends on the ambient temperature and composition of the medium. In deeper layers, the temperature is higher as compared to upper layers; hence sound waves travel faster in deeper layers, because the square of the sound speed is proportional to temperature, and is given by the expression:-

$$s = \sqrt{\left(\frac{\partial p}{\partial \rho}\right)} = \sqrt{\left(\frac{\gamma kT}{m_o \mu}\right)} = const.x\sqrt{\left(\frac{T}{\mu}\right)}, \tag{9.1}$$

where p is the gas pressure, ρ is the gas density, k is the Boltzmann's constant and m_o atomic mass, γ is the adiabatic index and μ is mean molecular weight, given by $\mu = 2/(1+3X+0.5Y+Z)$, where X, Y and Z are concentration by mass of hydrogen, helium and heavier elements respectively, $X+Y+Z=1$. For fully ionized hydrogen gas $\mu=1/2$, and fully ionized helium $\mu=2/1.5$ or $4/3$.

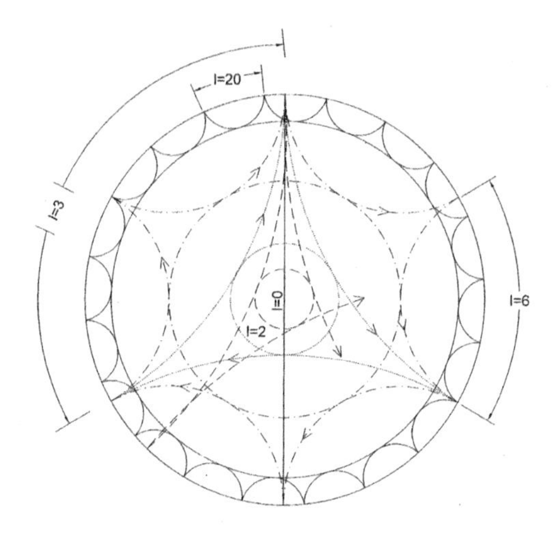

Figure 9.2 Schematic showing the trajectories of sound waves which are normally bent inside the Sun and circle the solar interior within a spherical shell or a resonant cavity. Each shell is bounded at the top by a large density drop near the photosphere and at the bottom is bounded by the increase in sound speed and corresponding increase in temperature with depth, which refracts the downward propagating wave back towards the photosphere. The bottom turning point is shown by dotted line, and the cavity is between the dotted line and the solar surface. How deep a wave penetrates and how far it goes around the Sun depends on the degree, '*l*'.

As shown in Figure 9.2, the waves travel in series of arcs, turning around at some depth and returning to the surface to be reflected back. The waves are trapped in spherical shells or resonant cavities. The bottom turning point occurs along the dotted circles as shown in Figure 9.2. How far deep a wave may penetrate in the solar interior and how far around the Sun it may go depends on the harmonic degree '*l*', which is discussed in Section 9.1.3. From this figure it will be noticed that short wavelength (high frequency) waves travel through shallow upper layers of the convection zone and near the photosphere, while long waves travel through deeper layers. Even in this simplified diagram, one can see that waves can reflect near the same point on the surface. In the Sun, many waves with slightly different frequencies and traveling in slightly different directions can be reflected from almost near the same point. The depth to which a wave can penetrate in the Sun also depends on the angle of incidence of the wave to the solar surface, for example, waves traveling near the surface start their journey at small angles, while waves traversing the Sun's center may be normal to the solar surface, as shown in Figure 9.2. Waves with large l values say 300 or 400 confine to the upper layer of the solar atmosphere, while intermediate degree modes limit to convection zone and l degree modes of 0,1,2, or 3, penetrate to deep solar interior. A typical wave may take about 5 days to circle the Sun. Waves with shorter period than 3 minutes escape into the overlying atmosphere.

9.1.1 *l-ν Diagram*

To test the theory proposed by Ulrich and by Stein and Leibacher, that the observed solar oscillations are in fact manifestations of standing waves in the solar interior; Deubner (1975) made very precise observations of the solar oscillations and obtained a '*diagnostic diagram*' by plotting the individual modes of oscillation, according to their mean frequency and the horizontal wavelengths. In Figure 9.3 is a plot of the mean frequency as a function of degree l obtained using 360 days of observations with SOHO/MDI instrument. It is noticed that all the nodes, along a solar radius n, fall on the same discrete ridge. The ridge at the lower right corner corresponds to n = 0, while the ridge at the upper left to n = 40. This diagnostic diagram confirmed the theory, which predicted that the frequencies and the horizontal wavelengths of the trapped waves in a cavity should form a family of distinct curves. The doted lines in the diagram refer to values obtained theoretically from a Standard Solar Model (which will be discussed later), while the crosses with error bars are the actual values of frequencies obtained from observations. It is clearly seen that both the observed and those calculated frequencies are in good agreement. The spread in the plotted values is due to errors in measurements, with improvement in frequency measurements this spread has remarkably reduced to less than 5 μHz or 0.15%.

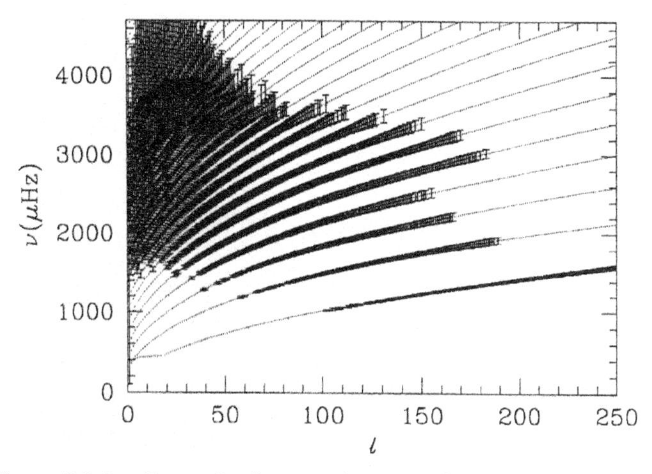

Figure 9.3 l-ν diagnostic diagram, the mean frequencies plotted as a function of degree, l up to 250, obtained from a solar model is shown as dots merging into as a line. The observed frequencies obtained from 360 days of observations from SOHO/MDI instrument are plotted as crosses with error bars. The error bars correspond to 5000σ error level. The lowest line refers to f-mode, or the surface wave, while the other lines are p-modes.

9.1.2 *Solar Standard Model (SSM)*

What is a Standard Solar Model (SSM)? This term might imply that the '*Standard Model*', is the one against which other models are to be compared. Actually, it is not so. The term Standard Model is used to distinguish between the models based on simplest possible assumptions and those computed using the best available physical inputs, such as the opacity, hydrogen-helium abundance, neutrino deficit estimates. The theory of stellar structure has been used to construct theoretically solar models, which provides a physical picture of the solar interior. This is done with the help of a set of mathematical equations governing its mechanical and thermal equilibrium, and nuclear generation, supplemented by observed boundary conditions. As described earlier in Chapter 3, the main problem of solar structure is to determine the run of the physical quantities with depth in the solar atmosphere, with the help of *equations of state*, governing the mechanical and the thermal equilibrium. The mechanical equilibrium in a gaseous envelope, like the Sun is established when the pressure gradient balances the gravitational forces, and is given by the following equations:-

$$\frac{dP(r)}{dr} = -\frac{Gm(r)}{r^2}\rho(r), \tag{9.2}$$

$$\frac{dm(r)}{dr} = 4\pi r^2 \rho(r), \tag{9.3}$$

where P(r) is the gas pressure, $\rho(r)$ is the density, and m(r) the mass inside radius r in the Sun.

For maintaining thermal equilibrium, the energy radiated by the Sun, as measured by its luminosity, should be balanced by thermonuclear energy generated in the solar interior and is given by:-

$$\frac{dL(r)}{dr} = 4\pi r^2 \rho(r)\varepsilon, \tag{9.4}$$

where ε is the energy generation per unit mass, and $L(r) = 4\pi r^2(F_{rad} + F_{conv})$ is the luminosity and F_{rad} & F_{conv} are the radiative and convective energy fluxes.

From these basic *equations of state*, supplemented by auxiliary input physics which describes the thermodynamic state of the matter, and assuming the Sun to be spherically symmetrical object with negligible effects due to rotation, magnetic field, mass loss and tidal forces on its global structure, and certain opacity, chemical abundance and nuclear energy generation rate and the age of the Sun; several authors have numerically integrated these equations to construct theoretical solar models. In recent times, Christensen-Dalsgaard *et al.*,(1996), Brun, Turck-Chièze and Zahu (1999), Bahcall, Pinsonneault and Basu (2001) and others have developed several solar models, based on minimum assumptions and the best available input physics. In all these models, it is considered that the energy generation takes place in solar core by thermonuclear reactions converting hydrogen to helium, through p-p reaction. Further the energy is transported outward from the core by radiative processes, but in the outer 2/3 of the solar radius, it is carried largely by convection and there is no mixing of the nuclear products outside the convection zone. Gravitational settling of helium and heavy elements by diffusion process in the radiative zone has been also proposed by Christensen-Dalsgaard *et al.*, (1996). The theoretically calculated SSM of solar interior is then matched to the atmospheric model given by Vernazza et al., (1981) at the photospheric level. As an example of a SSM, in Figure 9.4 is shown a typical solar model proposed by Brun *et al.*, (1999) giving the temperature, density, pressure, sound speed, adiabatic index, and hydrogen-helium abundance profiles as a function of the radial distance from the center of the Sun. In Table 9.1 are given values of sound speed, density, pressure, adiabatic index and temperature, at various depths in the solar interior.

Table 9.1 In this table the radial distance r/R, c – sound speed in arc/sec, density ρ in g/cm^3, pressure–p in dyn/cm^2, Γ- adiabatic index, and temperature in K in the solar interior are given as per Solar Standard Model by Christensen-Dalsgaard et al., (1996).

r/R	c (cm/sec)	ρ(g/cm^3)	p (dyn/cm^2)	Γ	T (K)
1.0000000	7.8925512e+05	1.9979759e-07	7.6084760e+04	1.6357894	5.7775075e+03
0.9003548	1.1595567e+07	2.6096106e-02	2.1039224e+12	1.6677463	5.9663708e+05
0.8002121	1.7598439e+07	9.1494512e-02	1.7020884e+13	1.6647968	1.3642083e+06
0.7003509	2.2959873e+07	2.0791810e-01	6.5847466e+13	1.6645322	2.3222985e+06
0.6002312	2.6511852e+07	5.0575827e-01	2.1346353e+14	1.6653266	3.1251830e+06
0.5007668	2.9865536e+07	1.3425548e+00	7.1863049e+14	1.6663530	3.9714613e+06
0.4001115	3.3881913e+07	3.9262490e+00	2.7035183e+15	1.6671872	5.1187994e+06
0.3006636	3.8951314e+07	1.1963180e+01	1.0882393e+16	1.6678864	6.7782516e+06
0.2005158	4.5470161e+07	3.4755322e+01	4.3076512e+16	1.6681449	9.3653365e+06
0.1001302	5.0774982e+07	8.7280708e+01	1.3488894e+17	1.6681745	1.3082521e+07
0.0901552	5.0965428e+07	9.4922247e+01	1.4779687e+17	1.6682230	1.3472348e+07
0.0804767	5.1071934e+07	1.0274942e+02	1.6064997e+17	1.6682584	1.3840839e+07
0.0700227	5.1104517e+07	1.1155292e+02	1.7463664e+17	1.6682615	1.4221622e+07
0.0602529	5.1066516e+07	1.1993604e+02	1.8748278e+17	1.6682487	1.4554533e+07
0.0502321	5.0973584e+07	1.2842233e+02	2.0001925e+17	1.6682422	1.4865436e+07
0.0402330	5.0846170e+07	1.3640550e+02	2.1139150e+17	1.6682489	1.5136635e+07
0.0301075	5.0706290e+07	1.4351664e+02	2.2118825e+17	1.6682606	1.5362637e+07
0.0200194	5.0582078e+07	1.4909599e+02	2.2866111e+17	1.6682725	1.5530562e+07
0.0108799	5.0502474e+07	1.5243240e+02	2.3304156e+17	1.6682810	1.5627274e+07

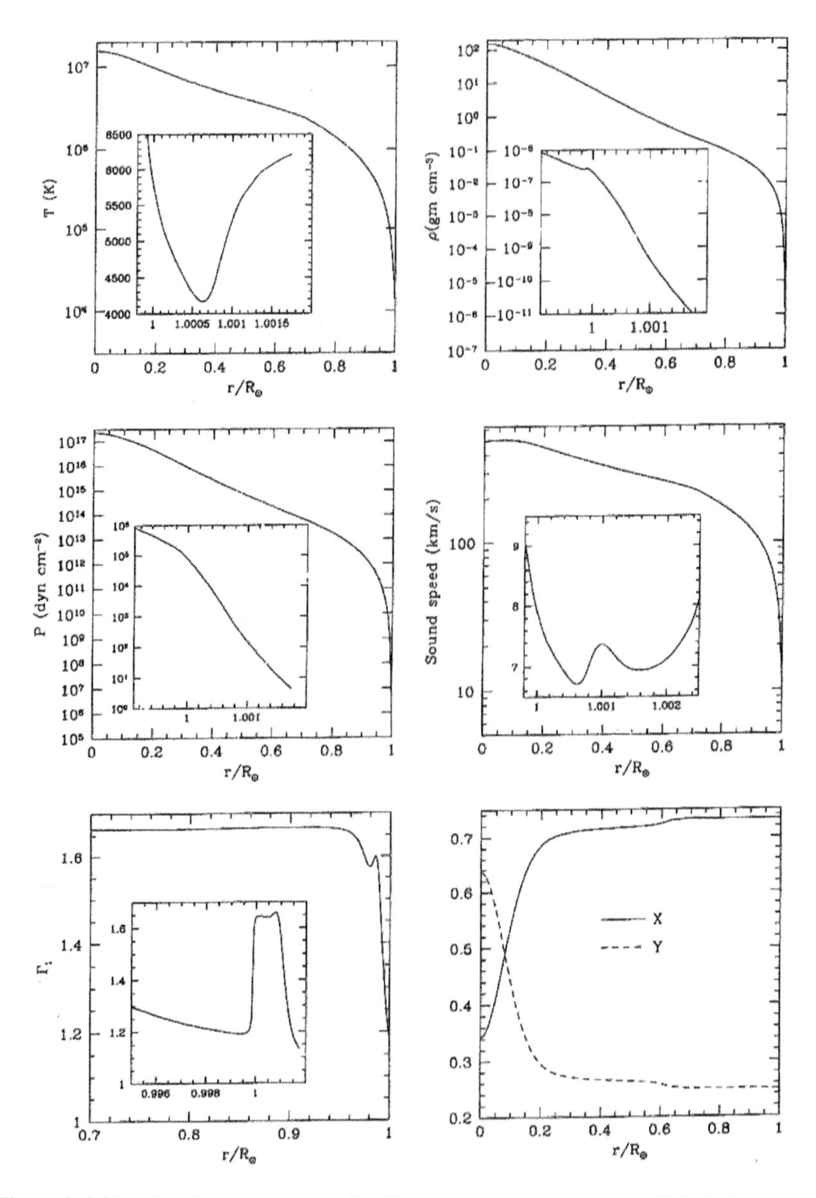

Figure 9.4 Showing the temperature, density, pressure, sound speed, adiabatic index-gamma, hydrogen and helium abundance profiles as function of depth in the solar interior, as per a solar model given by Brun et al., (1999). The inset shows a blowup of the region near the solar surface.

9.1.3 *Observations of Solar Oscillations*

The basic tool of helioseismology is the observations of solar oscillations from such observations the solar frequencies are derived. By comparing the observed frequencies with those predicted from theoretically worked out Standard Solar Model (SSM), one can decide whether the 'model' needs to be modified or not. It is well established now that solar models are fairly robust in its predictions; therefore the frequency measures must also be very accurate, to at least one part in 10,000. The basic method to measure the frequency is to count the number of cycles in a certain period of time. For 10,000 cycle of 5-minute period, it would require at least 35 days of continuous observations. Unfortunately it is difficult to observe the Sun continuously for this long a time, without the day–night gap.

To achieve uninterrupted solar oscillation observations, astronomers have devised several ingenious methods, such as observations from the South pole, where for 6 months the Sun does not set. Gerard Grec, Eric Fossat and Pomerantz (1980a, b) obtained in 1980 interesting results from their 5 continuous days of observations from the South pole, and detected lowest meridional oscillations degree modes *(l=0.1.2,3)* in the Sun. Jack Harvey and colleagues {Harvey et al., (1986)} made 3 successful expeditions in 1981, 1983 and in 1990 to the South pole to observe the intensity oscillations over a 2-dimensional grid on the solar surface. The longest continuous observations they could achieve was for 65 days. From these observations they could construct l-ν diagnostic diagram for degree l between 4 and 140.

The other approach for continuous observations was to establish chain or network of observatories around the globe with identical equipment. The first group to try this idea in 1979 was from the University of Birmingham and they obtained the oscillation spectrum of the integrated sunlight from two stations, using a resonance absorption cell as narrow band filter. This network has been extended to six stations and is now known as BISON (**BI**rmingham **S**olar **O**scillation **N**etwork). Following this effort the French group established six station network called IRIS (**I**nternational **R**esearch on the **I**nterior of the **S**un). Taiwan group also established 3-station network called TON (**T**aiwan

Oscillation **N**etwork) to observe the brightness oscillations using a K-line filter.

The most elaborate 6-station ground based network called GONG (**G**lobal **O**scillation **N**etwork **G**roup) was established through International collaboration in 1995, by the National Solar Observatory, and supported by the National Science Foundation of USA, for 2-dimensional velocity oscillation observations. These 6 stations use high precision and identical equipment, and are located at Learmonth in Australia, at Udaipur in India, at Tenerife in Canary Islands, in Chile at Cerro Tololo, at Big Bear in California and at Mauna Loa in Hawaii. By combing observations from all the 6-GONG stations, the Sun is under observation for more than 93 per cent of time.

9.1.3.1 *Observations from Space*

Using satellites and spacecrafts, astronomers have been able to obtain continuous observations of the total irradiance from very precise and stable instrument in space, called the Active Cavity Radiation Intensity Monitor (ACRIM), built by Robert Williams of Jet Propulsion Laboratory in Pasadena, USA. Using continuous 137 days ACRIM data, during 1980-82, Martin Woodard and Hugh Hudson found low degree oscillation modes in the solar irradiance.

On 2 December 1995 a very sophisticated spacecraft called SOHO (**SO**lar and **H**eliospheric **O**bservatory) was launched by NASA and the European Space Agency (ESA) to reach its permanent position around the L1 Lagrangian point, a location in space where the gravitational force of the Earth and Sun just balances. From this vantage point SOHO can see the Sun continuously for months and years at a time, unhindered by the terrestrial weather.

For helioseismic studies, three experiments are on board on SOHO:-

1. The Stanford University's experiment - SOI (**S**olar **O**scillation **I**nstrument), which uses a narrow band Michelson Doppler Imager (MDI), for 2-D full disk oscillation observations, with 2 arc sec pixel resolution, at a cadence of 1 minute. Full disk longitudinal magnetograms are also made through this instrument,

2. GOLF (**G**lobal **O**scillation of **L**ow **F**requency) is a resonance absorption cell to observe the low degree velocity oscillations, over the integrated image of the Sun,

3. VIRGO (**V**ariability of Solar **IR**radiance and **G**lobal **O**scillations) measures the lowest degree brightness oscillations.

Both MDI/SOHO and GONG projects complement each other and are taking minute-by-minute Dopplergrams and magnetograms with high precision. They have acquired enormous data during the last 9 years. The GOLF/VIRGO/MDI /SOHO is providing superb data for all modes from l = 0 to 4500. High precision data for low degree modes are especially important as they reveal the physical conditions near the solar core.

9.1.4 *Spherical Harmonic Quantum Numbers l, m, and n*

The solar oscillations, first observed by Leighton and colleagues are in fact manifestation of standing sound waves. These waves are formed in the Sun in a similar manner as standing waves on a violin string; it is free to vibrate up and down between fixed positions- '*nodes*'. Standing waves can be considered as a summation of constructive interference of two waves traveling in opposite directions. The destructive interference filters out all the other oscillations. The nodes remain stationary in space while the medium between them oscillates.

In the case of the Sun, a sound wave is free to travel in any direction; in or out, north or south, east or west or in any arbitrary direction. But only those waves of particular wavelengths and direction which can form standing waves can survive random collisions with other waves. To form standing wave pattern or 'mode', the wavelength of each twin wave must fit into the circumference of the Sun a whole number of times (an integer) and also fit into the depth of the spherical shell, a whole number of times without an *overlap,* only then the 'twin' waves will produce fixed nodes in 3-dimemsions. The Sun can be considered as a 3-dimensional perfectly spherical oscillating body, to which the principle of spherical harmonics can be applied.

In the case of the Sun, three spherical harmonic functions are required to represent the *trapped* modes (standing wave pattern). A single mode is designated by a set of three integer numbers *l, m, and n.* The spherical harmonic degree *l*, indicates the total number of nodes along a meridian on the solar disk, and it may range from zero to thousands in number. It is also a measure of the spatial scale of the mode, which extends from the size of a small granule to the entire Sun; from a thousand meters to hundreds of million meters. For $l=0$, the whole solar globe resonates across its radius of about 696×10^6 meters.

The second parameter is the *'azimuthal'* order '*m*', which describes the number of nodal lines that cross the equator and ranges from zero to $\pm\, l$. It represents the concentration of modes near the equator. Modes with $m = 1$, are located close to the equator.

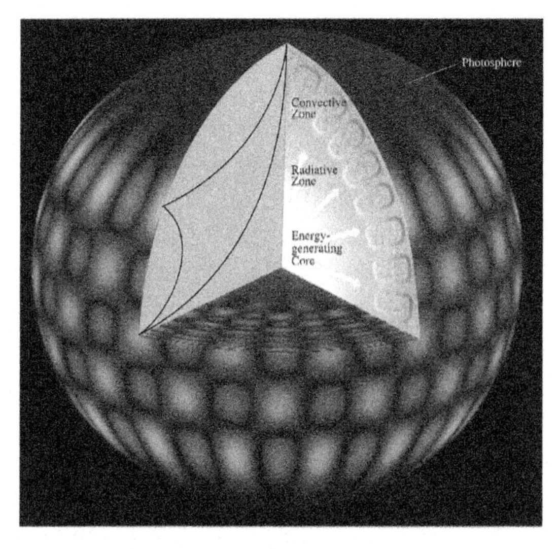

Figure 9.5 Computer generated picture shows that the sound waves inside the Sun make the visible solar surface to move in and out, creating a throbbing motion, these are due to superposition of millions of oscillations, as represented by white and grey regions. The black lines, shown in the cutaway indicate the path of sound waves resonate through the Sun, which are produced by turbulent motions in the convection zone, just about the radiative zone.

The third parameter is the radial order *'n'*, it indicates the number of nodal surfaces in the radial direction inside the Sun, on a line from the center of the Sun to the photosphere and has been observed from zero to 50.

To estimate the total number of nodes vibrating in the Sun, let us consider the following:

That all the standing waves on a single trajectory have the same number of nodes (stationary points) along a solar radius, some may have as many as 50 radial nodes *n*. Each curve represents a string of individual wave pattern or the 'modes', each of these modes has different number of nodes or degrees along the meridian of the Sun. Waves with as many as 500 meridian nodes have been observed. Further each wave can be classified by the number of nodes it has along the solar equator, waves with 1000 of these nodes have been observed. Any particular wave can be labeled with the number of nodes it has along the solar radius, meridian and the equator. Combining all the possible modes, there could be at least 25 million (50x500x1000) different modes, vibrating in the Sun at any instance of time!

9.2 Salient Results from Helioseismology

To extract information about the physical parameters; e.g., temperature, density, pressure, sound speed and chemical composition profile in the deep solar interior, we have two main methods; the '*forward*' method and the '*inverse*' method. In the case of the forward method, the oscillation frequencies and the amplitudes are predicted from the Standard Solar Model and compared with the observed ones. Any difference if found will require modifying the input physics, that is, temperature, density, pressure, chemical composition opacity. In the case of 'inverse' method, it is considered that different modes are trapped in different layers (or cavities) of the Sun. If one combines the observations of several waves in the same layer, one can obtain an *empirical* (although through observations) measure of the sound speed, resulting the temperature of that particular layer. In this manner one builds up an empirical temperature profile by examining deeper and deeper layers.

The accuracy of measurement of solar frequencies has tremendously increased in recent past; nearly one part in a million, due to which the predicted (from a solar model) and the observed frequencies match to within 5 microhertz.

9.2.1 *Tachocline*

A large team of investigators analyzing the oscillation data from the MDI instrument on the SOHO spacecraft, made a startling discovery of a distinct 'bump' in the sound speed (corresponding to temperature also) profile at the base of the convection zone. In Figure 9.6(a) is shown a plot of the relative differences found between the empirically derived values of the square of sound speed ($\delta c^2/c^2$) and those obtained from a Standard Solar Model, verses the distance from the center of the Sun (r/R_0), and in Figure 9.6(b) is plotted the density variation. The maximum discrepancy of about 0.4% appears as a 'bump' just below the base of the convection zone between 0.6 and $0.7R_0$.

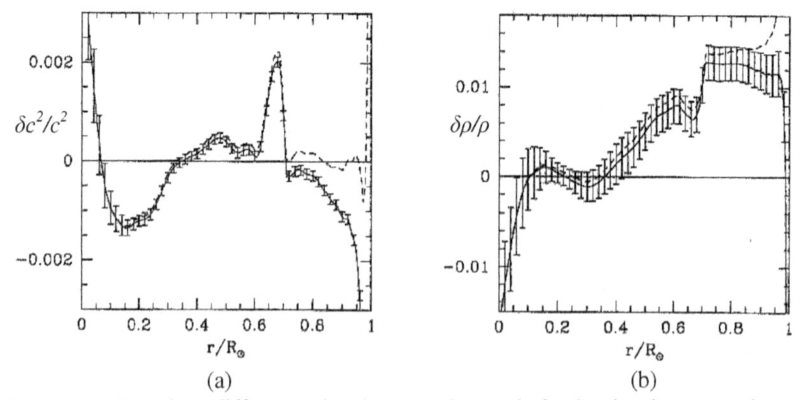

(a) (b)

Figure 9.6 Showing difference in the sound speed & density between the theoretically predicted values from SSM (is shown by solid lines with error bars) with those from observations, along a solar radius in the Sun. Here ($\delta c^2/c^2$) is plotted instead of c, because c^2 is proportional to the ambient temperature. Note a distinct 'bump' around $0.67R_0$, from the Sun's center, and a sharp decrease near the core of the Sun at about $0.25R_0$.

In a corresponding plot Figure 9.6(b) between the density and radial distance, a distinct discontinuity in density profile is also noticed around $0.7R_0$. Due to the sudden increase in sound speed and corresponding increase in temperature and density, a 'shear' region near $0.67R_0$ develops, which is kwon as ***'tachocline'***. This region has important consequence related to the dynamo theory for generation of magnetic fields in the Sun. Some scientists believe that this 'bump' in the sound speed and temperature profile is due to sharp gradient in the helium abundance profile in the solar model. In this plot a sharp decrease in sound speed is also seen in the radiative zone near the solar core at about $0.25R_o$ from the center of the Sun. No proper explanation has come forward to explain this decrease. Christensen-Dalsgaard *et al.*, (1996) had proposed a solar model with gravitational settling of helium and heavy elements, which substantially improved the theoretical model of the radiative interior. Perhaps the observed decrease in sound speed near $0.25R_o$ may be due to over abundance of helium at this level, as per Christensen-Dalsgaard's model.

9.2.2 *Helium Abundance*

Scientist have been debating for years about the amount of helium in the Sun. Different methods and different data sets give values ranging from 20% to 40% of the Sun's mass. Part of the problem is because helium emits relatively weak lines in the visible spectrum, and very strong lines in the EUV region, but it is difficult to interpret these lines. Helium atoms are also detected in the solar wind, which shows variation on the order of at least two. The LOWL and the SOHO data indicate that helium is more abundant in the core than the best theoretical models have assumed. The best estimates of the average abundance of helium in the Sun lies now between 23.2% and 25.2 % of the solar mass, which agrees well with the recent SSM, yielding a value of 24%.

9.2.3 *Temperature and Frequency Variation with Solar Cycle*

From ACRIM (**A**ctive **C**avity **R**adiometer **I**rradiance **M**onitor) irradiance data taken from 1980 to 1985, Woodard and Noyes (1985) found that the frequencies of the low degree modes change with solar activity cycle. The frequencies of the low degree showed a decrease during the declining phase of solar cycle from maximum (1980) to minimum (1985) by 0.4 microhertz or one part in 10,000. This effect has been confirmed by IRIS and BISON groups also. The frequencies again rose at the next maxima in 1991. This change in frequency with solar activity has been attributed to change in temperature in the deep solar interior. Bhatnagar *et al.*, (1999) have found frequency change also in the case of intermediate and higher modes (l= 20 to 150) with solar cycle. A decrease in frequency is noticed during the minimum period while an increase during the maximum period. Since then several authors have confirmed these findings.

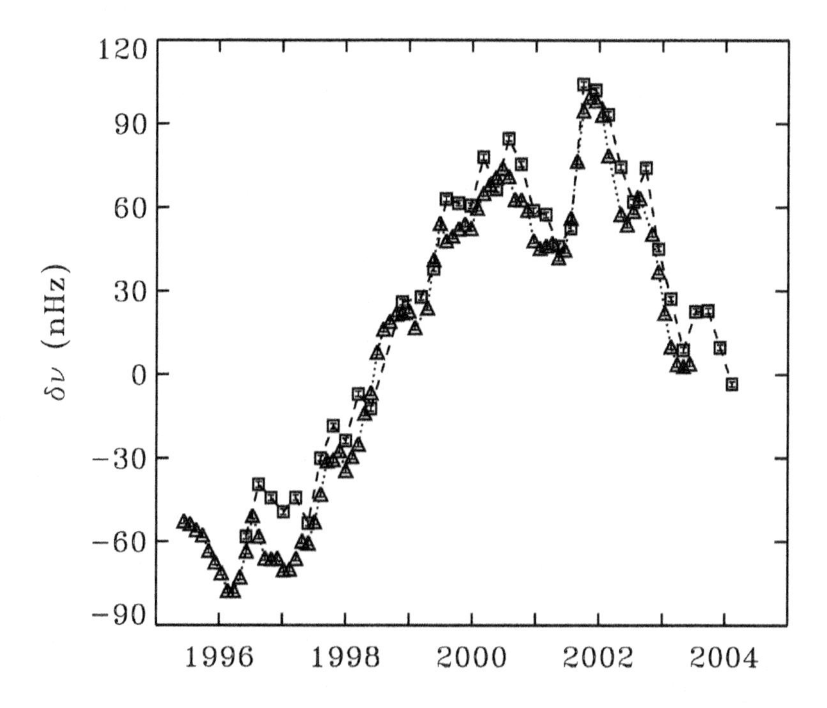

Figure 9.7 p-mode frequency change during the period 1995.5 to 2004, data from GONG, shown by triangles joined by dotted lines and MDI data is shown by squares joined by dashed lines. This plot shows distinct frequency shifts during solar activity cycle. (After Jain & Bhatnagar (2003), this plot is extended to 2004, by Jain & Tripathi from the earlier data available up to 2002).

9.2.4 *Back Side View of the Sun*

Besides the interesting results obtained on the global properties of the Sun from helioseismology, scientists have used the solar oscillations to derive the structure and properties of isolated features on the Sun. Braun *et al.*, (1988 a, b) developed a method to investigate the fate of traveling sound waves in and around sunspots. They found that half of the acoustic power that enters the sunspot is absorbed, although the absorbed power is too small to heat the spot, but the question remains as to what happens to the sonic power in the spot. From the MDI/SOHO, and recently from the GONG data also, a new technique called *helioseismic holography* has been developed. It has been used to *see through* the Sun and image its 'back side'. With this new technique, the images show structure, evolution and eruption of solar activity on the far side of the Sun, even many days before they rotate on the side facing the Earth.

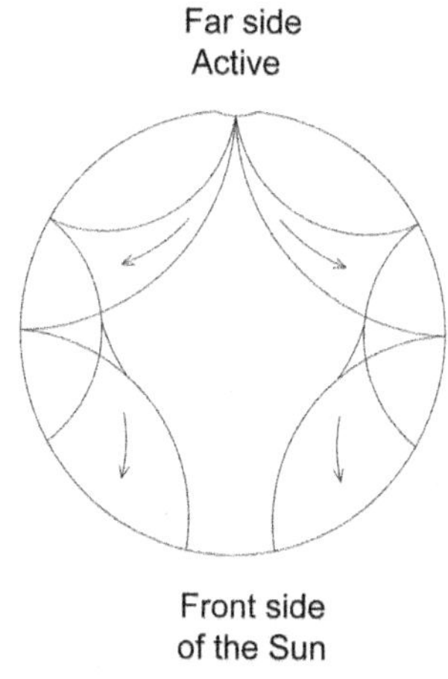

Figure 9.8 A schematic showing wave propagation from the back side of the Sun, which are reflected internally once before reaching the front side, where they can be observed by SOHO/MDI instruments.

In Figure 9.8 is shown a schematic of sound waves emanating from an active region on the far side of the Sun and reaching the near side that faces the Earth. The idea behind this technique is that, when a large active region is present on the back side of the Sun its intense magnetic field compresses the gases and slightly lowers them, and make them denser as compared to the surrounding material. A sound wave that would normally take 6 to 7 hours to travel from the near side to the far side and back again, takes approximately 12 seconds less, when it bounces off the compressed active region on the far side. When scientists examined the near-side oscillations obtained from MDI instrument, they could detect quicker return of these sound waves. Solar astronomers can now use this technique to monitor the evolution of large regions of magnetic activity as they cross the far side of the Sun, and may even warn the world of impending appearance of strong activity on the Sun. In Figure 9.9 is shown a map of active regions on the far side of the Sun, computed by the Stanford group.

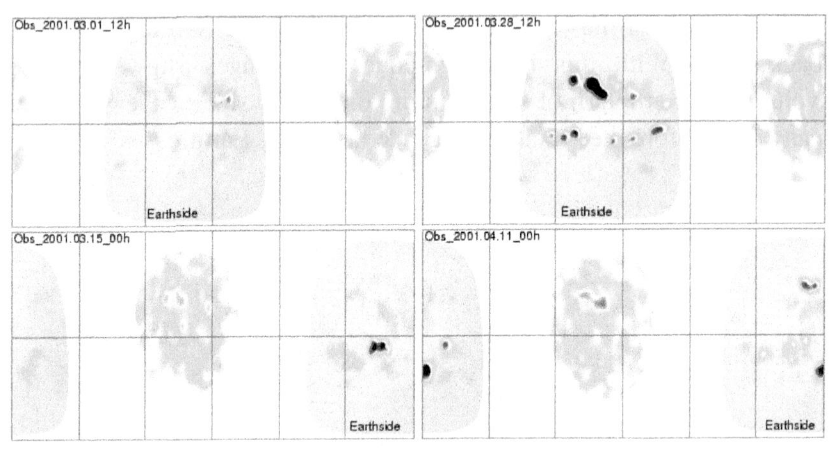

Figure 9.9 Showing location and structure of active regions on four sample images of the Sun taken from SOHO during 1 March to 11 April 2001, over a span of two solar rotations. A very large active region AR9393 at Carrington longitude 154 deg, and latitude 17N deg was observed during this time. These computed Carrington maps show the magnetic flux on the Earth side and inferred flux on the far side of the Sun was obtained from helioseismic holographic technique at 13.5 days interval.

9.2.5 *Sunquakes*

Solar flares are known to occur on the Sun since long. These are very powerful explosions releasing energy equivalent to billions of nuclear bombs in the solar atmosphere. Scientists have speculated that these explosions can produce powerful shocks, which in turn can make the Sun vibrate inside. The MDI/SOHO scientists (Kosovichev and Zharkova, 1998) had detected a powerful sunquake generated by an exploding solar flare which occurred high up in the photosphere on 9 July 1996, and observed circular seismic waves moving across the photosphere from their helioseismic data, like ripples spread out in all directions when a stone is thrown into a water pond. This was the first observation of seismic waves produced by a solar flare. The observed seismic waves

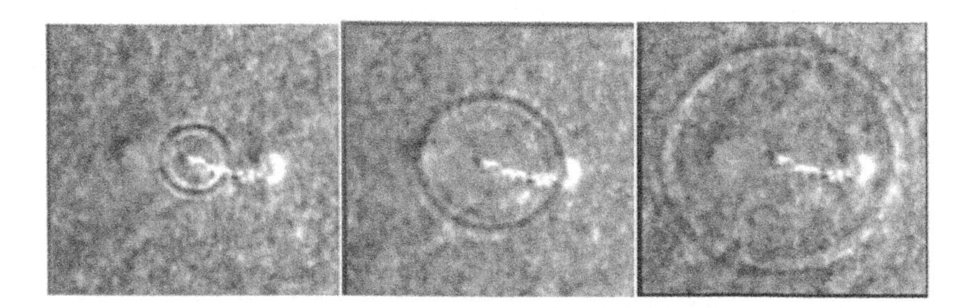

Figure 9.10 Sunquake - sequence of seismic waves produced by a X2.6/1B class solar flare on 9 July 1996, observed by MDI/SOHO instrument showing seismic waves spreading out from the flaring region towards the photosphere as ripples on a water pond.

moved out to a distance of at least 120,000 km from the flare region and were accelerated from about 30 km/s to 100 km/s, before fading out in the photospheric background. Seismic waves produced by sunquakes can really shake the Sun to its very center, like earthquakes can make our entire planet to vibrate. This 9th July sunquake was produced by relatively moderate X2.6/1B flare; however it was equivalent to an earthquake of magnitude 11.3 on Richter scale, about 40,000 times the earthquake that devastated San Francisco in 1906. From MDI/SOHO data, Kosovichev and Zharkova (1999) have reported observing rapidly propagating *magnetic waves* also, associated with X-class flare of 2nd May 1998. This effect is not the same as that observed on 9th July 1998, wherein actual seismic waves were identified. The 9th July event is the one and the only observation of sunquake waves made by any helioseismic instrument. Scientists do not understand how such a modest flare can give rise to such powerful seismic waves. In Figure 9.10 is shown a sequence of seismic waves spreading from the flare region.

9.2.6 *Velocity Structure & Rotation in the Solar Interior*

Some of the most interesting results emerging out from helioseismic studies concern the velocity structure of the solar interior. In Chapter 3 we have already discussed in detail the rotational and meridional flows in the Sun. One of the most striking result that helioseismology brought forward is that the radiative zone below the convective zone, rotates rigidly as a solid body down to 0.4 R_0 with an angular velocity between the highest and the lowest surface velocity. While in the convection zone, the solar interior rotates *differentially*, that is, the rotation rate varies with latitude and depth as seen in Figure 3.25. From this diagram, it is obvious that a strong *shearing* motion exits, between the radiative and the convective zones. What this *sheared layer* inside the Sun means in terms of physical conditions, is not yet well understood. Some scientists believe that the sheared layer may be a prime site for the generation of solar magnetic fields, but how exactly fields are generated is not known.

9.2.7 *The Neutrino Puzzle*

Helioseismology's biggest contribution to solar and particle physics had been to finally resolve the long standing question of solar neutrino deficit, those obtained from observations and theory. In Chapter 3 we have already discussed in detail the neutrino problem; here we shall mention only that the Standard Solar Models are accurate enough to match within 0.01% with the most sensitive and highly precision measurements of solar frequencies obtained from helioseismology. Thus it is established without doubt that the problem was not with the solar models but it was the neutrino physics which needed to be understood better and now with neutrino oscillation concept being accepted, this problem has been solved.

Chapter 10

On the Joy of Observing the Sun – A *Personal Experience*

Dawn on a summer day at Mount Wilson in California. The year could be 1900, 1950, or 2000; it doesn't matter.

The chirping of chickadees and sparrows awaken us. Reaching for our shoes we shake them out in case a mouse has nested overnight. Strolling up from the monastery along the path to the telescopes, we are pleased that the sky is clear, the air balmy. As is usual for this time of year, fog hides the valley below us. Millions reside there, unseen, unknowable, in complete silence so far as we are concerned. To the east, the red orb of the Sun is elevated a few degrees just south of Mt. Waterman in the San Gabriel Mountains.

The languid, soft air suggests the seeing might be exceptional. Our steps quicken as we approach the 150-foot solar tower. An open elevator bucket takes us to the top with questionable safety. There we open the dome and set the coelostat to track the Sun for the days work.

Down in the observation room two big questions occupy us: how good is the image and what is the Sun offering in the way of sunspots? If the seeing is really excellent almost any solar structure can be intriguing. We move the image to inspect the east limb, hoping solar rotation has brought a new region onto the disk. Such regions invariably form on the backside of the disk.

The best seeing usually is found about an hour after sunrise. It is a compromise between air mass and heating of the atmosphere. We must be prepared for this optimal interval: film holders loaded (1900-1980), computers up and running (2000), telescope focus determined.

During the past night, while wakeful, we considered possible experiments we might conduct on the morrow if conditions became just right. Such nighttime thoughts are often surprisingly original. It is as though our thinking mechanism is cleared of trash and an uninhibited slate is there to write on.

Fortunately there are always new discoveries to be made. Sometimes this follows from instrumental advances. Other times the Sun itself surprises us with her astonishing quintessence: umbrae, penumbrae, light-bridges, flare activity, faculae, prominences, to name only major categories. From these circumstances flows our joy of observing.

Appendix I

Basic Units, Conversion Factors, Physical and Astronomical Constants

A. Basic Units – Length

1 micron (μ) = 1 micrometer = 10^{-6} meter μm or μ
1 Ångstrom (Å or A) = 10^{-10} meters = 10^{-8} cm, 0.1 nm or Å

B. Conversion Factors

1 joule = 10^7 ergs	1 erg = 10^{-7} joule
1 electron volt (eV) = 1.60207 x 10^{-19} joules	1 joule = 6.2419 10^{18} eV
1 watt = 1 joule/sec	
1 cm = 0.3937 inch	1 inch = 25.4 mm
1 m = 1.0936 yards	1 yard = 0.9144 m
1 km = 0.6214 mile	1 mile = 1.6093 km
1 gm = 0.0353 oz	1 oz = 28.3 gm
1 kg = 2.2046 lb	1 lb = 0.4536 kg

C. Basic Physical Constants

Speed of light in vacuum	c = 299 792 458 m/s
Constant of gravitation	G = $(6.672\ 59 \pm 0.00085) \times 10^{-11} m^3/kg.s^2$
Planck's constant	h = $(6.626\ 0755 \pm 0.000\ 0040) \times 10^{-34}$ J.s
Boltzmann's constant	k = $(1.380\ 658 \pm 0.000\ 012) \times 10^{-23}$ J/K
Wien displacement constant	$\lambda_{max}T$ = 0.289 789 cm.K = 28.978 9×10^6 Å.K

Stefan-Boltzmann constant	σ = (5.670 51 ± 0.000 19)x10^{-8}W/m^2.k^4
Mass of hydrogen atom (kg)	m_H = (1.673 534 0 ± 0.000 0010)x10^{-27}
Mass of neutron (kg)	m_O = (1.674 928 6 ± 0.000 0010)x10^{-27}
Mass of Proton (kg)	mp = (1.672 623 1 ± 0.000 0010)x10^{-27}
Mass of electron (kg)	m_e = (9.109 389 7 ± 0.000 005 4)x10^{-32}
Rydberg's Constant	R = (1.097 373 153 4 ± 0.000 000 001 3)x10^7/m

D. Units used in Astronomy

For	Standard Units	Other
Energy	joule (J) = kg.m^2 / s^2	ergs, eV
Power	watts (W) = J/s	joule/sec
Frequency	hertz (Hz)	Cycles/sec

E. Astronomical Constants

Astronomical Unit	1 AU	= 1.495 978 70 x 10^{11}m
Solar Parallax	Π_O	= 8.794 148 arc sec
Parsec	1 pc	= 3.086x10^{16} m
		= 206 264.806 AU
		= 3.261 633 light years
Tropical year (Equinox to equinox)		= 365.242 190 78 days
Julian century		= 365 25 days
Mean solar day		= 864 00 sec
Sidereal year		= 365.256 366 days
		= (3.155 815)x10^7 sec
Mass of Sun	M_O	= (1989 1)x10^{30}kg
Radius of Sun	R_O	= 696 000 km
Luminosity of Sun	L_O	= 3.827x10^{26}J/sec
Solar Constant	S	= 1368 W/m^2
Mass of Earth	E_E	= (5.974 2)x10^{24} kg
Equatorial radius of Earth	R_E	= 6 378.130 km
Earth- Moon distance (mean)		= 384 403 km
Radius of Moon	R_M	= 1738 km
Mass of Moon	M_M	= 7.35x10^{22} kg

Appendix II

Glossary

Absolute luminosity: Denoted by L, the amount of energy radiated per unit time, measured in units of watts.

Absolute temperature: The temperature as measured on a scale whose zero point is absolute zero. The unit of absolute temperature is the degree Kelvin, denoted by K.

Absorption line: A "dark" line of less intensity than the continuum centered at a particular wavelength in the electromagnetic spectrum. It is formed when a cool, tenuous gas located between a hot, radiating source and the observer, absorbs electromagnetic radiation its wavelength.

ACRIM: Acronym for the Active Cavity Radiometer Irradiance Monitor, a space experiment to measure the total radiative output of the Sun which is called Total Solar Irradiance (TSI).

Active region: The magnetized region in, around, and above sunspots.

Alfvèn velocity: The velocity of propagation of disturbances in a magnetized plasma, propagating along the magnetic field lines and is given by $B/(4\pi\rho)^{1/2}$ for a magnetic field strength of B and a gas mass density ρ.

Alfvèn wave: A wave motion occurring in magnetized plasma in which the magnetic field oscillates, transverse to the direction of propagation, without a change in magnetic field strength.

Alpha particle: The nucleus of a helium atom, consisting of two protons and two neutrons.

Alt-azimuth: A two-axis telescope mounting in which motion around one of the axis is vertical which provides movement in azimuth, while motion in perpendicular axis provides up-down movement in altitude.

Altitude: Elevation in angular measure above the horizon.

Ångstrom: A unit of length equal to 10^{-10} m, or10^{-8} cm or 0.1 nm.

Angular momentum: The product of angular velocity and mass.

Angular velocity: The rate at which a body rotates or revolves, expressed as the angle covered in a given time.

Annular eclipse: A solar eclipse in which a ring (annulus) of solar photosphere remains visible.

Aperture: The diameter of the objective of the lens or mirror of a telescope.

Aphelion: The point in a planet's orbit that is farthest from the Sun.

Apparent magnitude: A measure of the relative brightness of a star or other celestial object, as perceived by an observer on Earth.

Arc degree: A unit of angular measure in which there are 360 arc degrees in a full circle.

Arc second: Abbreviated arc sec. A unit of angular measure in which there are 60 seconds in 1 arc minute and therefore 3600 arc seconds in 1 arc degree

Arcade: A series of arches formed by magnetic loops that confine hot gas.

Astronomical unit (AU): Parameter describing the Earth's orbit around the Sun, corresponding to the mean distance between the Sun and the Earth, approximately 1.495978×10^8 km.

Astrophysics: The science of astronomy in which the laws of physics are applied.

Atmosphere: The gases surrounding the surface of a planet, or natural satellite, or the outermost gaseous layers of a star held by their gravity.

Atmospheric window: A wavelength band in the electromagnetic spectrum that is able to pass through the Earth's atmosphere with relatively little attenuation through absorption, scattering or reflection.

Atom: The smallest particle of an element that has the characteristics of the element and is composed of a central nucleus, containing protons and neutrons, and a cloud of orbiting electrons equal to the number of protons.

Atomic mass unit: Abbreviated a.m.u., the unit of atomic mass equal to $1.660\ 54 \times 10^{-27}$ kg. A proton and neutron have masses of 1.007 276 5 and 1.008 664 9 a.m.u respectively.

Atomic number: Denoted by Z, the number of protons in the nucleus of an atom. The atomic number of an element indicates its place in the periodic table of elements

Atomic weight: The number of protons and neutrons in an atom.

Autumnal equinox: Positions in the sky where the ecliptic crosses the celestial equator, and where the Sun passes each year when moving from northern to southern declinations. It occurs on 21-22 September.

Azimuth: The angular distance around the horizon counted from North-East–South-West

Baily's beads: Beads of light visible around the rim of the moon at the beginning and end of a total solar eclipse. They result from the solar photosphere shinning through valleys at the edge of the moon. First described by the English astronomer Francis Baily in 1836.

Balmer continuum: Radiation associated with the recombination of free electrons to the second bound level of the hydrogen atom.

Balmer series: Hydrogen lines seen in visible wavelengths. The set of spectral absorption or emission lines resulting from a transition down to or up from the second energy level (first excited level) of hydrogen.

Bartel's rotation number: A cumulative sequence of 27-day rotation periods of solar and geophysical parameters, assigned by Julius Bartel to begin in January 1833.

Beta decay: Emission of an electron (or a positron) and a neutrino by a radioactive nucleus.

Bipolar region: A photospheric region where the magnetic fields show opposite magnetic polarities.

Black-body: A hypothetical body which is perfect absorber and emitter of all radiations incident on it. The intensity of the radiation emitted by a perfect black-body and the way it varies with wavelength depends only on the temperature of the body and can be predicted by quantum theory.

Black-body radiation: The radiation that would be emitted by a perfectly black-body.

Boltzmann's constant: The constant of proportionality between energy per degree of freedom and temperature, denoted by the symbol k. Boltzmann's constant has the value $k = 1.380\ 66 \times 10^{-23}$ j K^{-1}.

Bremsstrahlung: Radiation that is emitted when an energetic electron is deflected by an ion. It is also called a free-free transition because the electron is free both before and after the encounter, and remaining in an unbound hyperbolic orbit without being captured by the ion, to form an atom. Generally, it is a type of radiation emitted when high-energy electrons are accelerated

Butterfly diagram: A plot of the heliographic latitudes of sunspots with time during the 11-year sunspot cycles. The plot resembles butterfly wings, which gives the diagram its popular name.

Calcium H and K lines: Spectral lines of singly ionized calcium denoted by CaII, in the violet part of the spectrum at 3968 Å (H) and 3934 Å (K). The H and K designations were given by Fraunhofer and are still commonly used.

Calcium network: A pattern of emission features seen in the lines of ionized calcium.

Carbon-Nitrogen-Oxygen cycle: A sequence of nuclear reactions, abbreviated as the CNO cycle, which accounts for the production of energy inside main-sequence stars which are more massive and hotter than the Sun.

Carrington longitude: A system of fixed longitudes rotating with the Sun introduced by Richard C. Carrington in the 19th century.

Cassegrain telescope: A type of reflecting telescope in which the light is focused by the primary mirror is intercepted by a secondary mirror and reflected back and focused through a hole in the center of the primary mirror.

CCD: Charge Coupled Device, as solid-state imaging device.

Celestial equator: The intersection of the celestial sphere with the Earth's equator.

Celestial poles: The intersection of the celestial sphere with the axis of rotation of the Earth.

Celestial sphere: The hypothetical sphere in the sky centered at the center of the Earth

CGS: Centimeter-Gram-Second (abbreviated cm-gm-sec or cm-g-s).

Charged particles: Fundamental components of sub-atomic matter, such as protons and electrons that have electrical charge.

Chromosphere: The layer of the solar atmosphere lying between the photosphere and the corona.

Chromospheric network: A large-scale cellular pattern visible in spectroheliograms taken in the calcium H and K lines.

CME: Acronym for Coronal Mass Ejection..

Continuous spectrum: A spectrum with radiation at all wavelengths displaying neither absorption nor emission lines.

Continuum: That part of a spectrum that has neither absorption nor emission lines but only a smooth wavelength distribution of intensity.

Convection: The transfer of energy by material mass motion. Physical lifting of hot matter, thus transporting energy from a lower, hotter region to a higher cooler region.

Convection zone: The subsurface zone in which convection dominates energy transfer.

Convection: The method of energy transport in which the rising motion of masses from below carries energy upward.

Coordinate system: Method of assigning positions of objects with respect to suitable axes.

Core: The central region of a star or other celestial object.

Corona: The outermost, high-temperature region of the solar atmosphere above the chromosphere and transition region.

Coronagraph: An instrument used for observing the faint solar corona outside a solar eclipse.

Coronal green line: An emission line due to Fe XIV at 5303Å. The strongest line visible in the solar corona.

Coronal hole: Region of the corona that appears to be dark when observed in the Extreme Ultraviolet and soft X-ray region of the electromagnetic spectrum.

Coronal loop: A loop structure that traces magnetic lines of force and passes through the corona.

Coronal mass ejection: Abbreviated CME, a transient ejection of plasma and magnetic fields at a high speed from the Sun's corona into interplanetary space.

Coronal streamer: A magnetically confined, loop-like coronal structure in the low corona straddling a magnetic neutral line on the solar photosphere.

Coronium: Before the highly ionized coronal lines were identified, the observed coronal lines were suppose to originate from a hypothetical unknown chemical element called 'Coronium'.

Corpuscular radiation: Charged particles (mainly protons and electrons) emitted by the Sun.

Cosmic rays: Nuclear particles or nuclei traveling through space at high velocity and originating outside the solar system.

Coulomb barrier: The electric field repulsion experienced when two charged nuclear particles approach each other. When the barrier is overcome by the tunnel effect, nuclear reactions take place that produce the Sun's energy.

Current sheet: The two-dimensional surface within a magnetosphere that separates magnetic fields of opposite polarities.

D lines: A pair of lines from sodium that appear in the yellow part of the spectrum at 5890Å and 5896Å.

Decametric radiation: Radio radiation of approximately 10-meter in wavelength.

Decimetric radiation: Radio radiation of approximately 0.1-meter in wavelength.

Declination: Celestial latitude, measured in degrees north or south of the celestial equator.

Density: The amount of mass or number of particles per unit volume. In cgs units density is given by gm cm^{-3}.

Deuterium: The heavy isotope of hydrogen containing both a proton and a neutron in its nucleus.

Diamond ring effect: The last of the photosphere glowing brightly just at the beginning and just after the total phase of a solar eclipse,

Differential rotation: The rotation of a gaseous body, such as the Sun, where the rotation rate varies with latitude.

Diffraction grating: A metal or glass surface on which equidistant parallel lines have been ruled, typically at 100 to 1000 lines per millimeter. Light striking a grating is dispersed by diffraction into a spectrum.

Disk: The visible part of the Sun.

Doppler Effect: Change in the observed wavelength or frequency of electromagnetic radiation, due to the relative motion between the observer and the emitter.

Dynamo: A mechanism that generates electricity through the effect of motion of conducting matter in the presence of a magnetic field. **$E = mc^2$:** Einstein's formula (special theory of relativity) for the equivalence of mass and energy.

Eccentricity: Measure of the departure of an orbit from a perfect circle.

Echelle grating: A diffraction grating with lines ruled to produce spectra at high orders.

Eclipse: The partial or total obscuration of the light from a celestial body as it passes behind another body.

Ecliptic plane: The plane of the Earth's orbit around the Sun. It is inclined to the plane of the celestial equator by about 23.5 degrees, and intersects the celestial equator at the two equinoxes.

Ecliptic: The projection of the Earth's orbit around the Sun on the celestial sphere, marking the Sun's apparent yearly path against the background stars.

Effective temperature: The temperature of an object emitting black-body radiation. The effective temperature is denoted by the symbol T_e, and for the Sun it is equal to 5780^0 K.

Electric field: A force field set up by an electric charge.

Electromagnetic force: One of the four fundamental forces of nature, giving rise to electromagnetic radiation.

Electromagnetic radiation: Radiation that carries energy through space at the speed of light and generated by changing electric and magnetic fields.

Electromagnetic spectrum: The entire range of all kinds of radiations including Gamma rays, X-rays, ultraviolet radiation, visible, infra red and radio waves.

Electron: A negatively charged sub-atomic particle.

Electron flux: The rate of flow of electrons through a reference surface. In cgs units, measured in electrons s^{-1}, or simply s^{-1}.

Electron neutrino: A type of neutrino that interacts with the electron. It is the only kind of neutrino that is produced by the nuclear reactions in the Sun.

Electron volt (eV): The energy necessary to raise an electron through a potential of one volt.

Element: A kind of atom characterized by a certain number of protons in its nucleus. All atoms of a given element have similar chemical properties.

Ellerman bombs: Also called moustaches, these are short-lived very bright points of light in outer penumbra edge of spots observed in hydrogen-alpha wings.

Emerging flux region: A region on the solar surface where new magnetic flux appears from below.

Emission line: A bright spectral feature at a particular wavelength in a spectrum.

Energy flux: The rate of flow of energy through a reference surface. In cgs units, it is measured in erg s^{-1}, or in watts, where 1 watt = 1×10^7 erg s^{-1}.

Energy level: A state corresponding to an amount of energy that an atom is allowed to have by the laws of quantum mechanics.

Ephemeral region: A short-lived, tiny active region.

Ephemeris: A listing of astronomical positions and other data.

Equator: An imaginary line around the center of a body, where every point on the line is an equal distance from the poles. The equator defines the boundary between the northern and southern hemispheres.

Equatorial mount: A type of telescope mounting in which one axis, called the polar axis, points toward the celestial pole and the other axis is perpendicular. Motion around the polar axis can counter the effect of the Earth's rotation.

Equinox: An intersection of the ecliptic and the celestial equator.

Erg: A unit of energy equal to the work done by a force of 1 dyne acting over a distance of 1 centimeter. Energy of one joule is equal to 10^7 ergs.

Escape velocity: The velocity that an object must have to escape the gravitational pull of a mass.

Evershed effect: The radial flow of gases within the penumbra of a sunspot.

Excitation: An atomic process in which an atom or ion is raised to a higher-energy state by taking one of its electrons from one orbit to a higher one.

Extreme–ultraviolet: Abbreviated EUV, a portion of the electromagnetic spectrum from approximately 100 to 1000 Å.

Eyepiece: The lens at the eye end of a telescope used to examine the image formed by the objective.

F corona: A component of the white-light corona caused by sunlight scattered from solid dust particles in interplanetary space.

Faculae: Bright regions of the photosphere associated with magnetic fields seen in white light. They are visible only near the limb of the Sun.

Fibrils: Dark elongated features seen in hydrogen-alpha spectroheliograms, occurring near sunspots, plages or filament channels.

Filament channel: A broad pattern of fibrils in the chromosphere, marking the place that a filament may soon form or where a filament recently disappeared.

Filament: A feature of the solar surface seen in hydrogen-alpha as a dark wavy structure. It is in fact a prominence projected on the solar disk.

Filigree: Fine structure visible in the photosphere in far wings of Hα filtergrams or the G-band, first reported and named by Dunn and Zirker from Sac Peak Observatory, USA.

Five-minute oscillations: Vertical oscillations of the solar photosphere with a well-defined period of five minutes, usually interpreted in terms of trapped sound waves in the interior of the Sun.

Flare: A sudden and violent release of energy and matter within a solar active region in the form of electromagnetic radiation, energetic particles, wave motions and shock waves, lasting minutes to hours.

Flash spectrum: The emission-line spectrum of the solar chromosphere seen for a few seconds just before and after totality during the total eclipse of the Sun.

Flavors: A way of distinguishing quarks; up, down, strange, charmed, truth (top), beauty (bottom).

Flux: The rate of flow of any thing, e.g., energy, through a reference surface. The flux density is the flux measured per unit area.

Flux tube: A magnetic torus through which particles circulate.

Focal length: The distance between a lens or curved mirror and its optical focus.

Focal ratio: The ratio of the focal length of a lens, or curved mirror, to its diameter.

Focus: (pl; foci) (a) A point to which radiation is made to converge by an optical system; (b) the point at which an image is formed of a distant source lying on the axis of a lens or mirror, (c) in an ellipse, one of the two points, the sum of the distances to which remains constant.

Foot point: Intersection of a magnetic loop with the photosphere, the lowest visible portion of the magnetic loop.

Forbidden lines: Emission lines not normally observed under laboratory conditions because they have a low probability of occurrence, often resulting from a transition between a metastable excited state and the ground state. Under astrophysical conditions, including the solar corona and highly-rarefied nebulae, the metastable state can last long enough for the "forbidden" lines to be emitted.

Forbush Decrease: An abrupt decrease in the background cosmic-ray intensity observed at the Earth, by at least 10 percent, a day or two after a solar flare. This phenomenon is named after Scott Forbush who first noted it in 1954.

Force: In physics; something that can or does cause a change of momentum, measured by the rate of change of momentum with time.

Fraunhofer lines: The dark absorption features or lines in the solar spectrum, caused by absorption at specific wavelengths, in the cooler layers of the Sun's atmosphere, including the photosphere and chromosphere. Although first observed in 1802 by William Hyde Wollaston, they were first carefully studied in 1814 by Joseph von Fraunhofer, who also labeled some of the most prominent lines with letters of the English alphabet.

Free electron: An electron that has broken free of its atomic bond.

Frequency: The number of crests of a wave passing a fixed point each second, usually measured in units of Hertz (one oscillation per second).

Fusion: Joining of two or more lighter nuclei to produce a heavier nucleus, releasing energy as the result. Fusion powers hydrogen bombs and the Sun.

GALLEX: Acronym for the Gallium Experiment to measure solar neutrinos of low energy from the proton-proton reaction.

Gamma: A unit of magnetic field strength equal to 10^{-5} gauss, 10^{-9} tesla, or 1 nanotesla.

Gamma rays: Electromagnetic radiation with wavelengths shorter than approximately 1 A.

Gamma-ray radiation: The most energetic form of electromagnetic radiation, with the highest frequency and the shortest wavelength. Gamma rays have photon energies in excess of 100 KeV or 0. 1 MeV and wavelengths less than 1 A.

G-band: A wavelength region at 4304 A useful for the photography of magnetic features in the high photosphere. Typically the window is 12 A wide and is due to the CH molecule.

Gas pressure: The outward pressure caused by the motions of gas particles and increasing with their temperature.

Gauss: The c.g.s. (centimeter-gram-second) unit of magnetic field strength.

Geocentric: Earth-centered.

Geomagnetic activity: Disturbances in the magnetized plasma of a magnetosphere associated with fluctuations of the Earth's magnetic field, auroral activity, strong Ionospheric currents, and particle precipitation into the ionosphere.

Geomagnetic field: The magnetic field in and around the Earth. The intensity of the magnetic field at the Earth's surface is approximately 3.2 x10^{-5} T (Tesla), or 0.32 G(Gauss) at the equator and 6.2 x10^{-5} T, or 0.62 G, at the North Pole.

Geomagnetic storm: A rapid, world-wide disturbance in the Earth's magnetic field, typically of a few hours duration, due to the passage of a high-speed stream in the solar wind or caused by the arrival in the vicinity of the Earth of a coronal mass ejection.

Gleissberg period: A period of approximately 80 years in the number of sunspots, the frequency of terrestrial auroras, and in the terrestrial radiocarbon variability.

g-modes: Gravity oscillation mode of solar acoustic waves.

Gnomon: A device like a sundial that casts shadows for measurement.

GOES 1, 2,..... 8, 9: Acronym for a series of Geostationary Operational Environmental Satellites numbered 1 through 9 in order of launch. These satellites have provided, and continue to provide, information about terrestrial cloud cover, temperatures, water-vapor content, and other meteorological data relayed from central weather facilities to regional stations, as well as space environment monitoring systems to measure proton, electron and solar X-ray fluxes and magnetic fields.

GOLF: Acronym for the Global Oscillations at Low Frequency instrument on SOHO.

GONG: Acronym for the Global Oscillation Network Group of six Earth-based telescopes that continuously monitor solar oscillations.

Granulation: A mottled, cellular pattern visible at high spatial resolution in the white light of the solar photosphere.

Granule: One of about a million bright cells, that cover the visible solar disk and that comprise the granulation.

Grating: A surface ruled with uniformly spaced lines,which through diffraction breaks up light into its spectrum.

Gravitational force: One of the four fundamental forces of nature, the force by which two masses attract each other.

Gravitational instability: A situation that tends to break up under the force of gravity.

Gravitational red shift: A red shift of light caused by the presence of mass, according to the general theory of relativity. At the solar surface it is equivalent to 0.6 kms^{-1}.

Gravity: The universal force of attraction between all particles of matter. The gravitational force of attraction increases with mass and falls off as the inverse square of the separation.

Grazing incidence: Striking at a low angle.

Gregorian calendar: The calendar in current use, with normal years except for years that are divisible by 100 but not by 400.

Ground state: The lowest-energy state of the electron orbital motion around the nucleus of an atom.

Gyrofrequency: Frequency of the circular motion of charged particle perpendicular to the magnetic field.

Gyroradius: Radius of the circular orbit of a charged particle-gyrating in a magnetic field.

H alpha (Hα): See hydrogen-alpha line.

H and K lines: The strongest lines in the visible spectrum due to ionized calcium, Ca II, at the wavelengths of 3934 Å (K) and 3968Å (H).

Hale's law: The leading or western most spots of any sunspot group in the northern hemisphere of the Sun, have the same magnetic polarity, while the following, or eastern most spots have the opposite magnetic polarity. The polarities of sunspots are reversed in the southern hemisphere. The spot's magnetic polarities reverses in 11 years and return to their original polarity every 22 years. This is known as 22 year-magnetic cycle, or the Hale's law.

Half–life: The length of time for half a set of particles to decay through radioactivity or instability.

Hard X-rays: Electromagnetic radiation with photon energies of between 10 keV and 100 keV and wavelengths between about 10^{-10} and 10^{-11} m.

Helicity: A measure of the "twist" an object or system has.

Heliocentric: Sun-centered.

Heliographic latitude: The angle in degrees from the solar equator to an object as measured along a great circle passing, through the poles of the Sun.

Heliographic longitude: The angle in degrees measured along the solar equator from the central meridian of the Sun, to the foot of the great circle that passes through an object (sunspot) and the poles of the Sun.

Heliopause: The outer boundary or edge of the heliosphere marking the interface of the solar wind with the interstellar medium. The pressure of the solar wind equals that of the interstellar medium at the Heliopause.

Helioseismology: The study of the interior of the Sun by the analysis of sound waves that propagate through the solar interior and manifest themselves as oscillations at the photosphere.

Heliosphere: A vast region carved out of interstellar space by the solar wind; the region of interstellar space surrounding the Sun where the Sun's magnetic field and the charged particles of the solar wind control plasma processes.

Heliospheric current sheet: A thin sheet of current that forms at the interface between oppositely directed magnetic fields in the solar wind from the North and South solar hemispheres.

Heliostat: A moveable flat mirror on an equatorial mount used to reflect sunlight into a fixed solar telescope.

Heliotail: The elongated tail of the heliosphere extending downwind of the approaching interstellar gas formed by solar wind escaping from the heliosphere.

Helium: After hydrogen, the second most abundant element in the Sun and universe.

Helmet streamer: Named after spiked helmets once common in Europe, helmet streamers form in the low corona, over the magnetic inversion, or neutral lines in large active regions with a prominence commonly embedded in the base of the streamer.

Hertz: A unit of frequency equal to one cycle per second abbreviated Hz.

High-speed stream: A stream within the solar wind having speeds of 600 km s^{-1} and higher.

Homestake: The South Dakota gold mine where the chlorine neutrino detector is located.

Homologous flares: Solar flares that occur repetitively in the same active region, with essentially the same position and pattern of development.

Hour angle: The difference between the Local Sidereal Time (LST) and the Right Ascension (RA) (H.A. = L.S.T. – R.A.) of a celestial object.

Hour circle: The great circles passing through the celestial poles.

Hydrogen burning: The thermonuclear process in which a star (Sun) shines by converting hydrogen nuclei into helium nuclei.

Hydrogen: The lightest, simplest element.

Hydrogen-alpha line: The spectral line of neutral hydrogen in the red part of the visible spectrum, denoted by Hα. Light emitted or absorbed at a wavelength of 6563 Å.

Hydromagnetic wave: A wave in which both the plasma and magnetic field oscillate.

Hydrostatic equilibrium: The condition of stability in an atmosphere or stellar interior that exists when the inward gravitational force of the overlying material, is exactly balanced by the outward force of the gas and radiation pressure.

Hyperfine level: A subdivision of an energy level caused by such relativity minor effects as changes resulting from the interactions among spinning particles in an atom or molecule.

Hz: Abbreviation for Hertz, a unit of frequency equivalent to cycles per second. See hertz.

Hα: Light emitted at a wavelength of 6563 A° from an atomic transition in hydrogen, the lowest energy transition in its Balmer series.

Ice age: A period of cool, dry climate causing a long-term buildup of extensive ice sheets far from the poles. The major ice ages last for about 100,000 years; they are separated by warmer interglacial periods that last roughly 10,000 years

IMF: Acronym for Interplanetary Magnetic Field.

IMP-1, 2, ..., 8: Acronym for a series of Earth-orbiting Monitoring Platforms.

Impulsive flare: The most common type of solar flare, lasting minutes to hours.

Inclination of an orbit: The angle of the plane of the orbit with respect to the ecliptic plane.

Infrared Radiation: Beyond the visible red, about 7000 Å to 1 mm.

Insolation: The amount of radiative energy received from the Sun per unit area per unit time at any given location on the Earth's surface.

Interference: The property of radiation in which waves in phase can add (constructive interference) and waves out of phase can subtract (destructive interference.

Interferometer: A device that uses the property of interference to measure properties of radiation and objects.

Interferometry: Observations using an interferometer.

Interplanetary magnetic field: The Sun's magnetic field carried into interplanetary space by the expanding solar wind, abbreviated as IMF has a magnetic field strength of about 6×10^{-9} tesla, or 6×10^{-5} gauss at the Earth's orbital distance

Interplanetary medium: Gas and dust between the planets.

Interplanetary scintillation: Fluctuations in the signal received from a distant radio source observed along a line of sight close to the Sun. The scintillation is caused by irregularities in the solar wind.

Inverse-square law: A reduction in the intensity of radiation in proportion to the square of the distance from its source. Also a reduction in gravitational attraction by the same factor.

Ion: An atom that has gained or lost (more usually) one or more electrons, thus having a net electrical charge.

Ionization potential: The amount of energy required to remove the least tightly bound electron from a neutral atom or molecule is called the first ionization potential and is usually measured in electron volts, or eV.

Ionization: The process in which a neutral atom or molecule is given a net electrical charge.

Isotope: One of two or more forms of the same chemical element whose atoms have the same number of protons in their nucleus but a different number of neutrons, and therefore a different mass.

Isotropy: Being the same in all directions.

Julian calendar: The calendar with 365-day years and leap years every fourth year without exception; the predecessor to the Gregorian calendar.

Julian day: The number of days since noon on January 1, 713 B.C. used for keeping track of astronomical events. January 1, 2000, noon, will begin as Julian day 2,451,545.

K corona: The electron - scattered component of the white-light coronal intensity.

K line: The spectral line of ionized calcium at 3933 Å.

Kamiokande: A massive underground neutrino detector in Japan filled with water, replaced by the Super Kamiokande detector.

Kelvin: A unit of absolute temperature abbreviated K. Zero degree Celsius is equal to 273.16 Kelvin.

KeV: Abbreviation for kilo-electron volt, or one thousand electron volts. A unit of energy with 1 keV = 1.6022×10^{-16} J. The wavelength of radiation with a photon energy in keV is 1.24×10^{-9} meters (energy in keV).

Kinetic energy: The energy that an object possesses as a result of its motion.

Kirchhoff's law: The ratio of the emission and absorption coefficients of a black-body is equal to its brightness.

Lagrangian point: Also known as the inner Lagrangian point and designated L_1, the point about one one-hundredth of the way from the Earth to the Sun, where the gravitational pull of the Earth and Sun balance in such a way as to give an orbit of exactly one Earth year. The Lagrangian point is located at a distance from the Earth of about 1.5×10^9 m towards the Sun. The ACE and SOHO spacecrafts are both near this point, thus observing the Sun continuously.

LASCO: Acronym for the Large Angle Spectroscopic Coronagraph on SOHO.

Latitude: Number of degrees north or south of the equator measured from the center of a coordinate system.

Leader spot: The leader spot is the western preceding part of a magnetically bipolar or multi-polar sunspot group. Since the Sun rotates from East to West, the leader spot precedes the other members of the local group as the Sun rotates.

Lens: A device that focused waves by refraction.

Light: The kind of radiation to which the human eye is sensitive, in the wavelength between 3850 and 7000Å.

Limb darkening: The edge of the visible disk of the Sun appears darker as compared to the center of the Sun; this effect is called Limb darkening.

Limb: The apparent edge of a celestial object which is visible as a disk, such as the Sun or the Moon.

Line profile: The intensity variation of radiation verses wavelength for a spectral line.

Longitude: Angular distance around a body measured along the equator from some particular point; for a point not on the equator, it is the angular distance along the equator to a great circle that passes through the poles and through the point.

Lorentz force: Total electromagnetic force on a charged particle in the presence of an electric field due to its motion across a magnetic field.

Luminosity: Absolute brightness or luminosity of a glowing body, denoted by L. The amount of energy radiated per unit time by an object, measured in units of watts. One joule per second is equal to one watt of power, and to ten million ergs per second.

Lunar eclipse: Passage of the Moon into the Earth's shadow.

Lyman alpha: The spectral line (1216 Å) that corresponds to a transition between the two lowest energy levels of a hydrogen atom.

Magnetic braking: A process proposed to account for the slow rotation of the Sun and some other stars, in which angular momentum is transferred from the star to the surrounding plasma through its magnetic field.

Magnetic field: A magnetic force around the Sun, planets, and any other magnetized body, generated by electrical currents.

Magnetic field lines: Imaginary lines that indicate the strength and direction of a magnetic field. **Magnetic mirror:** A situation in which magnetic lines of force meet in a way that they reflect charged particles.

Magnetic pressure: A type of pressure inherent in magnetic plasma.

Magnetic reconnection: A change in the topology of the magnetic field where the magnetic field lines re-orient themselves by new connections. A process by which magnetic field lines are broken and then rejoined into a new configuration.

Magnetic storm: A disturbance in the Earth's magnetic field observed all over the Earth due to the passage of a high speed stream and/or a coronal mass ejection in the solar wind See geomagnetic activity, and geomagnetic storm.

Magnetism: One aspect of electromagnetism, a fundamental force of nature, whereby a magnetized object can affect the motion and direction of a charged particle

Magnetograph: An instrument used to map the strength, direction, and distribution of magnetic fields across the solar surface.

Magnetogram: A picture or map of the strength, direction and distribution of magnetic fields across the solar surface.

Magnitude: A measure of the brightness in the sky of a celestial object.

Major axis: The longest diameter of an eclipse that passes through the foci.

Mass number: Denoted by the capital letter A, the total number of protons and neutrons in a nucleus.

Maunder Minimum: The period roughly between 1645 and 1715 when few sunspots were observed.

Maxwellian distribution: Distribution of particle velocities of a gas in thermal equilibrium.

MDI: Acronym for the Michelson Doppler Imager on SOHO, for measuring solar oscillations.

Mean solar day: A solar day for the "mean Sun", assumed to move at a constant rate during the year.

Meridian: The great circle on the celestial sphere that passes through the celestial poles and the observer's zenith.

Metonic: Period of 235 lunar months.

MeV: A unit of energy equal to one million electron volts and 1.6022×10^{-13} J.

MHD: Abbreviation for magneto-hydro-dynamics.

Microflares: Also called nanoflares, these are small brightening on the Sun, formed in chromosphere and low corona, each lasting for a few minutes, which can be observed at extreme-ultraviolet, radio, ultraviolet and X-ray wavelengths.

Microwaves: Electromagnetic radiation with wavelengths between 0.001 and 0.06 meters, or frequencies from 5 to 300 GHz.

Molecule: A tightly knit group of two or more atoms, bound together by nuclear forces among the atoms.

Momentum: A measure of tendency of a moving body to keep moving. The momentum in a given direction (linear momentum) is equal to the mass of the body, times its component of velocity in the direction.

MSW effect: The transformation of a neutrino of one type or "flavor" into a neutrino of another kind while traveling through matter.

Nanoflares: Low-level flares, also called micro-flares, are detected at extreme ultraviolet, radio, ultraviolet and X-ray wavelengths. See micro-flares.

Nanotesla: A unit of magnetic field strength abbreviated nT. It is equal to 10^{-9} T, or 10^{-5} G, and to 1 gamma.

Negative hydrogen ion: Hydrogen atom with an extra electron.

Network: Chromosphere and photosphere features arranged in a cellular structure

Neutrino: A spinning, sub-atomic particle with no electric charge and very little or rest mass. **Neutrino oscillation:** The change of one type or flavor of neutrino another while traveling through matter or a vacuum. See MSW effect, neutrino and solar neutrino problem.

Neutron: A sub-atomic particle with no electric charge found in all atomic nuclei except that of hydrogen. The neutron has slightly more mass than a proton and 839 times heavier than an electron. The mass of the neutron is 1.008 665 a.m.u. or 1.6749×10^{-27} kg.

Newtonian (telescope): A reflecting telescope where the beam from the primary mirror is reflected by a flat secondary mirror to the side.

NOAA: Acronym for the National Oceanic and Atmospheric Administration, United States.

NOAO: Acronym for the National Optical Astronomy National Optical Astronomy Observatory, United States.

Non-thermal particle: A particle that is not part of a thermal gas. These particles cannot be characterized by a conventional temperature.

Non-thermal radiation: The electromagnetic radiation produced by a non-thermal electrons traveling at a speed close to that of light in the presence of a magnetic field.

Nuclear energy: The energy obtained by nuclear reactions.

Nuclear fission: A reaction involving an atomic nucleus in which the nucleus splits into two or more simpler and lighter nuclei.

Nuclear force: The force that binds protons and neutrons within atomic nuclei.

Nuclear fusion: The amalgamation of lighter nuclei into heavier ones.

Nucleosynthesis: The production of chemical elements from other chemical elements by naturally occurring nuclear reactions. Fusion reactions inside stars create elements heavier than helium. Helium is also synthesized from hydrogen inside stars; most of the helium in the Universe was created by nucleosynthesis during the first few minutes following the big bang.

Nucleus: The small, massive center of an atom, made up of protons and (except for hydrogen) neutrons, bound together by the nuclear force.

Objective: In optics, the principle lens or mirror of an optical system.

Oblate: With an equatorial diameter greater than the polar.

Obliquity: The angle between an object's axis of rotation and the pole of its orbit.

Occultation: The hiding of one astronomical body by another.

Ohmic dissipation: Conversion of an electrical current to heat because of the resistance of the medium in which it travels.

Opacity: A measure of the ability of a gaseous atmosphere to absorb radiation and become opaque to it. A transparent gas has little or no opacity.

Optical astronomy: The study of objects in space using visible light.

Optical depth: A logarithmic measure of the radiation absorbed as it passes through a medium, or how far one could ‘see’ into a semi transparent medium. A transparent medium has an optical depth of zero. The medium is optically thin when the optical depth is less than unity, and optically thick when greater than unity.

Optical radiation: Electromagnetic radiation that is visible to the human eye, with wavelengths of approximately 3850 to 7000 Å.

Optical spectrum: Spectrum of a source that spans the visible wavelength range, approximately from 3850 to 7000 Å.

Pair annihilation: Mutual destruction of an electron and positron with the formation of gamma rays.

Parallax: An angular displacement of a nearby star with respect to distant ones. *See* annual parallax, solar parallax,

Particle physics: The study of elementary nuclear particles.

Penumbra: (a) The lighter periphery of a sunspot seen in white light, surrounding the darker umbra, (b) For an eclipse, the part of the shadow from which the Sun is only partially occulted;

Perihelion: For a planet, comet or other object orbiting the Sun, the point in the orbit that is closest to the Sun. Compare aphelion.

Period: The interval over which something repeats.

Photodissociation: The breakdown of molecules due to the absorption of radiation.

Photo ionization: The ionization of an atom by absorption of a photon of electromagnetic radiation. Ionization can take place only if the photon carries at least, the energy corresponding to the ionization potential of the atom, that is, the minimum energy required to overcome the force binding the electron within the atom. *See* ionization potential.

Photometry: The electronic measurement of the amount of light.

Photon: A discrete unit or quantity of electromagnetic energy. A photon can be described as a particle, a quantum of light.

Photon energy: The energy of radiation of a particular frequency or wavelength. Short-wavelength or high-frequency photons have more energy than long wavelength, or low-frequency photons. The amount of photon energy, E is equal to the product of the frequency, ν and Planck's constant, $h = 6.6261\times10^{-34}$ Js, ($E = h\nu$).

Photosphere: That part of the Sun from which visible light originates. The lowest layer of the Sun's atmosphere viewed in white light.

Plage: From the French word for 'beaches', that portion of the solar magnetic active region that appears much brighter in Hα and CaII lines than the surrounding chromosphere.

Planck's constant: The constant of proportionality h between the frequency of an electromagnetic wave and the energy of an equivalent photon. $E = h\nu = hc/\lambda$.

Planck's law: The formula that predicts for gas at a certain temperature, how much radiation there is at every wavelength.

Plasma: An ionized gas consisting of electrons and ions.

Plumes: Thin structures in the solar corona near the poles.

p-mode: An acoustic mode of oscillation of the Sun in which pressure is the restoring force.

Polar axis: The axis of an equatorial telescope mounting that is parallel to the Earth's axis of rotation.

Polarity: The direction of a magnetic field, being north- or south-seeking. According to convention, magnetic lines of force emerge from regions of positive north polarity and re-enter regions of negative south polarity.

Polarization: According to wave theory of light, those electromagnetic vibrations which are not randomly oriented but have a preferred direction.

Pore: Small, short-lived dark area in the photosphere out of which a sunspot may or may not develop.

Positive ion: An atom that has lost one or more electrons.

Positron: A positively charged anti-particle of the electron. A sub-atomic particle having the mass of an electron but an equal positive electric charge.

Post-flare loop: An arcade of loops, or a loop prominence system, often seen after a major two-ribbon flare, which bridges the ribbons.

Potential energy: The energy that an object possesses as a result of its position.

p-p chain: Abbreviation for proton-proton chain. *See* proton-proton chain.

Precession: The slow, periodic conical motion of the rotation of a spinning body, like a wobbling top or the rotating Earth on its axis. The precessional motion of the Earth is caused by the tidal action of the Moon and Sun on the spinning Earth. As a result, the earth's axis of rotation sweeps out a cone in space, centered around the axis of the earth's orbit, completing one revolution in about 26,000 years.

Pressure: Force per unit area.

Prime focus: The point at which the objective lens or primary mirror of a telescope brings light to a focus.

Principle quantum number: The integer n that determines the main energy levels in an atom.

Prominence: A region of cool (10^4 K), high-density gas embedded in the lower part of the hot (10^6 K), low-density solar corona. A prominence is a filament viewed on the limb of the Sun in the light of the hydrogen-alpha line, or as bright protrusions at the limb seen during total solar eclipses or with a coronagraph.

Proton: A positively charged, sub-atomic particle located in the nucleus of an atom or set free from it. The nucleus of a hydrogen atom is a proton. The proton has a mass of 1.672623×10^{-27} kg and is 1836 times more massive than an electron.

Proton flare: Any flare which significantly produce fluxes of energetic protons, with energies greater than 10 MeV.

proton-proton chain: Abbreviated p-p chain, a series of thermonuclear reactions in which hydrogen nuclei, or protons are transformed into helium nuclei. **Proto sun:** The Sun in formation.

Quiescent prominence: A long-lived and relatively stationary prominence.

Quiet Sun: The Sun when it is at the minimum level of activity in the 11-year solar cycle; the photosphere outside magnetic regions.

Radar: The acronym for "radio detection and ranging". A passive radio technique in which radio signals are transmitted and their reflections received.

Radial velocity: The velocity of an object along the line-of-sight.

Radian: A dimensionless unit of angular measures equal to 206265 sec of arc. There are 2π radians in a full circle of 360 degrees, where $\pi = 3.14159$.

Radiation: A process that carries energy through space.

Radiation pressure: The pressure exerted by electromagnetic radiation. Radiation pressure can compete with gas pressure in supporting giant stars, and it blows the dust tails of comets away from the Sun.

Radiative zone: An interior layer of the Sun, lying between the energy generating core and the convective zone, where energy travels outward by radiation.

Radio Burst: A sudden, transient increase in solar radio radiation during a solar flare, emitted by energetic electrons.

Radio radiation: The part of the electromagnetic spectrum whose radiation has the longest wavelengths and smallest frequencies, with wavelengths ranging from about 0.001 m to 30 m and frequencies ranging between 10 MHz and 300 GHz.

Radio telescope: A large radio antenna designed to concentrate radio waves and permit the detection of faint radio signals reaching us from the Sun (and other celestial objects).

Radioactivity: The spontaneous decay of certain rare, unstable, heavy nuclei into more stable lighter nuclei with the release of energy.

Radioheliograph: A radio telescope designed for mapping the distribution of radio emission from the Sun.

Recombination: The capture of an electron by a positive ion. It is the opposite process to ionization.

Red shifted: When a spectrum is shifted to longer wavelengths by the Doppler Effect.

Reflecting telescope: A telescope that gathers radiation and forms an image by the reflection of light from a primary concave or parabolic mirror.

Refraction: The bending of electromagnetic radiation as it passes from one medium to another.

Relativistic: Having a velocity which is a very large fraction of the speed of light so that the special theory of relativity must be applied.

Resolution: The ability of an optical or radio imaging system to distinguish fine details. Angular resolution $\theta = \lambda/D$ of a telescope is given in radians, where λ is the wavelength, and D the aperture.

Rest mass: The mass an object that it would have if it was not moving with respect to the observer.

Revolution: The orbiting of one body around another.

Right ascension: Celestial longitude, measured eastward along the celestial equator in hours of time from the vernal equinox.

Rotation: The spin of an object about its own axis.

SAGE: Acronym for the Soviet-American Gallium Experiment begun in 1990, an underground neutrino detector in the northern Caucasus.

Scattered Light: Light which interacts with matter and then is re-emitted in different directions.

Schwabe cycle: Historical term for the 11-year sunspot cycle, discovered by the amateur German astronomer Samuel Heinrich Schwabe in the early1840's.

Scintillation: The twinkling of light intensity (or radio waves) caused by atmospheric density variations (or solar wind inhomogeneities).

Second of arc: A unit of angular measure or mentioned as 'arc sec'. There are 60 seconds of arc in one minute of arc, and therefore 3600 seconds of arc in one degree.

Sector boundary: A place in the solar wind where the predominant direction of the interplanetary magnetic field changes direction, from towards the Sun to away from the Sun or *vice versa.*

Seeing: Fluctuations in a visible-light image due to turbulence and inhomogeneities in the earth's atmosphere. In conditions of good seeing, images are sharp and steady; in poor seeing they are extended and blurred and appear to be in constant motion.

Seismic waves: Waves traveling through a body from an earthquake or other impact.

Seismology: Science of earthquake. "Seismo-" comes from the Greek for the earthquake.

Semi major axis: Half of the major axis of an ellipse.

Shock wave: A sudden abrupt discontinuous change in density and pressure propagating in a gas or plasma at supersonic speed.

Sidereal day: A day with respect to the stars.

Sidereal period: The orbital or rotation period of a planet or other celestial body with respect to the background stars.

Sidereal time: The hour angle of the vernal equinox equal to the Right Ascension of objects on ones local meridian.

Sidereal year: An apparent circuit of the Sun with respect to the stars.

Skylab: An American manned space station launched in Earth orbit on 4 May 1973 for study of the Sun.

Slit: A long, thin gap through which light is allowed to pass especially in spectrographs.

SMM: Acronym for the Solar Maximum Mission.

SNO: Acronym for the Sudbury Neutrino Observatory, Canada.

SNU: Abbreviation for the Solar Neutrino Unit equal to 10^{-36} captures per target atom per second.

Soft X-rays: Electromagnetic radiation with photon energies of 1 to 10 Kev and wavelengths between about 10^{-9} and 10^{-10} m.

SOHO: Acronym for the **So**lar and **H**eliospheric **O**bservatory.

SOI MIDI: Acronym for the **S**olar **O**scillations **I**nvestigation **Mi**chelson **D**oppler **I**mager instrument on SOHO.

Solar activity cycle: A cyclical variation in solar activity with a period of about 11 years between maxima (or minima) of solar activity.

Solar and Heliospheric Observatory: Abbreviated SOHO, a joint project of ESA and NASA, was launched on 2 December 1995, and reached its permanent position on 14 February 1996. SOHO orbits the Sun at the Lagrangian point where the gravitational forces of the Earth and Sun are equal and orbits the Sun once a year.

Solar atmosphere: The outer layers of the Sun from the photosphere through the chromosphere, transition region and corona.

Solar constant: The total amount of solar energy, integrated over all wavelengths, received per unit time and unit area at the mean Sun-Earth distance outside the Earth's atmosphere. Its value is 1366.2 J s^{-1} m^{-2}, which is equivalent to 1366.2 Wm^{-2}. It is now also called as Total Solar Irradiance (TSI).

Solar core: The region at the center of the Sun where nuclear reactions take place.

Solar cosmic rays: Abbreviated SCR, a historical name for energetic charged particles, mainly protons and electrons, accelerated to energies greater than 1 MeV by explosive processes on the Sun. It is now better to use the term *solar energetic particles* to avoid confusion with cosmic rays that come from interstellar space.

Solar cycle: The approximately 11-year variation in solar activity and the number of sunspots.

Solar day: A full rotation of the Earth with respect to the Sun.

Solar dynamo: The generation of sunspots by the interaction of convection, turbulence, differential rotation and magnetic fields.

Solar eclipse: A blockage of light from the Sun when the Moon is positioned precisely between the Sun and the Earth observer.

Solar Energetic Particle: Abbreviated SEP, charged particles, mainly protons and electrons, accelerated to energies greater than 1 MeV by explosive processes on the Sun.

Solar flare: A sudden explosive release of matter and energy from an active region in the form of electromagnetic radiation, energetic particles, wave motions and shock waves, lasting minutes to hours in time

Solar Insolation: The amount of radiative energy received from the Sun per unit area per unit time, at any given location on the Earth's surface. *See* Insolation, TSI.

Solar limb: The apparent edge of the Sun as it is seen in the sky.

Solar mass: The amount of mass in the Sun, equal to 1.989×10^{30} kg.

Solar Maximum Mission: Abbreviated SMM, a NASA Satellite, launched on 14 February 1980, for studying the Sun during a period of maximum solar activity. It failed after few months, but repairs were successfully done by a Space Shuttle crew in 1984. **Solar maximum:** The peak of the sunspot cycle when the numbers of sunspots is greatest.

Solar minimum: The beginning or end of a sunspot cycle marked by the near absence of sunspots and the relatively low output of energetic particles.

Solar neutrino problem: Solar neutrino detectors found only one-third solar neutrinos of the theoretically predicted number, based on the Standard Solar Model (SSM). Two possible explanations for the solar neutrino problem were proposed; either that there is some problem with the SSM or we do not understand the neutrino physics. The latter explanation has been found likely, because neutrinos can change *flavor* on their way out of the Sun, and thus not all neutrinos are detected.

Solar neutrino unit: Abbreviated SNU, a unit of solar neutrino capture rate by subterranean detectors, with 1 SNU = 10^{-36} solar neutrino captures per second per target atom. *See* SNU.

Solar parallax: The angular size of the radius of the Earth at a distance of one astronomical unit, amounting to 8.794158 seconds of arc. *See* astronomical unit.

Solar probe: A planned NASA spacecraft, the first to fly through the atmosphere of the Sun, taking in-situ measurements down to 3 solar radii in the solar low corona, where the radiation temperatures will exceed 2×10^6 K. Solar probe will study the heating of the solar corona and origin and acceleration of the solar wind.

Solar rotation period: The time for a complete rotation of the Sun with respect to the stars or Earth.

Solar sectors: A region in the solar wind that has predominantly one magnetic polarity, pointed away from or toward the Sun. *See* sector boundary.

Solar system: The Sun, its planets and the smaller bodies orbiting the Sun, including asteroids and comets.

Solar time: A system of time keeping with respect to the Sun, such that the Sun crosses the meridian at a given location at 12 noon local time.

Solar wind: The expansion of the solar corona mainly consists of electrons and protons, to form supersonic plasma streaming in all directions away from the Sun with speed ranging from 300 to 1,000 km s^{-1}.

Solar year (tropical year): Planet's complete circuit of the Sun; a tropical year is reckoned as time elapsed between two successive vernal equinoxes.

Solar-B: A space mission of the Japanese Institute of Space and Astronautical Science (ISAS) which is tentatively scheduled for launch in 2005. It will include a coordinated set of optical, extreme-ultraviolet and X-ray telescopes that will investigate the interaction between the sun's magnetic field and its corona.

Solstice: The point on the celestial sphere of northern most or southern most declination of the Sun in the course of a year.

South Atlantic anomaly: A region over the South Atlantic Ocean where the lower Van Allen belt of energetic, electrically-charged particles is particularly close to the earth's surface and presents a hazard for artificial satellites.

Space-weather: Changing conditions in interplanetary space and Earth's magnetosphere controlled by the variable solar wind.

Special theory of relativity: Einstein's 1905 theory of relative motion.

Speckle interferometry: A method obtaining higher resolution of an image by analysis of a rapid series of exposures that freezes atmospheric blurring.

Spectral classification: The sequence of stellar spectral types arranged according to temperature as inferred from spectral lines, designated as O, B, A, F, G, K sequence, from the hot O and B cool M ones. The Sun is G2.

Spectral line: A radiative feature observed in emission or absorption at a specific frequency or wavelength.

Spectrograph: Also known as a spectrometer. An instrument that separates light or other electromagnetic radiation into its component wavelengths.

Spectroheliogram: A monochromatic image of the Sun produced by means of a spectroheliograph.

Spectroheliograph: A type of spectrograph used to image the Sun, in the light of one particular wavelength.

Spectrohelioscope: An instrument for viewing the Sun's image in a narrow band of wavelengths by the eye. A spectroheliograph adapted for visual use. *See* spectroheliograph.

Spectrometer: A device to make and measure electronically the intensity of a spectrum with respect to wavelength.

Spectrophotometer: A device to measure the intensity of given wavelength bands.

Spectroscope: A device to look at a spectrum.

Spectroscopy: The study of a spectrum, including the wavelength and intensity of emission and absorption lines

Spectrum: The distribution of intensity of electromagnetic radiation with wavelength.

Speed of light: The velocity $c = 2.997 \times 10^8$ ms^{-1} at which all electromagnetic radiation travels. **Spicule:** A small jet of gas seen at edge of the quiet Sun in the chromosphere, approximately 1,000 km in diameter and 10,000 km high, with a lifetime of about 15 minutes.

Spörer Minimum: A period of low sunspot activity during the 15th century (about A.D. 1420-1500), named after the German astronomer Gustav Friedrich Wilhelm Spörer who called attention to this minimum in solar activity, as early as in1887. *See* Little Ice Age, and Maunder Minimum.

Spörer's law: The appearance of sunspots at solar lower latitudes over the course of the 11-year solar activity cycle, drifting from mid-latitudes towards the equator as the cycle progresses.

Standard Solar Model: A theoretical model of the evolution and internal properties of the Sun, based on physical laws and constrained by an assumed initial composition and age of the Sun as well as its observed mass, radius and luminosity.

Stefan-Boltzmann constant: The constant of proportionality, denoted by the symbol-σ, relating the radiant flux per unit area from a black-body to the fourth power of its effective temperature. The constant $\sigma = 5.67051 \times 10^{-8}$ J m^{-2} K^{-4} s^{-1}. *See* Stefan-Boltzmann law.

Stefan-Boltzmann law: The radiation law that states that the energy emitted by a black body varies with the fourth power of the temperature.

Steradian: A unit of solid angle equal to $32400/\pi^2$ square radian or 32828 square degrees.

Stereo: Acronym for the **S**olar **Te**rrestrial **Re**lations **O**bservatory, scheduled for launch by NASA in 2004. STEREO will use two spacecraft, preceding and following the Earth in its orbit, to simultaneously observe coronal mass ejections (CME), to study three-dimensional structure of CMEs from their onset at the Sun to the Earth's orbit.

Streamer: Coronal structures at low solar latitudes. *See* coronal streamer and helmet streamer.

Sudbury Neutrino Observatory: Abbreviated SNO, a massive under ground neutrino detector in Sudbury, Ontario, Canada filled with heavy water.

Sun: The central star of the solar system, around which all the planets, asteroids and comets revolve in their orbits. The Sun is a dwarf star on the main sequence, of spectral type G2V with an effective temperature of 5780 K.

Sunspot: A dark, cooler region with strong magnetic fields in the Sun's photosphere.

Sunspot belts: The heliographic latitude zones where sunspots are found moving from mid-latitudes to the solar equator, during the 11-year sunspot cycle. *See* solar activity cycles and sunspot cycle.

Sunspot cycle: The recurring, 11-year rise and fall in the number and position of sunspots. At the commencement of a new cycle, sunspots erupt around latitudes of 35 to 45 degrees North and South. Over the course of the cycle, subsequent spots emerge closer to the equator, continuing to appear in belts on each side of the equator and finishing at around 7 degrees north and south. This pattern can be demonstrated graphically as a butterfly diagram.

Sunspot number: A daily index of sunspot activity R defined as $R = k\ (10g + s)$ where k is a factor based on the estimated efficiency of observer and the telescope, g is the number of groups of sunspots, irrespective of the number of spots each contains, and s is the total number of individual spots in all the groups.

Super Kamiokande: A massive underground neutrino detector in Japan filled with pure water, replacing the KAMTOKANDE detector.

Supergranulation cells: Large convective cells seen in Dopplergrams of the solar photosphere, having average dimensions of about 35,000 km and lasting for about 20 hours, and having horizontal velocity with amplitudes of about 20 to 400 m s^{-1}. The supergranulation pattern covers the entire photosphere except in plages and sunspots.

Supergranule: A large convection cell on the Sun that is approximately 35,000 km in diameter.

Supersonic: Moving at a speed greater than that of sound in a medium.

Surge: Sudden high-velocity upwelling, or jet, from active regions seen prominently in the light of the hydrogen alpha line. A surge originates in the chromosphere and reaches coronal heights.

SXT: Acronym for the **S**oft **X**-ray **T**elescope on Yohkoh- a Japanese spacecraft for the study of the Sun.

Synchronous rotation: A rotation of the same period as an orbiting body.

Synchrotron radiation: Electromagnetic radiation emitted by an electron traveling almost at the speed of light in the presence of a magnetic field.

Synodic period: The period of apparent rotation or orbital revolution as observed from the Earth.

Temperature: A measure of the heat of an object.

Tesla: The unit of magnetic flux density or a measure of the strength of a magnetic field, named after Nikola Tesla. The cgs unit of magnetic field strength is the tesla. In astrophysics gauss is often used as the unit of strength of magnetic field; 1 tesla = 10000 gauss = 10^4 gauss.

Thermal Bremsstrahlung: Emission of radiation by energetic electrons in a hot gas moving in the field of a positive ion. *See* Bremsstrahlung.

Thermal diffusion: Heat transport resulting from a temperature gradient.

Thermal energy: Energy associated with the motions of the molecules, atoms, or ions.

Thermal equilibrium: Equilibrium attained by a system that can be characterized by the same constant temperature at all points, and a single temperature.

Thermal gas: A collection of particles that collide with each other and exchange energy, giving a distribution of particle energies that can be characterized by a single temperature.

Thermal pressure: Pressure generated by the motion of particles that can be characterized by a temperature.

Thermal radiation: Radiation whose distribution of intensity over wavelength can be characterized by a single number (the temperature). Black-body radiation, which follows Planck's law, is called thermal radiation.

Thermonuclear fusion: The combination of atomic nuclei at high temperature to form more massive nuclei with the simultaneous release of energy. Thermonuclear fusion is the main power source in the core of the Sun.

Thermosphere: The uppermost layer of the atmosphere of the Earth and some other planets, which is heated by absorption of high-energy radiations from the Sun.

Torsional oscillations: Zones of alternating fast and slow rotation appearing in the photosphere and below, moving from higher latitudes to the equator.

Transit: The passage of one celestial body in front of another celestial body, or when a celestial body crosses an observer's meridian.

Transition region: A tenuous region of the solar atmosphere, thickness less than about 100 km, between the chromosphere and corona. It is characterized by a large rise of temperature from 10^4 to 10^6 K.

Transition zone: The thin region between a chromosphere and a corona.

Transparency: Clarity of the sky.

Transverse velocity: Velocity along the plane of the sky or normal to the line of sight.

Tropical year: The length of time between two successive vernal equinoxes.

Tunnel effect: A quantum mechanical effect that permits two colliding protons to overcome the electrical repulsion between them, enabling their nuclear fusion and the release of energy.

Turbulence: The chaotic mass motions associated with convection.

Two-ribbon flare: A solar flare that has developed as a pair of bright, hydrogen-alpha strands (ribbons) on both sides of the main inversion, or neutral line, of the photospheric magnetic field in an active region.

Type I radio burst: Short duration of the order of seconds, narrow-band burst detected at meter wavelengths (frequencies 300 to 50 MHz) that may continue for hours, caused by solar flares.

Type II radio burst: Narrow-band emission that drifts slowly in tens of minutes, from high to low frequencies. These radio bursts begin in the meter range (300 MHz in frequency) and sweeps toward decameter wavelengths (10 MHz in frequency). Type II bursts are thought to be caused by shock waves, moving outwards at velocities of about 1000 km s^{-1}, exciting the local plasma frequency in the corona.

Type III radio burst: Narrow-band emission characterized by its brief duration of seconds, and rapid drift from decimeter to decameter wavelengths (frequencies of 500 to 5 MHz). They are produced by electrons accelerated to energies of 1 to 10 keV in solar active regions, and then moving outward through the corona at speeds of 0.05 to 0.2 times the velocity of light.

Type IV radio burst: A smooth continuum of broad-band bursts primarily in the meter range of wavelengths, or at frequencies of 300 to 30 MHz. These bursts are associated with some major flare events, generally begin 10 to 20 minutes after the maximum phase, and can last for hours. Type IV radio bursts are emitted by magnetically 'trapped" high-energy electrons.

U burst: A radio burst that has an U-shaped appearance in an intensity-frequency plot.

Ultraviolet radiation: Electromagnetic radiation with a higher frequency and shorter wavelength than visible blue light. Ultraviolet radiation has wavelengths between about 10^{-8} and 3.5×10^{-7} m (100-3500 Å) with the extreme ultraviolet (EUV) lying in the short-wavelength part of this range. **Ulysses:** A joint undertaking of ESA and NASA, *Ulysses* was launched by NASA's Space Shuttle Discovery on 6 October 1990, to study the interplanetary medium and the solar wind at different solar latitudes. It provided the first opportunity for measurements to be made over the poles of the Sun, using the gravity-assist technique to take it out of the plane of the solar system. After an encounter with Jupiter in February 1992, the spacecraft moved back towards the Sun, to pass over the solar South pole in September 1994 and the North pole in July 1995.

Umbra: The dark inner core of a sunspot visible in white light. In the eclipse context, the inner part of the shadow cast by the Moon during a total solar eclipse

Unipolar region: A large area with weak magnetic fields of a single polarity often located towards the pole-ward side of the sunspot belt.

Universal Time: Abbreviated UT. Formerly Greenwich Mean Time (GMT).

UV: Abbreviation for ultraviolet radiation. *See* ultraviolet radiation.

Vector magnetic field: Magnetic field giving both the magnitude and direction of the field in the photosphere.

Velocity: A quantity that measures the rate of movement and the direction of movement of an object.

Velocity of light: The fastest speed that anything can move is equals 2.997 924 58 $\times 10^8$ m s^{-1}, in vacuum.

Velocity of sound: Denoted by s, the velocity of sound is proportional to the square root of the gas temperature, T, and inversely proportional to the square root of the mean molecular weight of the medium through which sound waves travel.

Vernal equinox: The equinox crossed by the Sun as it moves to northern declinations.

Vignetting: In optics, having to do with obscuration because of inadequate size of elements.

Visible radiation: Radiation at the narrow range of wavelengths in the electromagnetic spectrum perceptible to the human eye; namely light. It extends roughly from violet wavelengths at 3850 Å to red wavelengths at 7000 Å.

VIRGO: Acronym for the **V**ariability of solar **Ir**radiance and **G**ravity **O**scillations instrument on SOHO.

Virial theorem: For a bound gravitational system, the long term average of the kinetic energy is one-half of the potential energy.

Visible light: The form of electromagnetic radiation that can be seen by human eyes. *See* visible radiation.

Wavelength: The distance between successive crests, or troughs, of an electromagnetic or other wave. Wavelength is inversely proportional to frequency. The product of the wavelength and the frequency of electromagnetic radiation is equal to the velocity of light.

White light: The visible portion of sunlight that includes all of its colors.

Wien's displacement law: The expression of the inverse relationship of the temperature of a black body and the wavelength of the peak intensity.

Wilson effect: The foreshortening of a sunspot umbra that appears displaced towards the Sun's center, when the sunspot is near the Sun's limb. It is accompanied by a widening of the penumbra on the side nearest the limb and a narrowing on the side farthest from the limb.

WIND: A NASA spacecraft launched on 1 November 1994, to investigate basic plasma processes occurring near the Earth because of the solar wind.

Winter solstice: For northern hemisphere observers, the southern most declination of the Sun and its date.

Wolf number: A historic procedure for computing the number of sunspots, proposed by Rudolf Wolf of Zurich in 1849. *See* sunspot number.

X: The mass fraction of hydrogen in the Sun. X has been estimated = 0.705 83 ± 0.025 for the solar material outside the solar core, i.e. about 70.5 % of the solar material is hydrogen.

X-ray bright point: A small X-ray emitting region in the corona associated with a bipolar magnetic region.

X-rays: Electromagnetic radiation between 1 and 100 Å.

X-ray flare: A solar flare emitting X-ray energy.

X-ray radiation: The part of the electromagnetic spectrum covering the wavelength range from about 10^{-8} to 10^{-11} m and the energy range between 0.1 and 100 keV. Soft X-rays have lower energy, between 1 and 10 keV and hard X-rays have higher energies ranging from 10 to 100 keV.

Y: The mass fraction of helium in the Sun. Y has been estimated = 0.2743 ± 0.026 for the solar material outside the solar core or about 27.4% of the solar material is helium.

Yohkoh: Meaning in Japanese 'Sun Beam' a satellite launched by the Institute of Space and Astronautical Science (ISAS), Japan on 30 August 1991 to study the Sun, particularly solar flares in soft and hard X-rays and gamma-rays.

Z: The mass fraction of elements heavier than hydrogen and helium. In the Sun, the fraction of other elements Z has been estimated = 0.01886 ± 0.0084 for the solar material outside the solar core or about 1.89% of the solar material, are other elements.

Zeeman components: The linearly and circularly polarized components of a line split in the presence of a strong magnetic field. *See* Zeeman Effect and Zeeman splitting.

Zeeman Effect: A splitting of a spectral line into components by a strong magnetic field. If the components cannot be resolved, there is an apparent broadening or widening of the spectral line. The amount of splitting measures the strength of the magnetic field and the direction of the magnetic field can be inferred from the polarization of the components.

Zenith: The point in the sky directly overhead an observer.

Zodiac: The band of constellations through which the Sun, Moon, and planets appear to move in the sky, in the course of the year. *See* ecliptic.

Zodiacal light: A faint conical glow in the night sky caused by sunlight scattering off interplanetary dust near the plane of the ecliptic. A luminous pyramid of light that appears brightest and widest in the direction of the Sun, stretching along the ecliptic or zodiac from the western horizon after evening twilight or from the eastern horizon before morning twilight. The zodiacal dust cloud probably originates from both matter ejected by the Sun and from the decay of comets and asteroids.

Appendix III

References

Abetti, G. 1929. 'Solar physics', Handbuch der Astrophysik, ed. G. Eberhard, A. Kohl-Schütter, and H. Ludendorff, Vol. **4**, p.57. Publ. Berlin, Springer.

Abetti, G. 1957. In The Sun. London, Faber and Faber.

Adams, M.G., 1959. MNRAS., **119**, 460.

Adams, W.S., 1911. *'An investigation of the Rotation Period of the Sun by Spectroscopic Methods'*. Carnegie Inst. Washington Publ. No. **138**, p. 1-132.

Agnellii G., Cacciani, A., Fofli, M. 1975. *'The magneto-optical filter. I - Preliminary observations in Na D lines'*. Solar Physics, **44**, 509.

Ahmad, Q. R., Allen, R. C., Andersen, T. C., Anglin, J. D., Barton, J. C., Beier, E. W., Bercovitch, M., Bigu, J., Biller, S. D., Black, R. A., and 169 co-authors. 2002. *'Direct Evidence for Neutrino Flavor Transformation from Neutral-Current Interactions in the Sudbury Neutrino Observatory'*. Phys. Rev. Lett., **89a**, p.1301.

Aldrich and Abbot, C., 1948. Smithsonian Inst. Mis. coll. Vol. **110**, No. 5, & No. 11.

Alfven, H. 1947. *'Magneto hydrodynamic waves and the heating of the solar corona'*. MNRAS., **107,** 211.

Alfven, H. 1950. In Cosmic Electrodynamics.

Ambastha, A., Bhatnagar, A., 1988. *'Sunspot Proper Motions in Active Region NOAA 2372 and its Flare activity during SMY period of 1980 April 4-13'*. J. Astrophys. & Astron., **9**, 137.

Ananthakrishnan, R. 1954. *'Prominence Activity 1905-1952'.* The Proc. of the Indian Academy of Sciences, Vol. **XL**, no.2, Sec A, p.72.

Anselmann, P., et al., 1999. Phys. Lett. B., **447**, 127.

Antia, H. M.; Basu, Sarbani. 2000. *'Temporal Variations of the Rotation Rate in the Solar Interior'*. Ap. J. 541, 442.

Antonucci, E., and Svalgaard, L. 1974. *'Rigid and Differential Rotation of the Solar Corona'*. Solar Physics, **34**, 3.

Athay, R.G. 1976. In The Solar Chromosphere and Corona, Publ. D. Reidel, New York.

Atherton, P.D., Reay, N.K., Ring, J. 1981. *'Tunable Fabry-Perot Filters'*. Opt. Eng., **20**, 806.

Atkinson, R.d'E., and Houtermanns, F.G., 1929. Z. Phys. **54,** 656,

Babcock, H.D. and Livingston, W.C. 1958. '*Changes in the Sun's Polar Magnetic Field*'. Science, **127**, 1058.

Babcock, H.W. 1953. '*The Solar Magnetograph*'. Ap. J. **118**, 387.

Babcock, H.W. 1961. '*The Topology of the Sun's Magnetic Field and the 22-YEAR Cycle*'. Ap. J., **133**, 572.

Babcock, H.W., and Babcock, H.D. 1952. '*Mapping the magnetic field of the Sun*'. PASP., **64**, 282.

Babcock, H.W., and Babcock, H .D. 1953. '*Mapping the Magnetic fields of the Sun*' in 'The Sun', ed. Kuiper, p. 704, University of Chicago Press.

Babcock, H.W., and Babcock, H.D. 1955. '*The Sun's Magnetic Field*, 1952-1954'. Ap. J., **121**, 349.

Bahcall, J.N., 1965. '*Observational Neutrino Astronomy*'. Science, **147**, 115.

Bahcall, J., and Ulrich, R., 1988. '*Solar models, neutrino experiments, and helioseismology*'. Rev. Modern Phys., **90**, 297.

Bahcall J., Pinsonneault, M., and Basu, S., 2001. '*Solar Models: Current epoch and Time Dependences, Neutrinos, and Helioseismological Properties*'. Ap. J., **555**, 990.

Bahng, J., and Schwarzschild, M. 1962. '*Hydrodynamic Oscillation of Solar Chromosphere*'. A J., **67**, 312.

Balthasar, H., Lustig, G., Woehl, H., and Stark, D. 1986. '*The Solar rotation elements i and omega derived from Sunspot groups*'. A & A., **160**, 277.

Basu, S., and Antia, H. M. 1997. '*Seismic measurement of the depth of the solar convection zone*'. MNRAS., **287**, 189.

Basu, S., and Antia, H.S., 2001. '*Seismic investigation of changes in the rotation rate in the solar interior*'. In Proceedings of the SOHO 10/GONG 2000 Workshop: Helio- and asteroseismology at the dawn of the millennium, 2-6 October 2000, Santa Cruz de Tenerife, Tenerife, Spain. Edited by A. Wilson, Scientific coordination by P. L. Pallé. ESA SP-464, Noordwijk: ESA Publications Division, ISBN 92-9092-697-X, 2001, p. 179.

Beckers, J. M. 1968a. '*Solar spicules*'. Solar Physics., 3, 367.

Beckers, J.M., 1968b. 'High-Resolution Measurements of Photosphere and Sun-Spot Velocity and Magnetic Fields using a Narrow-Band Birefringent Filter'. Solar Physics, **3**, 258.

Beckers, J. M., 2001. '*A Seeing Monitor for Solar and Other Extended Object Observations*'. Experimental Astronomy, **12**, 1.

Beckers, J.M., and Morrison, R.A. 1970. '*The Interpretation of Velocity Filtergrams. III: Velocities inside Solar Granules*'. Solar Physics, **14**, 280.

Beckers, J. M., Schröter, E. H. 1968a. '*The Intensity, Velocity and Magnetic Structure of a Sunspot Region. I: Observational Technique; Properties of Magnetic Knots*'. Solar Physics, **4**, 142.

Beckers, J. M., Schröter, E. H. 1968b. '*On the Relation between the Photospheric Intensity, Velocity and Magnetic Fields*'. Solar Physics, **4**, 165.

Beckers, J. M., and Schröter, E. H. 1968c. '*Intensity, Velocity and Magnetic Structure of a Sunspot Region. II: Some Properties of Umbral Dots*'. Solar Physics, **4**, 303.

Beckers, J.M., and Tallant, P.E. 1969. '*Chromospheric Inhomogeneities in Sunspot Umbrae*'. Solar Physics, **7,** 351.

Belvedere, G., Godali, G., Motta, S., Paternò, I., Zappalà, R.A. 1977. '*K faculae as tracers of the solar differential rotation*'. Astrophysical Journal, Part 2 - Letters to the Editor, vol. **214**, L91.

Bhatnagar, A. 1966. '*The Evershed Effect and line asymmetry in Sunspot Penumbrae*'. 1966. Kodaikanal Observatory, series A, no. **180**, pp A13-A50, 1966.

Bhatnagaer, A., and Rahim, K.C. 1970. '*On the Polar Coronal Rays of the Sun*'. Kodaikanal Observatory Bulletin, No. **204**, p. A189.

Bhatnagar, A, Ambastha, A, and Srivastava, N. 1992, '*Filament Eruptions, Flaring arches and Eruptive Flares*', in Eruptive Solar Flares, IAU Colloq. No.**133**, ed., Z. Švestka, B.V. Jackson, and M.E. Machado, p. 59.

Bhatnagar, A. 1996. '*Solar Mass ejections and Coronal holes*'. Astrophysics and Space Science, **243**, 105.

Bhatnagar, A., 2003. '*Instrumentation and Observational Techniques in Solar Astronomy*'. Lectures on Solar Physics, ed., H. M. Antia, A. Bhatnagar, P. Ulmschneider, vol. **619**, p. 27. Publ. Springer-Verlag.

Bhatnagar, A., Acton, L., Hudson, H., Kosugi, T., Strong, K., and Tripathi, S. C. 1996, private communication.

Bhatnagar, A., Jadhav, D.B., R.M., Jain, R. M., Shelke, R.N., and Purohit, S.P. 1981. '*Observations for Coronal velocity field and color movie of the flash spectrum during total eclipse of 16 February 1980*'. Proceedings of INSA, Part A, Physical Sciences, Vol. **48A**, Indian Academy of Science, New Delhi, ed. S. K. Trehan. p. 29.

Bhatnagar, A., Jain, K., and Tripathi, S.C. 1999. '*GONG p-mode frequency change with solar cycle activity*'. Ap. J., **521,** 885.

Bhatnagar, A., Mukerji, S., Babu, Y.S., Sehgal, N.K., Kamble, V.B.,Pandya, R.P., Pandya, N.P., Bhavsar, K.M, and Prajapati, R .P. 1997. '*Total Solar Eclipse observations MiG-25 at 80,000 ft*'. Kodaikanal Obs. Bull., **13**, 101.

Biermann, L. 1946. '*Zur deutung der chromospharischen turbulenz und des exzesses der UV-strahlung der sonne*'. Naturwissenschaften, **33**, 118.

Biermann, L. 1946. 'Zur Deutung der chromospatischen Turbulenz und des Exzesses der UV-Strahlung der Sonne'. Naturwiss., **33**,118.

Biermann, L. 1948. 'Über die Ursache der chromosphärischen Turbulenz und des UV-Exzesses der Sonnenstrahlung'. Z. Astrophys., **25**, 16.

Birge, R.T., 1942. Rep. Phys. Soc. Prog., **8**, 90.

Blackwell, D.E., Dewhirst, D.W., and Dollfus, A., 1959. '*The observation of solar granulation from a manned balloon. I. Observational Data and measurement of contrast*'. MNRAS., **119**, 98.

Braun, D. C., Duvall, T. L., Jr. and La Bonte, B. J. 1988a. *'The absorption of high-degree p-mode oscillations in and around sunspots'*. Ap . J., **335**, 1015.

Braun, D. C. La Bonte, B. J., and Duvall, T. L., Jr. 1988b. *'Tomography of Solar Active Regions'*. BAAS., **20**, 70.

Bray, R.J. and Loughhead R., 1958. *'Observations of Changes in the photographic granules'*. Australian Journal of Physics, **11**, 507.

Bray, R.J., and Loughhead, R.E., 1959. *'High resolution observations of the structure of sunspot umbrae'*. Australian Journal of Physics, **12**, 320.

Bray, R.J., and Loughhead, R.E. 1961. *'Facular granule life times determined with a seeing-monitored photoheliograph'*. Aust. J. Phys., **14,** 14.

Bray, R.J. and Loughhead, R.E., 1964. In 'Sunspots'. Wiley and Sons, N.Y.

Bray, R.J.,Loughhead, R.E. and Tappere, E.J., 1976. *'Convective velocities derived from granule contrast profiles in Fe I at 6569.2 A'*. Solar Physics, **49**, 3.

Bray, R.J., and Loughhead, R.E., 1977. *'A new determination of the granule/intergranule contrast'*. Solar Physics, **54**, 319.

Brisken, Walter F., and Zirin, Harold. 1997. *'New Data and Models of Running Penumbral Waves in Sunspots'*. Ap. J., **478**, 814.

Brun, A. S., Turck-Chièze, S., and Zahu, J.P. 1999. *'Standard Solar Models in the Light of New Helioseismic Constraints. II. Mixing below the Convective Zone'*. Ap. J., **525**, 1032.

Bruzek, A. 1967. *'On arch –filament systems in spot group'*. Solar Physics, **2**, 451.

Brynildsen, N., Maltby, P. Fredvik, T., and Kjeldseth-Moe, O. 2002. *'Oscillations above Sunspots'*. Solar Physics, **207**, 259.

Bumba, V. 1960. *'Results of the study of the Evershed Effect in single sunspot'*. Izv. Crim. Asytrophys. Obs., **23,** 253.

Burton, C. H., Leistner, A. J., Rust, D. M. 1987. *'Electro-optic Fabry-Perot Filter: Development for the Study of Solar Oscillations'*. Applied Optics, **26**, 2637.

Cacciani *et al.* 1998. *'A MOF-based full vector imaging magnetograph'*. Second Advances in Solar Physics Euro conference. ASP Conf. Ser., Vol. **155**, p. 265. Ed.,. C.E. Alissandrakis & B. Schmieder.

Chandrasekhar, S., 1950, Radiative Transfer, p.248, Publ. Oxford: Clarendon Press.

Chaplin, William J., *et al.* 1996. *'BISON Performance'*. Solar Physics, **168**, 1.

Christensen-Dalsgaard, J., Däppen, W., Ajukov, S.V. et al., 1996. *'The Current State of Solar Modeling'*. Science, **272**, 1286.

Clark D.H., and Stephenson, F.R. 1978. *'An Interpretation of the Pre-Telescopic Sunspot Records from the Orient'*. Qr. J.R. Astro. Soc., **19**, 387.

Code, A. D. 1950. *'Radiative Equilibrium in an Atmosphere in which Pure Scattering and Pure Absorption both Play a Role'*. Ap. J., **112**, 449.

Cook, A. 1996. *'Halley and the Saros'*. Qr. J.R. Astro. Soc., **37**, 349.

Cox, A.N. 1999. In 4th edition of Allen's Astrophysical Quantities, Publ. Springer-Verlag.

Crammer, S.R., Field, G.B., and Kohl, J.L. 1999. *'Spectroscopic Constraints on Models of Ion Cyclotron Resonance Heating in the Polar Solar Corona and High-Speed Solar Wind'*. Ap. J., **518**, 937.

Crump, T. 1999. 'Solar Eclipse'. Publ. Constable, London.

Davis, R. Jr. *et al.*, 1968. Phys. Rev. Lett., **20**, 1205.

Deinzer, W. 1965. *'On the Magneto-hydrostatic Theory of Sunspots'*. Ap. J., **141**, 548.

Dobrowolski, J. A., 1962, *'Narrow Band Interference Filter'*, U.S. Patent 3,039,362, p. 1-3 June 19, 1962.

Dunn, R. B., and Zirker, J. B. *'The Solar Filigree'*. 1973. Solar Physics, **33**, 281.

Dunn, R.B. 1951. *'How to build a quartz monochrometer'*. Sky and Telescope, **10**, 2.

Dunn, R.B. 1969. *'Sacramento- Peak's New Solar Telescope'*. Sky & Telescope, **38**,

Dupree, A.K., and Henze, W, Jr., 1972. *'Solar Rotation as Determined from OSO-4 EUV Spectroheliograms'*. Solar Physics **23**, 271.

Ebert, H. 1893. Astronomy & Astrophysics. **12**, 209.

Eddy, J., 1977. *'The case of the Missing Sunspots'*. Scientific American, (May), **236**, 80.

Eddy, J., Gilman, P.A., Trotter, D.E. 1977. *'Anomalous solar rotation in the early 17th century'*. Science, **198**, 824.

Edmonds, F.N., Jr. 1960. *'On Solar Granulation'*. Ap. J., **131**, 57.

Ellerman, F. 1917. *'Solar Hydrogen "Bombs" '*. Ap. J. 46, 298.

El-Raey, M., Scherrer, P.H. 1972. *'Differential Rotation in the Solar Atmosphere Inferred from Optical, Radio and Interplanetary Data'*. Solar Physics, **26,** 15.

Engvold, O. 1976. *'The fine structure of prominences. I - Observations - H-alpha filtergrams'*. Solar Physics, **49**, 283.

Engvold, O. 1988, in E. Priest (ed.), 'Dynamics and structure of Quiescent solar prominences'. Kluwer Academic Publishers.

Evans, J. 1958. J. Opt. Soc. Am., **48**,142.

Evans, J.W., and Michard, R. 1962. *'Observational study of macroscopic inhomogeneities in the solar atmosphere III. Vertical oscillatory motions in the solar photosphere.'* Ap. J., **136**, 493.

Evershed, J. 1909. *'Radial movement in sun-spots'*. MNRAS., **69,** 454.

Feynman, J., and Martin, S. F. 1993. *'The Initiation of Coronal Mass Ejections'*. Bull.AAS., **25, 1203.**

Fossat, E., Grec, G., Pomerantz, M. 1980. *'Solar pulsations observed from the geographic South Pole - Initial results'*. Solar Physics, **74**, 59.

Fotheringham, J. K. 1920. *'A solution of ancient eclipses of the sun'*. MNRAS., **81**, 104.

Fox, P., 1921. Pub. Yerkes, Obs., **3**, 67.

Franz Deubner, F.-L. 1975. *'Observations of low wave number non-radial eigenmodes of the Sun'*. Astron., and Astrophys., **44**, 371.

Frazier, E. N. 1972. *'The relation between chromospheric features and the photospheric magnetic fields'*. Solar Physics, **24**, 98.

Frazier, Edward N. 1970. *'Multi-Channel Magnetograph Observations. II. Supergranulation'*. Solar Physics, **14**, 89.

Fukunday, et al., 1996. Phys. Rev. Lett., **7**, 1683.

Fukunday. et al., 1999. Phys. Rev. Lett. **82**, 1810.

Furst, E., Hirt, W., and Lantos, O. 1979. '*Prominences at centimetric and millimetric wavelengths. I-Size and spectrum of the radio filaments'*. Solar Physics, **63**, 257.

Gabrial, A. H. 1976. Phil. Trans. R. Soc., **281**, 399.

Georgakilas, A. A., Christopoulon, E. B., Zirin, H. 2000. '*Oscillations and running waves observed in Sunspots- Analysis of an extended sample of sunspots*', Bull. AAS., **32**, 1489.

Georgakilas, A. A.; Muglach, K. and Christopoulon, E. B. 2002. '*Ultraviolet Observations of Periodic Annular Intensity Fluctuations Propagating around Sunspots*'. Ap. J., **576**. 561.

Gilman, P., and Foukal, P.V. 1979. *'Angular velocity gradients in the solar convection zone'*. Ap. J., **229**, 1179.

Ginzburg, V. I. 1946. C.R. Dokl. Academy. Sci. USSR, **52**, 487.

Giovanelli, R.G. 1972. '*Oscillation and Waves in a Sunspot'*. Solar Physics, **27**, 71.

Golub, L. and Pasachoff, J.M. 1997. 'The Solar Corona'. Publ. Cambridge Univ. Press, Cambridge.

Golub, L., Vaiana, G. S. 1978. *'Differential rotation rates for short - lived regions of emerging magnetic flux'*. Ap. J. Lett., **219**, L55.

Grec, G., Fossat, E., and Pomerantz, M. 1980. *'Solar oscillations - Full disk observations from the geographic South Pole'*. Nature, **541**, 288.

Hale, G. E. 1908a. '*Solar Vortices and the Zeeman Effect*'. PASP., **20,** 203.

Hale, G. E. 1908b. '*On the Probable Existence of a Magnetic Field in Sun-Spots*'. Ap. J., **28,** 244.

Hale, G. E., Ellerman, F., Nicholson, S.B., and Joy, A.H. 1919. '*Magnetic Polarity of Sunspots*'. Ap. J., **49**, 153.

Hansen, R.T., Hansen, S.F., Loomis, H. G. 1969. *'Differential Rotation of the Solar Electron Corona'*. Solar Physics, **10**, 135.

Hart, A.B. 1954. '*Motions in the Sun at the photospheric level. IV. The equatorial rotation and possible velocity fields in the photosphere'*. MNRAS., **114**, 17.

Hart, A.B. 1956. '*Motions in the Sun at the photospheric level. VI. Large-scale motions in the equatorial region'*. MNRAS., **116**. 38.

Harvey, K. L., and Martin, S. F. 1973.*'Ephemeral Active Regions'*. Solar Physics, **32**, 389.

Harvey, J. W. Duvall, T. L., Jr. and Pomerantz, M. A. 1986. *'Helioseismology Results from South Pole Observations'*. Bull. AAS., **18,** 1011.

Hathaway, D. H., and Wilson, R.M. 1990. '*Solar rotation and the sunspot cycle'*. Ap. J., **357**, 271.

Heaviside, O. 1904. Nature, **69**, 342.

Herschel, W. 1795. '*On the Nature and Construction of the Sun and fixed stars*'. Phil. Trans., **85**, 46.

Hey, J.S. 1946. Nature, **157**, 47.

Heyvaerts, J., Priest, E.R., and Rust, D. M. 1977. *'An emerging flux model for the solar flare phenomenon'*. Ap. J., **216**, 123.

Hiei, E. 1963. *'Continuous spectrum in the Chromosphere'*. Publ. Astron. Soc. Jap., **15**, 115.

Holmes, J. 1961. *'A study of sunspot velocity fields using a magnetically undisturbed line'*. MNRAS., **122**, 301.

Horne, K., Hurford, G. J., Zirin, H., de Graauw, T. 1981. *'Solar limb brightening at 1.3 millimeters'*. Ap. J., **244**, 340.

Houtgast, J. & Sluiters, A. Van. 1948. *'Statistical investigations concerning the magnetic fields of Sunspot- I'*. Bull. Astron. Inst. Netherlands, **10**, 325.

Howard, R., Gilman, T.S., Gilman, P.I. 1984. *'Rotation of the Sun measured from Mount Wilson white-light images'*. Ap. J., **283**, 373.

Howard, R.F. 1972. *'Polar Magnetic Fields of the Sun: 1960-1971'*. Solar Physics, **25**, 5.

Howard, R.F. 1974, *'Studies of Solar Magnetic fields'*. Solar Physics, **38**, 283.

Howard, R.F. and Harvey, J. 1970. *'Spectroscopic Determinations of Solar Rotation'*. Solar Physics, **12,** 23.

Howard, R.F., 1984. *'Solar Rotation'*. Ann. Rev. Astron. Astrophys., **22**, 131.

Howard, R.F., and Bhatnagar, A. 1969. *'On the Spectrum of Granular and Intergranular Regions'*. Solar Physics, **10**, 245.

Howard, R.F., and Harvey, J. W. 1964. *'Photographic magnetic fields and Chromospheric features'*. Ap. J., **139**, 193.

Howard, R.F., and LaBonte. B. 1980. *'A search for large-scale convection cells in the solar atmosphere'*. Ap. J. Lett., **239**, L33.

Hudson, H. S. 1994. *'The Yohkoh Context for High-Energy Particles in Solar Flares'*, in 'High-Energy Solar Phenomena-a New Era of Spacecraft Measurements', Proc., Workshop Held in Waterville Valley, New Hampshire, March 1993. American Institute of Physics: New York. Ed. by J. Ryan and W. T. Vestrand. AIP Conference Proceedings, Vol. **294**, p.151.

Jain, K., Tripathi, S.C., and Bhatnagar, A. 2001. *'On the solar rotation rate in the upper convection zone'*. Proc. SOHO 10/GONG 2000 Workshop, Helio- and Asteroseismolgy at the Dawn of the Millennium', Santa Cruz de Tenerife, Spain, 2-6 October 2000, ESA-SP-464, January 2001, p. 641.

Jones, H.S. 1940. *'Eros, on the suitability of, for the accurate determination of the solar parallax'*. MNRAS., **100**,422.

Kahler, S. W. 1992. *'Solar flares and coronal mass ejections'*, in Annual Review of Astron., and Astrophys. Vol. **30,** p. 113

Kane, S.R. 1974. In *'Coronal Disturbances'*, IAU Sym. No. 57, ed. G.A. Newkirk, Publ. Reidel, Dordrecht, Boston.

Kanno,M. 1966. 'A model of the upper chromosphere with the spicule structure'. Publ. Astron. Soc. Jap., **18**, 103.

Karpinsky, V.N. 1980. Sol. Dannye Bull., **7**, 94.

Kawaguchi, I. 1980. '*Morphological study of the solar granulation - The fragmentation of granules*'. Solar Physics, **65**, 207.

***Kaye**, G. R. 1918.* 'The Astronomical Observatories of Jai Singh', *published by The Superintendent, Government Printing Press, Calcutta, India, 1918.*

Keil, S.L., 1977. '*A new measurement of the center-to-limb variation of the RMS granular contrast*'. Solar Physics, **53**, 359.

Keller, C.F, Montoya, J.A, Strait, B.G., and Tabor, J.E. 1980. 'Airborne Photometry of the Corona from 1.1 to 20 R_0 during the February 16, 1980 Solar Eclipse'. Bull. AAS, **12**, p. 917.

Keller, C.F. 1982. '*Airborne Eclipse Experiment: A description of five Experiments to determine Temperature, Density and Structure in the Corona*'. A special Supplement to Proceedings of INSA, Part A, Physical Sciences, Vol. **48A**, Indian Academy of Science, New Delhi, ed. S. K. Trehan, p. 33.

Keller, C. U. 1992. '*Resolution of magnetic flux tubes on the Sun*'. Nature, **359,** 307.

Khan, J., Uchida. Y., McAllister, A., and Watanbe, Ta. 1994. '*YOHKOH Soft X-Ray Observations Related to a Prominence Eruption and Arcade Flare on 7 may 1992*', in 'X-ray solar physics from Yohkoh'. Frontiers Science Series, Proc., International Symposium on the Yohkoh Scientific Results, held February 23-25, 1993, Sagamihara, Kanagawa, Japan, Universal Academy Press, ed. by Yutaka Uchida, Tetsuya Watanabe, Kazunari Shibata, and Hugh S. Hudson, 1994., p. 201

Kiepenheuer, K. O. 1953. '*Photoelectric Measurements of Solar Magnetic Fields*'. Ap. J. **117**, 442.

Kiepenheuer, K.O. 1964. '*Solar Site Testing*'. In Site Testing, ed. J. Rösch, IAU Sym. **19**, 193

Kinman, T. D. 1952. '*Motions in the Sun at Photospheric level- III. The Evershed effect in Sunspots of different sizes*'. MNRAS., **112,** 425.

Kippenhahn, R., Schlüter, A. 1957. '*Eine Theorie der solaren Filamente*'. Zeitschrift für Astrophysik, **43**, 36.

Kneer, F.; Wiehr, E. 1998. '*The Gregory-Coudé- Telescope at the Observatorio Del Teide Tenerife*', in **'**Solar and Stellar Granulation'**,** Proc. 3rd International Workshop of the Astronomical Observatory of Capodimonte (OAC) and the NATO Advanced Research Workshop on Solar and Stellar Granulation, June 21-25, held at Capri, Italy, Dordrecht: Kluwer, 1989, ed. by Robert J. Rutten and Giuseppe Severino. NATO Advanced Science Institutes (ASI) Series C, Vol. **263**, p.13.

Kosovichev, A. G., and Zharkova, V. V. 1998. '*X-ray Flare Sparks Quake inside the Sun*'. Nature, **393,** 317.

Kosovichev, A. G., and Zharkova, V. V. 1999. '*Variations of Photospheric Magnetic Field Associated with Flares and CMEs*', Solar Physics, **190,** 459.

Koval, A.N. 1965. '*The position of moustache in spot group relative to the magnetic field*'. Izv. Crim. Astrophys. Obs., **37**, 62.

Krieger, A. S., Timothy, A. F., Roelof, E. C. 1973. 'A Coronal Hole and its Identification as the Source of a High Velocity Solar Wind Stream'. Solar Physics, **29**,505.

Kuiper, G.P. 1938. *'The Magnitude of the Sun, the Stellar Temperature Scale, and Bolometric Corrections'*. Ap. J., **88**, 429.

Kuperus, M.; Raadu, M. A.1974. *'The Support of Prominences Formed in Neutral Sheets*'. Astro. & Astrophys., **31**, 189.

Kuperus, Max, and Tandberg-Hanssen, Einar. 1967. *'The Nature of Quiescent Solar Prominences'*. Solar Physics, **2**, 39.

LaBonte, B., & Howard, R.F. 1982a. *'Torsional waves on the Sun and the activity cycle'*. Solar Physics, **75**, 161.

LaBonte, B., & Howard, R.F. 1982b. *'Are the high-latitude torsional oscillations of the Sun real'*. Solar Physics, **80**, 373.

LaBonte, B., Simon, G.W., and Dunn, R. B. 1975. *'A Phenomenological Study of High Resolution Granulation Photography'*. Bull. American Astron. Society, **7**, 366.

Labs. 1957. Heidelbeige Sym. Problems der Spectra photo. Springer Verlag. Heidelberg.

Leibacher, J. and Stein, R. F. 1970. *'A New Description of the Solar Five-Minute Oscillation'*. Ap. J. Lett., **7**, 191.

Leighton, R.B., Noyes, R.W., Simon, G.W., 1962. *'Velocity Fields in the Solar Atmosphere. I. Preliminary Report'*. Ap. J., **135**, 479.

Lewis, D. J., Simnett, G. M., Brueckner, G. E., Howard, R. A., Lamy, P. L. Schwenn, R. 1999. *'LASCO observations of the coronal rotation'*. Solar Physics, **184**, 297.

Livingston, W.C., 1968. *'Magnetograph Observations of the Quiet Sun. I. Spatial Description of the Background Fields'*. Ap. J., **153**, 929.

Livingston, W.C., 1969. *'On the Differential Rotation with Height in the Solar Atmosphere*'. Solar Physics, **9**, 448.

Loops, R.H., and Billings D.E. 1962. Z. Ap., **55**, 24.

Lyot, B. 1930. Compt. Rend. Acad. Sci. Paris, **101**, 834.

Lyot, B. 1933. Compt. Rend. Acad. Sci. Paris, **197**, 1593.

Lyot, B. 1939. *'The study of the solar corona and prominences without eclipses (George Darwin Lecture, 1939)*'. M.N.R.S., **99**, 586.

Lyot, B. 1944. *'Le filtre monochromatique polarisant et ses applications en physique solaire'*. Ann. Astrophys., **7**, 31.

Macris, C.J. 1978. *'Size variation of photospheric granules in the vicinity of sunspots'*. Academie des Sciences (Paris), Comptes Rendus, Serie B - Sciences Physiques, vol. **286**, no. 22, June 5, p. 315, 316. In French. Astro. Astrophys., **78**, 186.

Makarov, V.I., Tlatov, A.G., Callebaut, D.K., Sivaraman, K.R. 2001. *'Pole-ward migration rate of the magnetic fields and the power of solar cycle.*' SOHO 10/GONG 2000 workshop. 'Helio-and Astroseismology at the Dawn of Millennium'. Santa Cruz de Tenerife, Spain. 2-6 October 2000 (ESA SP-464, January 2001, p.115.

Makita, M. 1963. *'Physical States in Sunspots'*. Pub. Astr. Soc., Japan, **15**, 145.

Makita, M., and Morimoto, M. 1960. '*Photoelectric study of Sunspots*'. Publ. Astron. Soc., Japan, **44**, 63.

Maltby, P. 1960. '*Note on the Evershed Effect in Sunspots*'. Ann. Astrophys., **23**, 983.

Malville, *M. 1998.* 'Megaliths and Neolithic astronomy in southern Egypt'. *Nature,* ***392****, 488.*

Marsh, K. A., Hurford, G. J., Zirin, H. '*High resolution interferrometric observations of the solar limb at 4.9 and 10.7 GHz during the solar eclipse of October,1977*'. Astron. Astrophys., **94**, 67.

Martin, S. F., Livi, S.H.B., and Wang, J. 1985. '*The cancellation of Magnetic Flux. II in decaying Active* region'. Aust. J. Phys., **38**, 929.

Martin, S.F., and Harvey, K.L. 1979. '*Ephemeral active regions during solar minimum*'. Solar Physics, **64**, 93.

Martres, M. J. and Soru-Escaut, I. 1971.'*Chromospheric absorbing features promising the appearance and development of active center*'. Solar Physics, **21**, 137.

Mathew, S. K., Bhatnagar, A., Debi Prasad, C., Ambastha, A. 1998. '*Fabry-Perot filter based solar video magnetograph*'. Astron. & Astrophysics, Suppl. Ser., **133**, 285.

Mattig, W. 1958. '*Beobachtungen randnaher Sonnenflecken in Hα*'. Z. Astrophys., **44**, 280.

Maunder, E.W., and Maunder A.S.D. 1905. '*Sun, rotation period of the, from Greenwich sun-spot measures, 1879-1901*' . MNRAS., **65**, 813.

McCready, L. L., Pawsey, J. L., Payne-Scott, Ruby. 1947. Proc. R. Soc. London, A., **190**, 357.

Meeus, Jean. 2000. '*Where Eclipses Come Thrice*'. Sky and Telescope, vol. **99**, No. 4, p 63.

Mehltretter, J.P. 1974. '*Observations of photospheric faculae at the center of the solar disk*'. Solar Physics, **38**, 43.

Mehltretter, J.P. 1978. '*Balloon-borne imagery of the solar granulation. II - The lifetime of solar granulation*'. Astron. Astrophysics, **62**, 311.

Michalitsanos, A.G., and Bhatnagar, A. 1975. '*Observations of Large–Scale Moving Magnetic Features near Sunspots*'. Astrophys. Lett., **16**, 43.

Michard, R. 1951. '*Remarques sur l'effet Evershed*'. Ann. Astrophys., **14,** 101.

Michard, R. 1953. '*Contibution à l'ètude physique de la photosphère et des taches solarires*'. Ann. Astrophys., **16**, 217.

Montesinos, B., and Thomas, J.H., 1997. '*The Evershed effect in Sunspots as a siphon flow along a magnetic flux tube*'. Nature, **390**, 485.

Moreton, G. E. 1960. '*Hα Observations of Flare-Initiated Disturbances with Velocities ~1000 km/sec*'. A. J, **65**, 494.

Moreton, G. E., Ramsey, H. E. 1960. '*Recent Observations of Dynamical Phenomena Associated with Solar Flare*'. PASP., **72**, No. 428, 357.

Muglach, K. 2003. '*Dynamics of solar active regions. I. Photospheric and chromospheric oscillations observed with TRACE*'. Astron., and Astrophys., **401,** 85.

Müller, R. 1954. *'Über die Rotation der Sonne in Polnähe. Mit 3 Textabbildungen'*. Z. Ap., **35**, 61.

Müller, R. 1977. *'Morphological properties and origin of the photospheric facular granules'*. Solar Physics, ***52**, 249.*

Namba, O., and Diemel, W.E. 1969. *'A Morphological Study of the Solar Granulation'*. Solar Physics, **7**, 167.

Nash, A. G., Sheeley, N. R., Jr., and Wang, Y.-M.1988. *'Mechanisms for the rigid rotation of coronal holes'*. Solar Physics. **117**, 359.

Newkirk, G., Jr. 1971. *'Large Scale Solar Magnetic Fields and Their Consequences'*, in IAU Symposium No. 43 on the Solar Magnetic Fields, ed. Howard, D. Reidel, Dordrecht, p. 54.

Newton, H.W., and Nunn, M.L. 1951. *'The Sun's rotation derived from sunspots 1934-1944 and additional results'*. MNRAS., **111,** 413.

Nicholson, S. B. 1933. *'Area of Sunspot and the Intensity of its magnetic field '*. PASP., **45**, 51.

November, L. J., Toomre, J., Gebbie, K.B. Simon, G.W., 1981. *'The detection of mesogranulation on the sun'*. Ap. J., **245**, L123.

Ohman, Y. 1938 . *'A new Monochromator'*. Nature, **141**, 157.

Osterbrock, D. E. 1961. 'The Heating of the Solar Chromosphere, Plages, and Corona by Magnetohydrodynamic Waves'. Ap. J., **134**, 270.

Paul, H.E. 1953. Amateur Telescope Making, Vol. **3**, ed., A. G. Igalls, p. 376.

Payne, T.W. 1993. *'A Multi-wavelength study of Solar Ellerman bombs'*. Thesis, New Mexico State University, Las Cruces, N.M. Pecker, J. C. 1950. Ann. d'Ap., **13**, 294 and 319.

Penn, M.J., Cao, W. D., Walton, S. R., Chapman, G. A., Livingston, W. 2003. *'Weak Infrared Molecular Lines Reveal Rapid Outflow in Cool Magnetic Sunspot Penumbral Fibrils'*. Ap. J., **590**, L119.

Petit, E. 1953. Amateur Telescope Making, Vol. **3**, ed., A. G. Igalls, p. 413

Piddington, J. H. 1978. *'The flux-rope-fibre theory of solar magnetic fields'*. Astrophysics and Space Science**,** **55**, 401.

Pierce, A.K. 1964. *'The McMath Solar Telescope of the Kitt Peak National Observatory'*. Applied Optics, **3**, 1337.

Plaskett, H.H. 1916. *'A Variation in the Solar Rotation'*. Ap. J., **43**, 145.

Plaskett, J. S. 1915.*'The Spectroscopic Determination of the Solar Rotation at Ottawa'*. Ap. J., **42**, 373.

Pravdjuk, C.H.B., Karpinsky, V.N., Andrelko, A.V. 1974, Solnechnye Dannye Bull. 70.

Priest, E. R., Foley, C. R., Heyvaerts, J., Arber, T. D., Mackay, D., Culhane, J. L., and Acton, L. W. 2000.*'A Method to Determine the Heating Mechanisms of the Solar Corona'*. Ap. J., **539**, 1002.

Rabe, E. 1950. *'Derivation of fundamental astronomical constants from the observations of Eros during 1926-1945'*. A.J., ***55,*** 112.

Rast, M.P., Fox, P.A., Lin, H., Lites, B.W., Meisner, R.W., and White, O.R. 1999. *'Bright rings around sunspots'*. Nature, **401**, 678.

Rhode, E.J. Jr., Deubner, F.-L., Ulrich, R.K. 1979. *'A new technique for measuring solar rotation'*. Ap. J., **227**, 629.

Richardson, R.S., Schwarzschild, M. 1950. *'On the Turbulent Velocities of Solar Granules'*. Ap. J., **111**, 351.

Ringnes, T.S., & Jensen, E. 1960. *'On relation between magnetic fields and the areas of sunspots in the interval 1917-1956'*. Astrophysica Norvegica, **7**, 99.

Rösch, J. 1962. *'La détérioration des images solaires par l'atmosphère'*. Trans. of the International Astronomical Union, vol. **11B**, ed. D.H. Sadler, p. 197. London: Academic Press.

Rösch, J., 1959. *'Observations sur la photosphère solaire: II. Numération et photométrie photographique des granules dans le domaine spectral 5900-6000 Å'*. Ann. d'Astrophysique, **22**, 584.

Royds, T. 1920. *'Some features of Hα dark markings on the Sun'*, Kodaikanal Observatory Bulletin, No. 63.

Rust, D. M. 1985. *'New Materials Applications in Solar Spectral Analysis'*. Austr. J. Phys., **38,** 781.

Rutten, R. 1999. *'The Dutch Open Telescope: History, Status, Prospects'*, in High Resolution Solar Physics: Theory, Observations, and Techniques, ASP Conference Series #183, eds. T. R. Rimmele, K. S. Balasubramaniam, and R. R. Radick, p.147.

Ryle, M., and Vonbery, D. D. 1946. Nature, **158,** 339.

Scharmer, G.B., Gudiksen, B.V., Kriselman, MD., Lofdahl, M.G. & Rouppe van der Voort, L.H.M. 2002. *'Dark cores in Sunspot penumbral filaments'*. Nature, **420**, 151.

Schlichenmaier, R. 2002. *'Penumbral fine structure: theoretical understanding'*. Astronomische Nachrichten, **323**, No.3/4, 303.

Schlichenmaier, R., Jahn, K., Schmidt, H. U. 1998. *'A Dynamical Model for the Penumbral Fine Structure and the Evershed Effect in Sunspots'*. Ap. J., **493**, L121.

Schröter, E, H., Soltau, Wohl, H., and Vazquez. 1978. *'An attempt to compare the differential rotation of the Ca/plus/-network with that of the photospheric plasma'*. Solar Physics, **60**, 181.

Schröter, E.H. 1962. Z. f. Astrophys., **56**, 183.

Schröter, E.H., Wohl, H. 1975. *'Differential rotation, meridional and random motions of the solar Ca+ network'*. Solar Physics, **42**, 3.

Schröter, E. H.; Soltau, D.; Wiehr, E. 1985. *'The German solar telescopes at the Observatorio del Teide'*. Vistas in Astronomy, vol. **28**, p. 519.

Schwarzschild M., 1959. *'Photographs of the Solar Granulation taken from Stratosphere'*. Ap. J., **130**, 345.

Schwarzschild, M. 1948. *'On noise arising from the Solar Granulation'*. Ap. J., **107**, 1.

Scudder, Jack D. 1994. *'Ion and electron supra-thermal tail strengths in the transition region: Support for the velocity filtration model of the corona'*. Ap. J., **427,** 446.

Secchi, A. 1875. Le Solelil, 2nd edn., Vol 1, Paris: Gauthier-Villars.

Semel, M. 1962. C.R., **254**, 3978.

Servajean, R. 1961. '*Contribution à l'etude de la cinematique de la matrière dans les taches et la granulation solaires*'. Ann. Astrophys., **24**, 1.

Severny, A. B. 1965. '*Fine Structure in Solar spectrum*'. Observatory, **76,** 241.

Seykora, E. J. 1993. '*Solar scintillation and the monitoring of solar seeing*'. Solar Physics, **145**, 389.

Simon, G.W., and Leighton, R.B. 1964. '*Velocity Fields in the Solar Atmosphere. III. Large-Scale Motions, the Chromospheric Network, and Magnetic Fields*'. Ap. J., **140**, 1120.

Simon, G. W., and Noyes, R.W. 1972. '*Solar Rotation as measured in EUV Chromospheric and Coronal lines*'. Solar Physics, **26**, 8.

Simon, G.W., and Zirker, J.B. 1974. '*A Search for the foot points of Solar Magnetic Fields*'. Solar Physics, **35**, 331.

Smartt, R. N. 1982. '*Solar corona photoelectric photometer using mica etalons*'. Instrumentation in astronomy IV; Proceedings of the Fourth Conference, Tucson, AZ, March 8-10, 1982 (A83-31976 14-35). Bellingham, WA, SPIE - The International Society for Optical Engineering, **331**, 442.

Snodgrass, H. B. 1983. '*Magnetic rotation of the solar photosphere*'. Ap. J., **270**, 288.

Sofaer Anna, *Zinser, Volker, and Sinclair, Rolf M. 1979.* 'A Unique Solar marking Construct'. *Science, Oct.19, vol.* ***206,*** *No. 4416, 283.*

Šolc, I. 1965. '*Birefringent Chain Filters*'. J. Opt. Soc. Am., **55**, 621.

Soonawala, *M. F. 1952.* 'Maharaja Sawai Raja Jai Singh II of Jaipur and his observations', *published by the Jaipur Astronomical Society, Jaipur.*

Spruit, H.C. 1982. '*The flow of heat near a starspot*'. Astron., and Astrophys., **108**, 356.

St. John, C. E. 1932. Trans. IAU **4**, 42.

St. John, Charles E. 1913. '*Radial Motion in Sun-Spots*'. Ap. J., **37**, 322.

Stenflo, J. O. 1976. '*Resonance-line polarization. I - A non-LTE theory for the transport of polarized radiation in spectral lines in the case of zero magnetic field*'. Astron., and Astrophys., **46**, 61.

Stepanian, N. N. 1994. '*Coronal Holes and Background Magnetic Fields on the Sun*'. Solar coronal structures. Proceedings of the 144th colloquium of the International Astronomical Union held in Tatranska Lomnica; Slovakia; September 20-14; 1993; on the occasion of the 50th anniversary of the Skalnate Pleso Observatory; Tatranska Lomnica: VEDA Publishing House of the Slovak Academy of Sciences; lc1994; edited by Vojtech Rusin, Petr Heinzel and Jean-Claude Vial, p. 61.

Stephenson, F.R. 1997. In 'Historical Eclipses and Earth's Rotation'. Publ. Cambridge Univ. Press, Cambridge.

Steskenko, N.V. 1960. Izvestiya Krymskoj Astrofizicheskoj Observatorii, **22**, 49.

Stoney, J. 1932. MNRAS., **92**, 737.

Stuart, F.E., and Rush, J.H. 1954. '*Correlation Analyses of Turbulent Velocities and Brightness of the Photospheric Granulation*'. Ap. J., **120**, 245.

Sturrock, P. 1980. In 'Solar Flares', ed., Sturrock, Colorado Ass. Uni. Press

Thomas, J.H. 1994. In 'Solar Surface Magnetism', Eds R.J.Rutten and C.J. Schrijver. Publ. Kluwer, Netherlands, p. 219.

Thomas, R.N. 1948. '*Phenomena in Stellar Atmospheres I- Spicules and the Solar Chromosphere*'. Ap. J., **108**, 13.

Thomas, R.N., and Athay, R.G. 1961. In 'Physics of the Solar Chromosphere'. Publ. Inter Science, New York.

Timothy, A. F., Krieger, A. S., Vaiana, G. S. 1975.'*The structure and evolution of coronal holes'*. Solar Physics, **42**,135.

Tomczyk, S., Schou, J., Thompson, M.J. 1995. '*Measurement of the Rotation Rate in the Deep Solar Interior'*. Ap. J., **448**, L57.

Topka, K., Moore, R., Labonte, B. J., and Howard, R. 1982. '*Evidence for a poleward meridional flow on the sun*'. Solar Physics, **79**, 231.

Tsuneta S., Takahashi T., Acton L. W., Bruner Marilyn E., Harvey K. L., Ogawara Y. 1992.'*Global restructuring of the coronal magnetic fields observed with the YOHKOH Soft X-ray Telescope'*. Publ. Astro. Soc. Japan, **44**, 299.

Tuominen,I., and Virtanen,H. 1987. '*Solar rotation variations from sunspot group statistics'*. In 'The internal solar angular velocity: Theory, observations and relationship to solar magnetic fields'. Proc. of the Eighth Summer Symposium, Sunspot, NM, Aug. 11-14, 1986 (A88-38601 15-92). Dordrecht, D. Reidel Publishing Co., p. 83.

Ulmschneider, P. 2003. '*The Physics of Chromosphere and Coronae'*. In 'Lectures on Solar Physics'. Ed. Antia, Bhatnagar, and Ulmschneider. 232. Publ. Springer-Verlag.

Ulrich, Roger K. 1970. '*The Five-Minute Oscillations on the Solar Surface'*. Ap. J., **162**, 993.

Valnicek, B. 1968. '*The 'de-twisted' Prominence of September 12, 1966'*. In 'Structure and Development of Solar Active Regions', IAU Sym. No. 35, ed. Kiepenheuer, p. 282.

Vernazza, J., Avertt, E., and Loeser, R. 1981.'*Structure of the Chromosphere'*. Ap. J. Suppl., **45**, 635.

Vrabec, D. 1973. '*Streaming magnetic features near sunspots'*, in 'Chromospheric fine structure', ed. R. Grant Athay, IAU Sym. 56, p. 201.

Wagner, W. J. 1984. '*Coronal Mass Ejections'*. Ann. Rev. Astro. Astrophys., **22**, 267.

Wahab Uddin, and Bondal, K.R., 1996. '*Eruption of a large quiescent prominence on January 14, 1993.* Bull. Astro. Soc. India, ***24**, 39.*

Waldmeier, M. 1953. Ergebnisse und Probleme der Sonnenforschung. 2nd ed. Leipzig, Geest u. Portig.

Waldmeier, M. 1961, '*The Sunspot Activity in the years 1610-1960*'. Quarterly Bulletin, **525**, 34.

Waldmeier, W. 1950. Z. Ap., **27**, 24.

Waldmeier,W. 1955. Z. Ap., **38**, 37.

Wang, John C. H. 1980. '*A note on sunspots records from China*'. Proc. Conf. Ancient Sun, ed. R.O. Pepin, J. A. Eddy, and R. B. Merrill, p.135.

Wang, Y. M. and Sheeley, N. R., Jr.1993. '*Understanding the rotation of coronal holes*'. Ap. J., **414**, 916.

Wang, Y.-M., Sheeley, N. R., Jr., Nash, A. G., and Shampine, L. R. 1988. '*The quasi-rigid rotation of coronal magnetic fields*'. Ap. J., **327**, 427.

Weart, S. R., and Zirin, H. 1969.'*The birth of active region*'. PASP., **81**, 270.

Weizsäcker, C.F., 1938, Phys. Z., **39,** 633.

Wilcox, J.M., and Howard, R.F. 1970. '*Differential Rotation of the Photospheric Magnetic Field*'. Solar Physics, **13**, 251.

Wilson, A., and Maskelyne, Nevil. 1774. '*Observations on the Solar Spots*'. Phil. Trans., **6**, 6.

Wilson, R.C., Hudson, H.S. 1988. '*Solar luminosity variations in solar cycle 21*'. Nature, **332**, 810.

Wittman, A. and Mehltretter, J.P. 1977. '*Balloon-borne imagery of the solar granulation. I - Digital image enhancement and photometric properties*'. Astron., and Astrophys., **61**, 75.

Woodard, M. F., and Noyes, R. W. 1985. '*Change of Solar Oscillation Eigen frequencies with the Solar Cycle*'. Nature, **318**, 44.

Worden, S. P. 1975. '*Infrared observations of Supergranule temperature structure*'. Solar Physics, **45**, 521.

Ye, B., and Livingston, W. 1998. '*Peering over the Sun's pole: behavior of its rotational vortex*'. Solar Physics, **179**, 1.

Yun, H.S. 1968. '*Model Sunspots*'. Astron. J., **72**, 838.

Zirin, H. 1971. '*Application of the Chromospheric magnetograph to active regions*'. In 'Solar Magnetic fields', ed. R. Howard, Publ. Reidel, Dordrecht Holland, p. 237.

Zirin, H. 1974. '*The Magnetic structure of plages*'. IAU Sym. No. 56, in Chromospheric fine Structure, ed. Grant R. Athay, p. 161.

Zirin, H., and Stein, A. 1972. '*Observations of Running Penumbral waves*'. Ap. J.**, 178**, p. L85.

Zirin, H., and Tanaka, K. 1973. '*The Flares of August 1972*'. Solar Physics, **32**, 173.

Zirker, J. B., Engvold, O., Martin, S. F. 1998. '*Counter-streaming gas flows in solar prominences as evidence for vertical magnetic fields*'. Nature, **396,** 440.

Zirker, J.B. 1984. In 'Eclipses'. Publ. Van Nostrand Reinhold, New York.

Appendix IV

Acknowledgement for Illustrations

Chapter 1. Ancient Solar Astronomy. Fig. 1.1 Robin Edgar. **Fig.1.2** © Stonehenge - Web site. **Fig. 1.3** Knowth.com.**Fig.1.4** Web site **Fig. 1.5** © Abu Simbal Web site. **Figs. 1.6, 1.7. and 1.8** © High Altitude Observatory. Fig.**1.9** Solstice project. **Fig.1.10** Bull. Astron. Soc India. **Fig.1.11** Samarkand © Web site. **Fig.1.12** Kshitij Garg.

Chapter 2. Modern Solar Observatories. Fig. 2.1 Big Bear Solar Observatory. **Fig. 2.2** San Fernando Solar Observatory. **Fig. 2.3** John Wilcox. **Fig. 2.4** A. Bhatnagar. **Fig. 2.5** National Solar Observatory. **Figs. 2.6** and **2.7.** Themes-Web site. **Fig. 2.8** Kiepenheuer Institute- Web site. **Fig. 2.9** Dutch Open Telescope -Web site. **Fig. 2.10** New Swedish Solar Telescope- Web site. **Fig. 2.11** Potsdam Observatory - Web site. **Fig. 2.12** Baikal Solar Observatory- Web site. **Fig. 1.3** A. Bhatnagar. **Fig. 1.14** National Astronomical Observatory Japan. **Fig. 2.15** Li Ting. **Fig. 2.16** Beijing Astronomical Observatory. **Fig. 2.17.** Dr. Park **Fig. 2.18** A. Bhatnagar-USO. **Fig. 1.19** Indian Institute of Astrophysics. **Fig. 2.20** Nobeyama Radio Observatory – Web site. **Figs. 2.21** and **2.22** Sebrian Solar Radio Observatory-Web site. **Fig. 2.23** SOHO/NASA/LMATC Web site **Fig. 2.24** TRACE/NASA/LMATC Web site. **Fig. 2.25** STEREO Web site.

Chapter 3. Structure of the Solar Atmosphere. Figs. 3.1 and **3.2** Anita Jain. **Fig.3.3** Bahcall and Ulrich. **Fig. 3.4** Hathaway. **Fig. 3.5** Joe Hickox, Mt. Wilson Observatory. **Fig 3.6** Drawing by Anita Jain. **Fig. 3.7...... Figs. 3.8, 3.9 and 3.12** Howard (1984). **Fig 3.10** Piddington. **Fig 3.11** Eddy © Science. **Fig. 3.13** Jain. Tripathi & Bhatnagar. **Fig. 3.14** Antia (2000). **Fig. 3.15** Basu & Antia (2001). **Fig. 3.16** H.M. Antia. **Fig. 3.17. (a)** Howard & La Bonte (1980), **Figs. 3.17. (b)** and **3.18** Basu & Antia (2001).

Chapter 4. The Quiet Sun. Fig. 4.1 Father Secchi – *Le Soleil* 187.7.. **Fig. 4.2** Project Stratoscope & New Swedish Solar Telescope. **Fig 4.3** New Swedish Solar Telescope. **Fig. 4.4** Neiss, Kiepenheuer Institute. **Fig. 4.5** R. Dunn/ National Solar Observatory/ Sac Peak. **Fig. 4.6** R. B. Leighton. **Fig. 4.7.** Drawing by Anita Jain.

Fig. 4.8 Father Secchi & R. Dunn/ national Solar Observatory/ Sac Peak. **Fig. 4.9** A. Bhatnagar/Udaipur Solar Observatory. **Fig. 4.10** Kodaikanal Observatory. **Fig. 4.11** ©Bray & Loughhead. **Fig. 4.12** New Swedish Solar Telescope- Web site. **Fig 4.13** © Vernazza *et al.* (1981). **Fig. 4.14** SOHO/TRACE Web site. **Fig. 4.15** Van de Hulst (1953). **Fig. 4.16** TRACE Web site. **Fig. 4.17.** High Altitude Observatory. **Fig. 4.18** SOHO/LASCO. **Fig. 4.19** SkyLab/ A S & E. **Fig. 4.20** Yohkoh/ISAS, Japan. **Fig. 4.21** © A. Gabrial *et al.* **Fig. 4.22** © M. R. Kundu. **Fig. 4.23** Stanford Radio Observatory. **Fig. 4.24** SOHO/ Nobeyama Radio Observatory/ Solar X-ray Imager/NOAA. **Fig. 4.25** Newkrik (197.1) and A S & E. **Fig. 4.26** SOHO/EIT/MDI.

Chapter 5. The Active Sun. Figs. 5.1 and **5.2** Dutch Open Telescope (DOT). **Fig. 5.3(a)** Rimmele/National Solar Observatory. **Fig. 5.3(b)** Father Secchi. **Fig. 5.5(a, b)** New Swedish Solar Telescope (NSST). **Fig. 5.6** St. John (1913). **Fig. 5.7** Bumba (1960). **Fig. 5.8** Larry Webster/ Mt. Wilson Observatory. **Fig. 5.9** Beckers/ National Solar Observatory/ Sac Peak. **Fig. 5.10** Big Bear Solar Observatory. **Figs. 5.12** and **5.14** W. Livingston. **Fig. 5.13** Severny. **Figs. 5.15** and **5.17** Drawing by Anita Jain. **Figs. 5.18** and **5.19** Hathaway. **Fig. 5.20** National Solar Observatory/ Kitt Peak. **Fig. 5.21** and **5.22** Big Bear Solar Observatory. **Fig. 5.23** V. Gaizauskas/ National Research Council, Canada. **Fig. 5.24** D. Vrabec/San Fernando Observatory/ Aerospace Corp. **Fig. 5.25** and **5.27** Harvey/ National Solar Observatory. **Fig. 5.26** New Swedish Solar Telescope. **Fig. 5.28** w. Livingston. **Fig. 5.29** SOHO/MDI. **Fig. 5.30** Ananthakrishnan(1954) and Makarov *et al.,* (2001). **Fig. 5.31** Drawing by Anita Jain. **Fig. 5.32** Big Bear Solar Observatory. **Fig. 5.33** (a,b) National Solar Observatory/Sac Peak. **Fig. 5.33** (c,d,e) Rompolt/Astronomical Observatory, Wroclaw, Poland. **Figs. 5.34 and 5.35** Udaipur Solar Observatory. **Fig. 5.36** High Altitude Observatory. **Fig. 5.37** TRACE/NASA. **Fig. 5.38** Udaipur Solar Observatory and Yohkoh/ISAS, Japan. **Fig. 5.40** S. R. Kane(1974). **Fig. 5.41** H. Zirin ©D. Reidel Publ. **Fig. 5.42** A. Ambastha/ Udaipur Solar Observatory. **Fig. 5.43** TRACE/ NASA. **Fig. 5.44** A. Bhatnagar & Naresh Jain/ Udaipur Solar Observatory. **Fig. 5.45** H. Zirin © Blaisell Publ. Co. **Fig. 5.46** EIT/SOHO. **Fig. 5.47** H. Rosenberg (1976). **Fig. 5.48** © Cambridge University Press. **Figs. 5.49** and **5.50** Yohkoh/ISAS, Japan/LMSAL/NASA. **Fig. 5.51** R.M. Jain/ Physical Research Laboratory. **Fig. 5.52, 5.53** and **5.54** Drawing by Anita Jain. **Fig. 5.55** LASCO/SOHO. **Fig. 5.56** Yohkoh and John Wilcox Solar Observatory. **Fig. 5.57** Yohkoh/ISAS, Japan. **Fig. 5.58** Schematic drawing.

Chapter 6. Observational Techniques. Fig. 6.1 O. Engvold. **Figs. 6.2** and drawing by Anita Jain. **Fig. 6..3** George Hale. **Fig. 6.4** Solar Geophysical Data/ NOAA.

Chapter 7. Solar Instrumentation. Fig. 7.1 A. Bhatnagar. **Fig. 7..2(a)** W. Livingston. **Fig. 7.2(b)** Steve Padilla, Mt. Wilson Observatory. **Fig. 7..3(a)** W. Livingston. **Fig. 7.3(b)** Drawing by Anita Jain. **Fig. 7.4** Kodaikanal Observatory. **Fig. 7.5** Schematic after Lyot. **Fig. 7.6** National Solar Observatory/Sac Peak. **Fig. 7.7.** Keith Pierce (1963). **Fig. 7..8** Schematic by Anita Jain. **Fig. 7..9** Kodaikanal Observatory. **Fig. 7.10** Schematic drawing. **Fig. 7.11** after Lyot. **Fig. 7.12.** Schematic drawing. **Fig. 7.13** After Alan Title. **Fig. 7.14** Schematic drawing. **Figs. 7.15** and **7.16** After Alan Title. **Fig. 7.17.** D. Rust (1987.). **Fig. 7. 18** Cacciani *et al.* (1998). **Fig. 7.19** GONG report. **Fig. 7.20** GONG/Udaipur. **Fig. 7.21** Schematic drawing. **Fig. 7.22** Schematic drawing.

Chapter 8. Solar Eclipse. Fig. 8.1 Drawing by Anita Jain. **Fig. 8.2** Captain Tupman. **Fig. 8. 3** W. Livingston. **Fig. 8.4** Fred Espenak/NASA. **Fig. 8.5** Koutchmy. **Fig. 8.6** High Altitude Observatory. **Fig. 8.8** Ms. Deepak Sundar. **Fig. 8..9** LASCO/ SOHO.

Chapter 9. Solar Interior and Helioseismology. Fig. 9.1 Howard *et al.*, (1968). **Fig. 9.2** Drawing by Anita Jain. **Fig. 9.3** J. Harvey/GONG. **Fig. 9.4** Brun *et at.*, (1999). **Fig. 9.5** J. Harvey/NSO. **Fig. 9.6** SOHO/SOI/MDI. **Fig. 9.7.** Jain, Bhatnagar and Tripathi (2002). **Fig. 9.8** Schematic drawing. **Fig. 9.9** SOHO/MDI. **Fig. 9.10** SOHO / After Kosovichev & Zharkova.

Appendix V

Index

CPSIA information can be obtained
at www.ICGtesting.com
Printed in the USA
BVOW07s1427250417
482096BV00022B/153/P